VOLUME EIGHTY NINE

CURRENT TOPICS IN MEMBRANES

Role of TRPV4 Channels in Different Organ Systems

CURRENT TOPICS IN MEMBRANES, VOLUME 89

Series Editor

IRENA LEVITAN
Division of Pulmonary and Critical Care Medicine
University of Illinois at Chicago
Chicago, IL, United States

VOLUME EIGHTY NINE

Current Topics in MEMBRANES

Role of TRPV4 Channels in Different Organ Systems

Edited by

SWAPNIL K. SONKUSARE
Robert M. Berne Cardiovascular Research Center;
Department of Molecular Physiology and Biological Physics,
University of Virginia, Charlottesville, VA,
United States

ELSEVIER
ACADEMIC PRESS
An imprint of Elsevier

Academic Press is an imprint of Elsevier
50 Hampshire Street, 5th Floor, Cambridge, MA 02139, United States
525 B Street, Suite 1650, San Diego, CA 92101, United States
The Boulevard, Langford Lane, Kidlington, Oxford OX5 1GB, United Kingdom
125 London Wall, London, EC2Y 5AS, United Kingdom

First edition 2022

ISBN: 978-0-323-99782-9
ISSN: 1063-5823

For information on all Academic Press publications
visit our website at https://www.elsevier.com/books-and-journals

Publisher: Zoe Kruze
Acquisitions Editor: Leticia Lima
Developmental Editor: Jhon Michael Peñano
Production Project Manager: Abdulla Sait
Cover Designer: Victoria Pearson

Typeset by STRAIVE, India

Contents

Contributors

Tusar Kanta Acharya
School of Biological Sciences, National Institute of Science Education and Research (NISER)-Bhubaneswar, Khurda, Odisha; Homi Bhabha National Institute, Mumbai, India

Ranabir Chakraborty
School of Biological Sciences, National Institute of Science Education and Research (NISER)-Bhubaneswar, Khurda, Odisha, India

Saurabh Chawla
School of Biological Sciences, National Institute of Science Education and Research (NISER)-Bhubaneswar, Khurda, Odisha, India

Brenan Cullimore
Department of Ophthalmology & Visual Sciences, University of Utah School of Medicine, Salt Lake City, UT, United States

Zdravka Daneva
Robert M. Berne Cardiovascular Research Center, University of Virginia, Charlottesville, VA, United States

Rashmita Das
School of Biological Sciences, National Institute of Science Education and Research (NISER)-Bhubaneswar, Khurda, Odisha; Homi Bhabha National Institute, Mumbai, India

Timothy L. Domeier
Department of Medical Pharmacology and Physiology, University of Missouri School of Medicine, Columbia, MO, United States

Nishant Kumar Dubey
School of Biological Sciences, National Institute of Science Education and Research (NISER)-Bhubaneswar, Khurda, Odisha; Homi Bhabha National Institute, Mumbai, India

Chandan Goswami
School of Biological Sciences, National Institute of Science Education and Research (NISER)-Bhubaneswar, Khurda, Odisha; Homi Bhabha National Institute, Mumbai, India

David Krizaj
Department of Ophthalmology & Visual Sciences, University of Utah School of Medicine; Department of Neurobiology; Department of Bioengineering, University of Utah, Salt Lake City, UT, United States

Satish Kumar
School of Biological Sciences, National Institute of Science Education and Research (NISER)-Bhubaneswar, Khurda, Odisha, India

Shamit Kumar
School of Biological Sciences, National Institute of Science Education and Research (NISER)-Bhubaneswar, Khurda, Odisha; Homi Bhabha National Institute, Mumbai, India

Maniselvan Kuppusamy
Robert M. Berne Cardiovascular Research Center, University of Virginia, Charlottesville, VA, United States

Monika Lakk
Department of Ophthalmology & Visual Sciences, University of Utah School of Medicine, Salt Lake City, UT, United States

Luka Lapajne
Department of Ophthalmology & Visual Sciences, University of Utah School of Medicine, Salt Lake City, UT, United States; Department of Ophthalmology, University Medical Centre, University of Ljubljana, Ljubljana, Slovenia

Victor E. Laubach
Department of Surgery, University of Virginia, Charlottesville, VA, United States

Carlene Moore
Division of Headache and Division of Translational Brain Sciences, Department of Neurology, Duke University School of Medicine, Durham, NC, United States

Vivian C. Onyali
Department of Medical Pharmacology and Physiology, University of Missouri School of Medicine, Columbia, MO, United States

Megan Elizabeth Perkins
Department of Neurological Sciences, The Larner College of Medicine, The University of Vermont, Burlington, VT, United States

Oleh Pochynyuk
Department of Integrative Biology and Pharmacology, the University of Texas Health Science Center at Houston, Houston, TX, United States

Kyrylo Pyrshev
Department of Integrative Biology and Pharmacology, the University of Texas Health Science Center at Houston, Houston, TX, United States

Sarah N. Redmon
Department of Ophthalmology & Visual Sciences, University of Utah School of Medicine, Salt Lake City, UT, United States

Tejas Pravin Rokade
School of Biological Sciences, National Institute of Science Education and Research (NISER)-Bhubaneswar, Khurda, Odisha; Homi Bhabha National Institute, Mumbai, India

Christopher N. Rudzitis
Department of Ophthalmology & Visual Sciences, University of Utah School of Medicine, Salt Lake City, UT, United States

Daniel Ryskamp
Department of Ophthalmology & Visual Sciences, University of Utah School of Medicine, Salt Lake City, UT, United States

Ram Prasad Sahu
School of Biological Sciences, National Institute of Science Education and Research (NISER)-Bhubaneswar, Khurda, Odisha; Homi Bhabha National Institute, Mumbai, India

Deep Shikha
School of Biological Sciences, National Institute of Science Education and Research (NISER)-Bhubaneswar, Khurda, Odisha; Homi Bhabha National Institute, Mumbai, India

Swapnil K. Sonkusare
Robert M. Berne Cardiovascular Research Center; Department of Molecular Physiology and Biological Physics, University of Virginia, Charlottesville, VA, United States

Anna Stavniichuk
Department of Integrative Biology and Pharmacology, the University of Texas Health Science Center at Houston, Houston, TX, United States

Viktor N. Tomilin
Department of Integrative Biology and Pharmacology, the University of Texas Health Science Center at Houston, Houston, TX, United States

Margaret A. Vizzard
Department of Neurological Sciences, The Larner College of Medicine, The University of Vermont, Burlington, VT, United States

Oleg Yarishkin
Department of Ophthalmology & Visual Sciences, University of Utah School of Medicine, Salt Lake City, UT, United States

Oleg Zaika
Department of Integrative Biology and Pharmacology, the University of Texas Health Science Center at Houston, Houston, TX, United States

Preface

Transient receptor potential vanilloid 4 (TRPV4) channels belong to the TRP family of cation channels at the cell membrane. Studies over the past two decades underscore the importance of TRPV4 channels in the vasculature, heart, lung, skin, kidney, bladder, brain, eye, musculoskeletal system, and immune cells. Global TRPV4 knockout mice did not show an obvious phenotype, perhaps due to the absence of TRPV4 channels from multiple cell types or compensatory mechanisms that have developed over time. However, recent studies using conditional, cell-specific knockout mice have made significant breakthroughs, identifying cell type-specific roles of TRPV4 channels in controlling physiological function and disease pathogenesis. Moreover, the availability of potent and specific pharmacological agonists and inhibitors has been extremely helpful in advancing our understanding of the functional roles of TRPV4 channels in different organ systems. It is anticipated that future studies will uncover new roles for TRPV4 channels in different cell types, the contribution of abnormal TRPV4 channel activity to disease pathogenesis, and meaningful ways of targeting TRPV4 channels for therapeutic benefit.

TRPV4 channel is a symmetric tetramer formed by four subunits, each with six transmembrane segments (S1–S6) and linking loops. Among the six transmembrane segments, S5, S6, and the interconnecting loop form the central cation-permeable pore. The intracellular N- and C-termini include multiple regions for protein–protein interactions that regulate the assembly and activity of TRPV4 channel. TRPV4 channel is a Ca^{2+} influx pathway and a crucial regulator of Ca^{2+} signaling mechanisms in multiple cell types. The physiological and pathological roles of TRPV4 channels have been attributed to Ca^{2+} influx through the channel, activation of other Ca^{2+}-sensitive signaling elements, effects on membrane potential, and interactions with structural proteins at the cell membrane. TRPV4 channels can also form functionally important heterotetramers with other TRP channels in some cell types. Depending on the cell type involved, excessive activation or impaired activity of TRPV4 channels can contribute to the pathogenesis of multiple disorders.

The presence of functional TRPV4 channels in multiple organ systems indicates their importance from a physiological point of view. Among the TRP family members, TRPV4 channels have been associated with the most naturally occurring mutations that result in human diseases. Most studies investigate the role of TRPV4 channels in a specific organ system. TRPV4 channels in neighboring cells can counterbalance each other's effects—for example, endothelial TRPV4 channels dilate the arteries, whereas smooth muscle TRPV4 channels constrict them. Similarly, it is conceivable that TRPV4 channels from different organ systems interact through paracrine signaling. This gap in our knowledge was the impetus for the current volume—to review detailed information on the TRPV4-dependent mechanisms in different organ systems by experts on TRPV4 channels in these organ systems. Understanding the behavior of TRPV4 channels in various organ systems may provide a big picture on their roles and therapeutic targeting. I hope this collection of articles will enhance our understanding of the role of TRPV4 channels at the whole-animal level and inform future studies.

Chapter 1 discusses the role of TRPV4 channels in vasculature and their involvement in vascular disorders. TRPV4 channels have been widely studied in endothelial cells, smooth muscle cells, and fibroblasts in the vascular wall. Chapter 2 addresses a well-studied role for endothelial TRPV4 channels in lung edema and injury. Recent findings on the physiological and pathological role of TRPV4 channels in cardiac myocytes are reviewed in Chapter 3. An important role played by TRPV4 channels in flow-sensitivity of the renal nephron is the focus of Chapter 4. Chapter 5 discusses the essential roles of TRPV4 channels in regulating the structure and function of the urinary bladder. Chapter 6 focuses on the role of epidermal TRPV4 channels in skin homeostasis. Chapter 7 addresses the expression, activity, and functional roles of TRPV4 channels in immune cells. TRPV4 channels in the eye are the topic of Chapter 8. Chapter 9 covers the physiological and pathological roles of TRPV4 channels in the musculoskeletal system. The chapter also discusses human disorders caused by naturally occurring TRPV4 mutations.

Although TRPV4 channels have been a topic of intense research over the past decade, our understanding of channel function, its physiological roles, signaling partners, interacting proteins, and involvement in disease pathogenesis continues to evolve. In many organ systems, a pathological role for TRPV4 channels is well established, but their physiological role remains unclear. It is expected that future studies will address these gaps in our knowledge. There is sufficient evidence in the literature linking abnormal TRPV4 channel activity to the pathogenesis of multiple disease conditions. Altering the activity of TRPV4 channels in such disease conditions may be therapeutically beneficial. However, with widespread expression in numerous cell types, a therapeutic strategy involving global targeting of TRPV4 may be associated with side effects. Investigations addressing the therapeutic targeting of TRPV4 channels and potential side effects are the need of the hour. One potential strategy is to identify the interacting proteins or signaling partners to modify the activity or effects of TRPV4 channels and target these proteins for a more specific therapeutic effect in disease conditions. Tissue-specific targeting of TRPV4 channels could be another promising therapeutic strategy. More studies of TRPV4-dependent mechanisms using state-of-the-art techniques are needed for identifying novel TRPV4 channel-dependent mechanisms and therapeutic targets.

Acknowledgments

I thank the authors of each chapter for their contribution to this volume and for their assistance with the review process. Special thanks goes to Dr. Irena Levitan, University of Illinois, Chicago, for coming up with the idea for this topic and for helping me at multiple points during the completion of this volume on TRPV4 channels.

Swapnil K. Sonkusare
Robert M. Berne Cardiovascular Research Center,
University of Virginia, Charlottesville, VA, United States
Department of Molecular Physiology and Biological Physics,
University of Virginia, Charlottesville, VA, United States

CHAPTER ONE

TRPV4-dependent signaling mechanisms in systemic and pulmonary vasculature

Zdravka Daneva* and Maniselvan Kuppusamy
Robert M. Berne Cardiovascular Research Center, University of Virginia, Charlottesville, VA, United States
*Corresponding author: e-mail address: zd9gj@virginia.edu

Contents

Abstract

The delicate balance between constrictor and dilator mechanisms is a vital determinant of blood pressure and blood flow. The maintenance of this balance requires constant communication between different cell-types in the vascular wall. In this regard, the transient receptor potential vanilloid type 4 (TRPV4) ion channel, a Ca^{2+}-permeable non-selective cation channel, has emerged as a crucial regulator of Ca^{2+}-mediated changes in vascular reactivity. Recent studies suggest that TRPV4 channels regulate vasoconstriction and arterial pressure in the systemic and pulmonary vasculature. New emerging data support a dilatory role of endothelial TRPV4 channels, and both constrictor and dilator roles of smooth muscle TRPV4 channels. Moreover, TRPV4 channel activity has been implicated in physiological functions of vascular support cells, such as fibroblasts and pericytes, to assist the sustenance of vascular reactivity in response to changes in intravascular pressure or external stimulation. Importantly, a growing body of evidence connects abnormal TRPV4 channel activity to multiple

Current Topics in Membranes, Volume 89
ISSN 1063-5823
https://doi.org/10.1016/bs.ctm.2022.07.003

vascular disorders. This chapter will review the current literature on the cell-type specific roles of vascular TRPV4 channels in regulating physiological function. Additionally, we summarize our understanding of the contribution of abnormal TRPV4 channel activity to various vascular disorders.

1. Introduction

The mammalian vasculature is a complex system. Circulating blood carries diverse signaling molecules to vascular beds throughout the body. The vasculature is organized into networks of arteries, veins, and interconnecting capillaries. Moreover, blood flow exerts mechanical sheer stress on vascular walls, which is met by an intrinsic resistance to flow to maintain intravascular pressure (White & Frangos, 2007). Such neurotransmitters and mechanical forces are detected by unique cellular mechanisms within vascular endothelial and smooth muscle cells and lead to distinct vascular reactivity, based on the vascular bed they occur in. In this regard, blood pressure regulation occurs at the level of resistance-sized arteries. These are small arteries (~50–100 μm in diameter (Daneva et al., 2021a; Ottolini et al., 2020a)) that regulate peripheral vascular resistance in response to changing intravascular pressure. Importantly, resistance-sized arteries regulate blood pressure via the opposing actions of smooth muscle cells (SMCs) and endothelial cells (ECs). SMCs from resistance-sized arteries contract in response to increases in intravascular pressure and ECs limit SMC contraction. Vascular reactivity can also be influenced by the cellular crosstalk between neurons and blood vessels. Disruption in intercellular communication can lead to regional impairment of cerebral blood flow, which, in turn, leads to serious neuronal damage (Kisler et al., 2017). Additionally, healthy vascular development and vascular reactivity can be affected by fibroblasts, which remain largely under-investigated in vivo due to their heterogeneity and plasticity (Di Carlo & Peduto, 2018). Thus, vascular reactivity to changes in intravascular pressure and blood flow depends on numerous signaling events within multiple cell-types in the vascular wall. Numerous ion channels, which allow the passage of different ions along their electrochemical gradients, play critical roles in regulating physiological functions of ECs, SMCs, and fibroblasts. In this regard, the transient receptor potential (TRP) channel family has recently emerged as an important regulator of vascular reactivity.

In mammals, there are 28 TRP channel homologs, distributed in different tissues. Based on amino acid homology, there are six subfamilies of TRP channels: canonical (TRPC), vanilloid (TRPV), melastatin (TRPM), ankyrin (TRPA), mucolipin (TRPML), and polycystin (TRPM) (Earley & Brayden, 2015). TRP channels are present virtually in every organ and play vital roles in both excitable and non-excitable cells to regulate sensory and signal transduction processes. A unifying feature that all TRP channels have is their significant contribution to the regulation and distribution of intracellular Ca^{2+} concentration ($[Ca^{2+}]_i$). Several TRP channels have been reported to regulate EC and SMC $[Ca^{2+}]_i$. Importantly, studies from the past two decades have identified the transient receptor potential vanilloid type 4 (TRPV4) ion channel as a key player in regulating vascular reactivity through its functions in endothelial and SMCs (Ottolini, Hong, & Sonkusare, 2019; Ottolini & Sonkusare, 2021).

In the human vasculature, TRPV4 channels are expressed, and are functional, in several major cell types regulating vascular reactivity: ECs, SMCs fibroblasts and pericytes. In these cells, TRPV4 channels have largely been thought to promote vasodilation, and therefore, reduce vascular resistance (Earley et al., 2005; Sonkusare et al., 2012, 2014; Tajada et al., 2017). Until the recent development of endothelial- and smooth muscle cell-specific TRPV4 ($TRPV4_{EC}$ and $TRPV4_{SMC}$, respectively) channel knockout mice, it has been challenging to determine with accuracy the physiological roles of TRPV4 channels in regulating vasodilation and blood pressure (Kwangseok Hong, Marziano, & Sonkusare, 2016; Zhang et al., 2009). Regarding vascular reactivity, TRPV4 channels have also been found in perivascular cells of the central nervous system, such as astrocytes and neurons (Benfenati et al., 2007; Dunn et al., 2013; Filosa et al., 2006), where they cause vasodilation of small cerebral arteries. Additionally, an emerging role of TRPV4 channel signaling in fibroblasts has attracted great interest in elucidating the channel's effects in different types of fibrosis (Zhan & Li, 2018). In this chapter, we will review the functional roles of TRPV4 channels in different cell types in the vasculature under normal and disease conditions.

2. TRPV4 channel structure

TRPV4 channels are Ca^{2+}-permeable non-selective cation channels at the plasma membrane (Ferrandiz-Huertas et al., 2014). They have been reported to have a permeability ratio (P_{Ca}/P_{Na}) of ~6, and single-channel conductance of ~90 pS (Ferrandiz-Huertas et al., 2014). TRPV4 channels consist of 871 amino acids with intracellular N- and C-terminals. TRPV4 channels are characterized by six transmembrane domains (TM1–6) (Fig. 1).

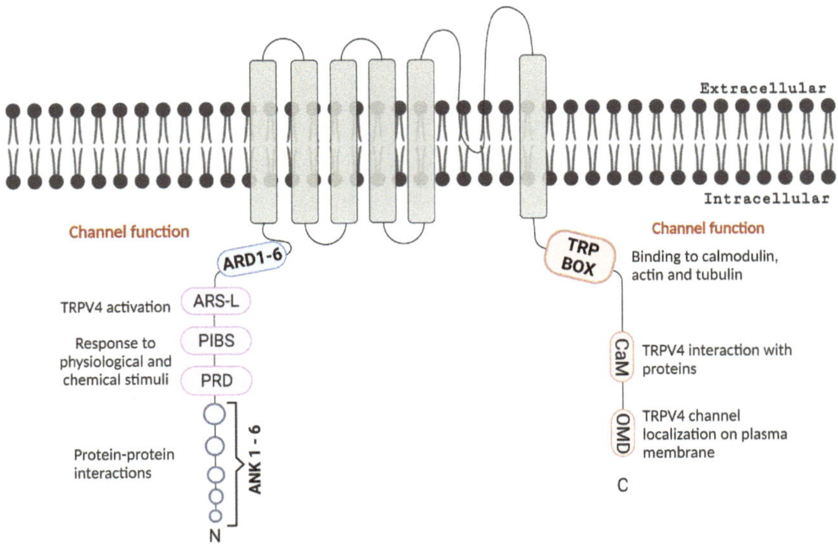

Fig. 1 Schematic overview of the TRPV4 channel structure. TRPV4 channels consist of 871 amino acids (AA) with intracellular N- and C-terminals, organized into six transmembrane domains (TM1–6). The cation conductivity pore is located between TM5 and TM6. The N-terminal contains: an arachidonate-like recognition sequence (ARS-L), a very short phosphoinositide-binding site (PIBS), a proline-rich domain (PRD), and six ankyrin repeats (ANK1–6). In the C-terminal tail, TRPV4 channels have a TRP box, a calmodulin binding domain (CaM), an oligomerization domain (OMD).

Their cation conducting pore, especially sensitive to Ca^{2+}, is formed between TM5 and TM6 (Tomilin et al., 2016). The N-terminal of TRPV4 channels has been widely studied and proven to be crucial for activation and protein–protein communication. Nilius et al. (Nilius, Watanabe, & Vriens, 2003) demonstrated that an arachidonate-like recognition sequence (ARS-L) is involved in the activation of TRPV4 channels (Fig. 1). Similarly, the presence of a short phosphoinositide-binding site (PIBS) and a proline-rich domain (PRD) are essential for channel activation by physiological and chemical stimuli (D'Hoedt et al., 2008; Garcia-Elias et al., 2013). Notably, ankyrin repeats (ANK1–6) on the TRPV4 N-terminal tail, responsible for protein–protein interactions, are directly involved in the pathogenesis of multiple diseases (Lamande et al., 2011; Li, Mahajan, & Tsai, 2006).

The C-terminal of TRPV4 channels consists of a different set of functionally important domains (Fig. 1). A calmodulin (CaM)-binding domain allows TRPV4 channels to interact with other proteins, the most common of which is inositol 1,4,5-triphosphate receptor (IP_3R) (Garcia-Elias et al., 2008). Moreover, an oligomerization domain (OMD) on the C-terminal facilitates the localization of TRPV4 channels at the plasma membrane (Nilius & Flockerzi, 2014). In recent years, Nilius and Flockerzi (Nilius & Flockerzi, 2014) discovered the presence of a TRP box on the C-terminal tail of TRPV4 channels for calmodulin, actin and tubulin binding (Fig. 1). Since then, single-particle cryo-electron microscopy (cryo-EM) studies on *Xenopus tropicalis* TRPV4 channels have revealed that inner helices form an intracellular gate in the ion-conduction pore. However, TRPV4 channels are unique in that they do not possess an extracellular gate in their selectivity filter, possibly explaining the less selective nature of the channel compared to other TRPV channels (Deng et al., 2018). The distinctive structural features of TRPV4 channels at least partially explain the differences in gating mechanisms and selectivity when compared to other ion channels in the TRP superfamily.

3. Ca^{2+} influx signals through TRPV4 channels and channel regulation by Ca^{2+}

Ca^{2+} signals regulate a plethora of biological processes in mammals, including muscle contraction, hormone secretion, and cell proliferation (Berridge, Bootman, & Roderick, 2003). The intracellular free concentration of Ca^{2+} is usually maintained at a low level (~100–300 nM) by membrane-bound and cytosolic proteins, as well as Ca^{2+}-pumping ATPases on the plasma membrane and other organelles. The intracellular Ca^{2+} concentration can be changed by influx of Ca^{2+} from extracellular spaces or release of Ca^{2+} from intracellular storage areas. Cells possess several different organelles that accommodate intracellular Ca^{2+} storage, but the most prominent is the endoplasmic/sarcoplasmic reticulum (ER/SR). Ca^{2+} signals additionally can be divided into two groups based on spatial feature: global, where Ca^{2+} changes occur throughout the whole cell, and local, where changes in Ca^{2+} are localized to certain regions of the cells.

TRPV4 channels play a crucial role in Ca^{2+} signaling mechanisms. While TRPV4 channels can conduct cations other than Ca^{2+}, functional effects of TRPV4 channels have mostly been attributed to Ca^{2+} influx through the channel (Bagher et al., 2012; Marziano et al., 2017; Mendoza et al., 2010; Sonkusare et al., 2012; Wu et al., 2009; Yin et al., 2008). The unitary conductance ranges of TRPV4 channels are 50–60 pS at −60 mV, and 90–100 pS at +60 mV (Strotmann et al., 2000; Watanabe et al., 2002a, 2003). TRPV4 channels display higher permeability for Ca^{2+} over Na^+ ($P_{Ca2+}/P_{Na+} = 6–10$) (Clapham et al., 2003), and can be activated by temperature, mechanical stimuli, and neurohumoral mediators. Elementary Ca^{2+} signals representing Ca^{2+} influx through single TRPV4 channels, called TRPV4 sparklets, have been recorded from airway (Zhao et al., 2014) and vascular SMCs (Mercado et al., 2014), isolated ECs (Sullivan et al., 2012), and intact endothelium (Chen et al., 2022; Sonkusare et al., 2012). Moreover, sophisticated technology now allows us to record Ca^{2+} influx through individual TRPV4 channels using total internal reflection fluorescence imaging (optical patch clamp) (Mercado et al., 2014; Sullivan et al., 2012). Thus, TRPV4 channel signaling is not only studied in the traditional whole-cell patch clamp models of isolated or cultured cells, but also in the intact tissue under physiological conditions, where cell–cell contacts and intercellular communications are maintained.

Cytosolic Ca^{2+} has been demonstrated to have a biphasic effect on TRPV4 channel activity. While low concentrations of intracellular Ca^{2+} facilitate TRPV4 channel opening, high concentrations limit the channel activity. Ca^{2+}-dependent activation/inactivation of TRPV4 channels occurs via Ca^{2+}-CaM binding on the C-terminal tail (CCaMB) of the channel (Plant & Strotmann, 2007; Strotmann, Schultz, & Plant, 2003; Watanabe et al., 2002a). Half-maximal CaM-binding affinity at CCaMB was observed at nanomolar concentrations of Ca^{2+} (150 nM). Experiments with mutations in CCaMB resulted in impaired Ca^{2+}-dependent potentiation of TRPV4 channel activity (Watanabe et al., 2002b). Importantly, Ca^{2+}-dependent inhibition of the TRPV4 channel at a higher concentration of Ca^{2+} (IC_{50} of 406 nM) was maintained in the mutants (Strotmann et al., 2003; Watanabe et al., 2002b). These findings suggested that Ca^{2+}-dependent inhibition of TRPV4 channel does not rely on CCaM binding to CCaMB. Strotmann and colleagues (Strotmann et al., 2010) proposed a mechanism whereby an interaction between the N- and C-terminal tails prevents channel activation. An increase in cytosolic Ca^{2+} enabled CaM binding to the CCaMB, which resulted in the displacement of N-terminal from the C-terminal tail and TRPV4 channel activation. Recently, a Ca^{2+}-CaM binding site was identified on the N-terminal tail (aa 132–383, NCaMB) (Phelps et al., 2010). Phelps and colleagues (Phelps et al., 2010) demonstrated that ATP interaction with N-CaMB increases TRPV4 channel activity in patch-clamp studies. It could be speculated that at higher cytosolic Ca^{2+} concentrations, Ca^{2+}-CaM competes with ATP for binding to N-CaMB, which leads to TRPV4 channel inhibition. Unfortunately, studies on Ca^{2+}-induced inactivation of the TRPV4 channel are scarce.

3.1 TRPV4 channel activity in arterial endothelial cells

In resistance-sized arteries, ECs send out projections to SMCs, called myoendothelial projections (MEPs). MEPs establish an electrical connection between the two cell types via myoendothelial gap junctions (MEGJs). Increase in EC Ca^{2+} activates several targets: (1) endothelial nitric oxide synthase (eNOS), (2) intermediate and small conductance Ca^{2+}-activated K^+ channels (IK and SK channels, respectively) and (3) release of endothelium-derived hyperpolarizing factors (EDHFs) (Fig. 2). Nitric oxide (NO) diffuses to the SMC layer, stimulates cyclic guanosine monophosphate (cGMP)-protein kinase G (PKG) signaling in SMCs, which activates myosin light chain phosphatase (MLCP). This signaling cascade results in a decrease in contractile force and causes vasodilation (Davis, Hill, & Kuo, 2008). Cytosolic Ca^{2+} activates IK and SK channels, thereby causing EC membrane hyperpolarization that is transmitted to SMCs via MEGJs (Fig. 2). This transmission triggers inactivation of SMC voltage-dependent Ca^{2+} channels (VDCCs) and induces vasodilation. Additionally, increases in intracellular Ca^{2+} concentrations lead to the release of several EDHFs, including but not limited to epoxyeicosatrienoic acids (EETs) and hydrogen peroxide (H_2O_2) (Ellinsworth et al., 2016). Such EDHFs hyperpolarize SMCs and cause vasodilation. Pressure myography and patch-clamp studies have shown that the $TRPV4_{EC}$—IK/SK channel linkage underlies endothelium-dependent vasodilation in systemic resistance-sized arteries (Bagher et al., 2012; Ottolini et al., 2020b; Sonkusare et al., 2012) (Fig. 2). However, in resistance-sized pulmonary arteries (PAs), the $TRPV4_{EC}$ channel—eNOS signaling axis appears to be the major dilatory pathway (Marziano et al., 2017; Ottolini et al., 2020b) (Fig. 3). Thus, $TRPV4_{EC}$ channels couple with distinct downstream targets in different vascular beds. The selective $TRPV4_{EC}$—IK/SK channel coupling in resistance-sized systemic arteries is to a large extent due to the localization of $TRPV4_{EC}$ channels and IK/SK channels at MEPs (Fig. 2) (Sonkusare et al., 2014). Recently, Ottolini et al. (Ottolini et al., 2020b) elucidated the mechanisms underlying the lack of $TRPV4_{EC}$—eNOS coupling in systemic arteries and the preferential $TRPV4_{EC}$—eNOS coupling in PAs (Figs. 2 and 3). Reportedly, the presence of the NO-scavenging protein, hemoglobin a Hbα, limits $TRPV4_{EC}$—eNOS signaling in mesenteric arteries (MAs), while spatial proximity of $TRPV4_{EC}$ channels with eNOS and the absence of Hbα favor $TRPV4_{EC}$—eNOS signaling in PAs (Figs. 2 and 3). However, the mechanisms underlying the lack of $TRPV4_{EC}$—IK/SK channel coupling in pulmonary arteries remain elusive.

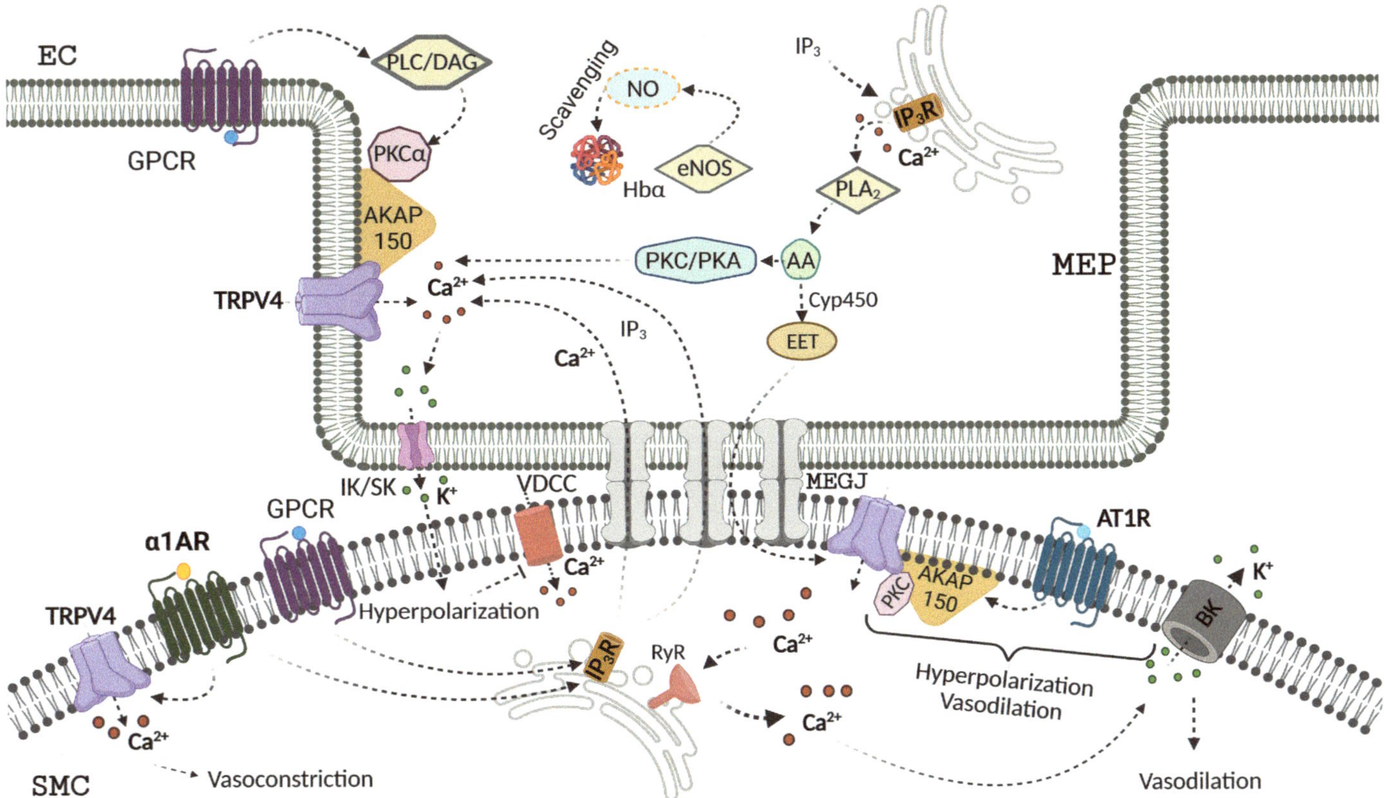

Fig. 2 See figure legend on opposite page.

3.1.1 Regulation of $TRPV4_{EC}$ channel activity by the formation of hetero-multimers with other channels

$TRPV4_{EC}$ channel activity can be altered by other proteins. Recent studies have provided compelling evidence that other ion channels, scaffolding and cytosolic proteins interact with TRPV4 channels to alter their activity. For example, Du et al. reported that in primary rat MA $TRPV4_{EC}$ channels form a heterotrimeric TRPV4–TRPC1–TRPP2 channel complex (Du et al., 2014). The authors further showed that the presence of this signaling complex is important for flow-induced influx of Ca^{2+} (Du et al., 2014). In congruence with this report, Ma and colleagues showed that in human umbilical vein $TRPV4_{EC}$ dimerizes with TRPC1 channels to promote flow-induced entry of Ca^{2+} (Ma et al., 2010). More specifically, flow-induced Ca^{2+} influx

Fig. 2 Schematic depiction of TRPV4 channel activity in endothelial and smooth muscle cells (ECs and SMCs, respectively) of resistance-sized systemic arteries. In resistance-sized systemic arteries, ECs send out projections to SMCs, called myoendothelial projections (MEPs). Within MEPs, activation of the endothelial TRPV4 channels ($TRPV4_{EC}$) elevates cytosolic Ca^{2+}, which activates several targets: (1) endothelial nitric oxide synthase (eNOS), (2) intermediate and small conductance Ca^{2+}-activated K^+ channels (IK and SK, respectively) and (3) release of endothelium-derived hyperpolarizing factors (EDHFs), such as epoxyeicosatrienoic acids (EETs). The presence of the NO-scavenging protein, hemoglobin α (Hbα), limits $TRPV4_{EC}$—eNOS signaling in systemic arteries. Endothelial inositol triphosphate (IP_3) and increased cytosolic Ca^{2+} concentration activate inositol triphosphate receptors (IP_3Rs) on the endoplasmic reticulum (ER) in ECs, which promotes arachidonic acid (AA)-mediated activation of phosphokinase C/phosphokinase A (PKC/PKA) and formation of epoxyeicosatrienoic acids (EETs). PKC and PKA stimulate further Ca^{2+} influx through $TRPV4_{EC}$ channels and promote vasodilation. EETs diffuse to SMCs, where they hyperpolarize the plasma membrane and cause vasodilation. Certain G protein-coupled receptors, such as purinergic P2Y2 receptors (P2Y2Rs), activate the phospholipase C—diacylglycerol (PLC—DAG) signaling mechanism to enable PKCα anchoring to scaffolding protein, AKAP150, in proximity to $TRPV4_{EC}$ channels and activate these channels. Elevation of IP_3 and Ca^{2+} via diffusion from SMCs to ECs at MEPs limits vasoconstriction by activating $TRPV4_{EC}$–IK/SK channel and IP_3R–IK/SK channel signaling. Stimulation of G_q protein-coupled receptors (G_qPCRs) on SMC membranes leads to the formation of IP_3 and DAG. DAG activates PKC, which phosphorylates voltage-gated Ca^{2+} (VDCC) channels, leading to an increase in SMC Ca^{2+} and vasoconstriction. SMC TRPV4 ($TRPV4_{SMC}$) channel activation by angiotensin type 1 receptors (AT1Rs) via PKC—AKAP150 anchoring causes large conductance Ca^{2+}-activated K^+ channel (BK) activation, SMC membrane hyperpolarization and vasodilation. Similarly, $TRPV4_{SMC}$ activation promotes Ca^{2+} influx, which stimulates ryanodine receptor (RyR)-mediated Ca^{2+} release from the SMC sarcoplasmic reticulum. This Ca^{2+}-induced Ca^{2+} release stimulates BK channel activity and induces vasodilation. $TRPV4_{SMC}$ channels also can be directly activated by α1 adrenergic receptors (α1ARs), which results in vasoconstriction of the smooth muscle layer.

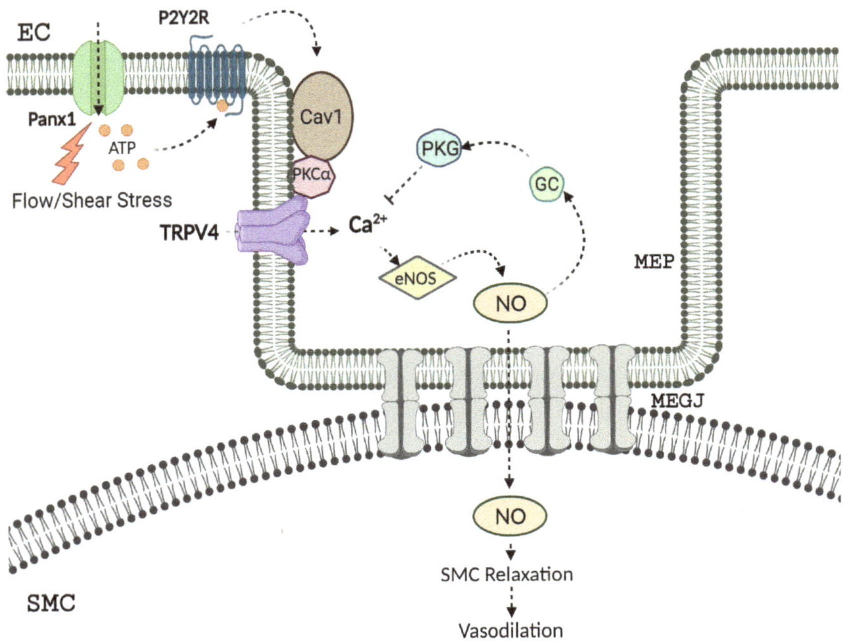

Fig. 3 Schematic depiction of TRPV4 channel activity in endothelial and smooth muscle cells (ECs and SMCs, respectively) of resistance-sized pulmonary arteries (PAs). In small PAs, elevated flow/shear stress activates the endothelial Pannexin 1 (Panx1) channels, which triggers a release of extracellular adenosine triphosphate (eATP) in the extracellular space. eATP binds to the P2Y2 purinergic receptors (P2Y2Rs) and activates endothelial TRPV4 ($TRPV4_{EC}$) channels via PKCα. Endothelial caveolin-1 ($Cav\text{-}1_{EC}$) provides a signaling scaffold that ensures spatial proximity among the elements of the $Panx1_{EC}$–$P2Y2R_{EC}$–PKCα–$TRPV4_{EC}$ pathway. Activated $TRPV4_{EC}$ channels promote endothelial nitric oxide synthase (eNOS) activity and nitric oxide (NO) release. Endothelium-derived NO then causes vasodilation through the activation of SMC guanylyl cyclase (GC)-protein kinase G (PKG) signaling and GC-PKG-independent mechanisms. Blue arrows indicate this pathway. NO also induces activation of endothelial GC-PKG signaling, which lowers Ca^{2+}-dependent cooperative opening of TRPV4 channels and limits TRPV4-mediated vasodilation.

and 4α-phorbol 12, 13-didecanoate (4α-PDD, selective TRPV4 agonist)-activated whole-cell current were enhanced in HEK cells co-expressing TRPV4-TRPC1 channels (Ma et al., 2010). Thus, the formation of hetero-multimers presents a novel target for studying the function of TRPV4 channels in ECs, and potentially other cell types.

3.1.2 $TRPV4_{EC}$ channel regulation by scaffolding/anchoring proteins

$TRPV4_{EC}$ channels have been found to interact with several anchoring proteins in systemic and pulmonary arteries. In systemic arteries, TRPV4 channels interact with AKAP150 (A-kinase anchoring protein 150), which anchors protein kinase C (PKC) and protein kinase A (PKA), respectively; Fig. 2) (Sonkusare et al., 2014). Protein kinases are crucial endogenous regulators of TRPV4 channel activity. In expression systems, PKC amplified TRPV4 channel activity by phosphorylating Ser162, Thr175, and Ser189 on the N-terminal tail; whereas, PKA increased TRPV4 channel activity by phosphorylating Ser824 on the C-terminal tail (Cao et al., 2018; Fan, Zhang, & McNaughton, 2009). Moreover, AKAP150, expressed only in systemic arteries, stimulates TRPV4 phosphorylation by PKC and PKA in expression systems (Cao et al., 2018; Fan et al., 2009). In this regard, co-localization studies demonstrated that endothelial AKAP150 ($AKAP150_{EC}$) stands in close proximity to $TRPV4_{EC}$ channels at MEPs, where it anchors PKC to enhance the activity of TRPV4 channels (Sonkusare et al., 2014). Further support for this finding was provided when the classical muscarinic agonist, acetylcholine, was used to induce $AKAP150_{EC}$—PKC—$TRPV4_{EC}$ channel-mediated dilation of systemic arteries. Sonkusare et al. showed that $TRPV4_{EC}$ channels at MEPs open in a coupled manner, indicating that simultaneous openings of multiple $TRPV4_{EC}$ channels can occur in a cluster. Coupled openings of $TRPV4_{EC}$ channels amplify the influx of Ca^{2+} through a small number of channels, enabling a small portion of $TRPV4_{EC}$ channels to exert a large functional effect. The channel coupling is lost in systemic arteries from global $AKAP150^{-/-}$ mouse models, implying that AKAP150 is crucial for $TRPV4_{EC}$ channel coupling and activity (Sonkusare et al., 2014). Sonkusare et al. also showed that the coupled openings of $TRPV4_{EC}$ channels are not dependent on PKC, suggesting that $AKAP150_{EC}$ can enhance TRPV4 channel activity in a PKC-dependent or -independent manner. More recently, Ottolini et al. demonstrated that inducible $TRPV4_{EC}$ channel or endothelial $AKAP150_{EC}$ knockout mice show higher resting blood pressures (Ottolini et al., 2020a). Impairment of $AKAP150_{EC}$—$TRPV4_{EC}$ channel signaling was reported to contribute to the loss of endothelial function and elevated blood pressure in a mouse model of obesity and obese patients. Ottolini and colleagues further showed that lowering the levels of the oxidant molecule peroxynitrite (PN), produced by upregulated endothelial nicotinamide adenine dinucleotide phosphate (NADPH) oxidase 1 (NOX1) and inducible nitric oxide synthase (iNOS), rescued $AKAP150_{EC}$—$TRPV4_{EC}$ channel signaling, vasodilation, and blood pressure in obesity (Ottolini et al., 2020a). PN causes cysteine oxidation of $AKAP150_{EC}$, resulting in the impairment of $AKAP150_{EC}$—$TRPV4_{EC}$ channel signaling in obesity. It remains to be elucidated whether the

PKC-independent $TRPV4_{EC}$ channel activation by $AKAP150_{EC}$ occurs via a PKA anchoring and/or channel regulation by direct binding to $TRPV4_{EC}$ channels.

Caveolin-1 (Cav-1) is an essential structural protein of the caveolae (membrane invaginations), which serves to localize key signaling elements (Minshall et al., 2003). In resistance-sized PAs, Cav-1 is an important scaffolding protein that interacts with eNOS (Bernatchez et al., 2005), acting by stabilizing eNOS expression and negatively regulating its activity. Earlier studies used pulmonary EC monolayer to show that $TRPV4_{EC}$ channels co-immunoprecipitate with Cav-1 (Saliez et al., 2008). Additionally, rat uterine arteries revealed a higher contribution of $TRPV4_{EC}$ channels in controlling vasodilation during pregnancy, an effect that was attributed to a higher expression of TRPV4 channels and their presence in caveolae (Senadheera et al., 2013). However, due to the lack of cell-specific animal models, the exact relationship between endothelial Cav-1 (Cav-1_{EC}) and TRPV4 channels was difficult to identify. Recently generated inducible $TRPV4_{EC}$ and Cav-1_{EC} knockout mice have proven invaluable to the investigations of Cav-1_{EC}—$TRPV4_{EC}$ signaling in relation to vasodilation and regulation of pulmonary arterial pressure (PAP) (Daneva et al., 2021a, 2021b). We showed that arterial Cav-1_{EC}–$TRPV4_{EC}$ channel signaling promotes pulmonary vasodilation and maintains low PAP (Daneva et al., 2021b) (Fig. 3). Furthermore, PKCα anchoring by Cav-1_{EC} enables PKCα activation of $TRPV4_{EC}$ channels in PAs (Daneva et al., 2021a, 2021b). These discoveries raise the possibility that disruption of PKCα anchoring by Cav-1_{EC} could impair the function of TRPV4 channels in different disease conditions.

3.1.3 TRPV4 channel regulation by NO and protein kinase G

Protein kinase G (PKG), a cGMP-dependent protein kinase, has also been demonstrated to regulate $TRPV4_{EC}$ channel activity. PKG inhibited flow-induced, TRPV4—TRPC1 channel-dependent Ca^{2+} influx in ECs, and flow-induced vasodilation in systemic arteries (Ma et al., 2010). Recently, Marziano and colleagues demonstrated that TRPV4 channel-dependent release of NO leads to vasodilation of small PAs. Simultaneously, NO also activates endothelial guanylyl cyclase (GC)-PKG-dependent negative-feedback loop that inhibits the coupled openings of $TRPV4_{EC}$ channels (Daneva et al., 2021a) (Fig. 3), thereby limiting Ca^{2+} influx through the channel. Whether PKG-modulation of TRPV4 channel activity requires a scaffolding protein has not been explored yet.

3.1.4 Role of $TRPV4_{EC}$ channels in flow/shear stress-induced vasodilation

Studies using genetic and pharmacological tools provide evidence that TRPV4 channels play an essential role in shear stress-induced activation of endothelial Ca^{2+} signaling in systemic and pulmonary vasculature. Earlier studies reported that the non-selective TRPV4 channel inhibitor ruthenium red inhibits shear stress-induced dilation of rat carotid artery and arteria gracilis (Kohler et al., 2006). Also, shear stress- and flow/reperfusion-induced vasodilation was significantly attenuated in carotid arteries from global $TRPV4^{-/-}$ mice (Hartmannsgruber et al., 2007). Furthermore, elevated luminal flow resulted in dilation of carotid artery via EET-endothelium derived hyperpolarization (EDH) pathway, a response that was blunted in the arteries from global $TRPV4^{-/-}$ mice. Mendoza and colleagues (Mendoza et al., 2010) reported that TRPV4 is critically important for flow-induced dilation of mouse MAs and showed that shear stress stimulates Ca^{2+} influx through TRPV4 channels in human coronary artery ECs. In mouse MAs, shear stress elicited a rapid Ca^{2+} influx through $TRPV4_{EC}$ channels and caused vasodilation via eNOS and EDH-dependent mechanisms (Mendoza et al., 2010). More recently, Darby et al. provided evidence that shear stress enhanced the sensitivity of $TRPV4_{EC}$ channel agonist and coupled $TRPV4_{EC}$ activation to muscarinic receptor-mediated, EC-dependent dilation of rat cremaster arteries (Darby et al., 2018).

The interaction between shear stress and TRPV4 channels has predominantly been studied in systemic arteries, partly due to the technical challenges involved in isolating and testing small, resistance-sized PAs. PAs are a high-flow, low-resistance circulation. Daneva et al. recently reported that in small PAs, shear stress activates the efflux of adenosine triphosphate (ATP) through endothelial Pannexin 1 ($Panx1_{EC}$) (Fig. 3). $Panx1_{EC}$—effluxed ATP, in turn, stimulates endothelial purinergic P2Y2 receptor ($P2Y2R_{EC}$) to increase $TRPV4_{EC}$ channel activity (Daneva et al., 2021b). These data supported the idea that flow-$Panx1_{EC}$ and $P2Y2R_{EC}$–$TRPV4_{EC}$ signaling may be involved in flow-induced dilation of PAs (Daneva et al., 2022) (Fig. 3). Future investigations assessing the role of $Panx1_{EC}$–$P2Y2R_{EC}$–$TRPV4_{EC}$ signaling axis in flow-induced dilation of small PAs are awaited.

3.1.5 Regulation of TRPV4 channel activity by G protein-coupled receptor signaling

The activity of $TRPV4_{EC}$ channels is also regulated by G-protein-coupled receptors. Predominantly, G_q protein-coupled receptor (G_qPCR) signaling has been associated with $TRPV4_{EC}$ channel activation via two mechanisms: phospholipase C (PLC)—diacylglycerol (DAG)—PKC-dependent activation of TRPV4 channels in systemic and pulmonary arteries (Daneva et al., 2021b; Sonkusare et al., 2014) (Figs. 2 and 3), and depletion of a negative regulator of TRPV4 channel activity, phosphatidylinositol 4,5-bisphosphate (PIP_2), in the capillary endothelium (Harraz et al., 2018). Bradykinin, an inflammatory mediator that acts on G_q-coupled bradykinin receptors, was shown to stimulate the activity of TRPV4 channels at MEPs in resistance-sized systemic arteries (Sonkusare et al., 2014). Similarly, Sonkusare et al. showed that the G_q-coupled muscarinic receptor agonist, acetylcholine, activated TRPV4 channels at MEPs via PLC—DAG—PKC signaling (Sonkusare et al., 2012, 2014). Interestingly, Marziano et al. showed that ATP dilates resistance-sized PAs via activation of endothelial TRPV4—eNOS—NO signaling (Marziano et al., 2017) (Fig. 3). Several purinergic receptor subtypes are expressed in the pulmonary vasculature, including P2YRs and P2XRs (Hennigs et al., 2019; Konduri et al., 2004; Syed et al., 2010). Recent studies in inducible EC-specific knockout mouse models showed that ATP released from $Panx1_{EC}$ selectively activates $P2Y2R_{EC}$ on the EC membrane in resistance-sized PAs (Daneva et al., 2021b). Stimulation of $P2Y2R_{EC}$ recruited PKCα, which anchors to the scaffolding protein $Cav\text{-}1_{EC}$ in proximity to $TRPV4_{EC}$ channels. Briefly, $Cav\text{-}1_{EC}$ provides a signaling scaffold that ensures spatial proximity among the elements of the $Panx1_{EC}$–$P2Y2R_{EC}$–PKCα–$TRPV4_{EC}$ pathway (Fig. 3). This enhances the activity of $TRPV4_{EC}$ channels, which causes vasodilation and lowers PAP (Daneva et al., 2021b). Thus, impairment of one or more of these elements could have detrimental effects in vascular diseases.

Recently, it was discovered that TRPV4 channels have an IP_3 binding domain, and that binding of IP_3 to this domain increases the activity of TRPV4 channels (Takahashi et al., 2014). To this effect, Hong et al. demonstrated that IP_3 generated in SMCs by the activation of SMC α1-adrenergic receptors (α1AR) or thromboxane A_2 (TXA_2) receptors could travel to ECs via MEGJs, and activate $TRPV4_{EC}$ channels (Hong et al., 2018) (Fig. 2). Interestingly, activation of α1ARs or TXA_2 receptors causes vasoconstriction, but the IP_3-dependent activation of $TRPV4_{EC}$ channels serves as a negative-feedback mechanism that limits vasoconstriction (Hong et al., 2018). IP_3Rs can also be activated by intracellular Ca^{2+}. In this context, Heathcote and colleagues recently demonstrated that Ca^{2+} influx through $TRPV4_{EC}$ channels activates Ca^{2+}-induced Ca^{2+} release through IP_3R ion channels

(Heathcote et al., 2019), amplifying the Ca^{2+} response that underlies a large-scale endothelial communication system (Heathcote et al., 2019). These findings suggest that TRPV4-mediated control of vascular reactivity requires Ca^{2+} release from internal stores.

3.1.6 Regulation of $TRPV4_{EC}$ channel activity by arachidonic acid metabolites

Arachidonic acid (AA) and its metabolites have been studied in systemic arteries in relation to arterial vasoreactivity. In human coronary arteries, exposure to AA activated TRPV4 channel-mediated Ca^{2+} influx and caused EC membrane hyperpolarization and direct activation of (Cao et al., 2018). Zhang et al. showed that the novel vasodilatory effect of AA requires PKA-mediated phosphorylation of TRPV4, activation of TRPV4 channels, and endothelial hyperpolarization (Zheng et al., 2013). Furthermore, phospholipase A_2 (PLA_2), via its main product, AA, was reported to increase $TRPV4_{EC}$ channel activity and endothelial Ca^{2+} levels in cerebral arteries (Marrelli et al., 2007; Watanabe et al., 2003). Vriens et al. further showed that CYP epoxygenase metabolizes AA to biologically active EETs, which increase the activity of $TRPV4_{EC}$ channels (Vriens et al., 2005) (Fig. 3). However, it appears not all EETs activate TRPV4 channels to cause vasodilation (Sonkusare et al., 2014; Thorneloe et al., 2012; Vriens et al., 2005; Watanabe et al., 2003). More investigations are needed to complete this signaling puzzle.

3.1.7 Differential coupling of $TRPV4_{EC}$ channels with their signaling targets

Accumulating number of reports support differential regulation and signaling targets of TRPV4 channels in distinct vascular beds, possibly contributing to endothelial heterogeneity. For instance, the activity of $TRPV4_{EC}$ channels is inhibited by a NO—GC—PKG-dependent negative-feedback pathway in resistance-sized PAs, but not in MAs (Marziano et al., 2017) (Fig. 3). AKAP150 increases the activity of $TRPV4_{EC}$ channels at MEPs in MAs, but is not expressed in PA endothelium (Sonkusare et al., 2014). Additionally, the physiological activator ATP activates $TRPV4_{EC}$ channels and dilates small PAs, but not MAs (Daneva et al., 2021b; Marziano et al., 2017). Further, $TRPV4_{EC}$ channels couple with distinct downstream signaling targets in different vascular beds. In small systemic arteries, TRPV4 channels exclusively couple with and activate IK/SK channels (Bagher et al., 2012; Sonkusare et al., 2012), whereas in PAs, they couple predominantly with eNOS (Marziano et al., 2017). Recently, Ottolini et al. showed that $TRPV4_{EC}$ channels co-localize with IK/SK channels in MAs, but not in PAs (Ottolini et al., 2020b), supporting the idea that differential co-localization may determine the signaling targets of TRPV4 channels in different vascular beds. Additionally, the presence of the NO-scavenging protein, hemoglobin α, limits $TRPV4_{EC}$—eNOS signaling in MAs (Fig. 2). Spatial proximity of $TRPV4_{EC}$ channels with eNOS and the absence of hemoglobin α favors $TRPV4_{EC}$—eNOS signaling in PAs (Ottolini et al., 2020b) (Fig. 3). Further research into the molecular mechanisms that underlie $TRPV4_{EC}$ channel signaling may provide exciting insights into endothelial cell heterogeneity.

3.2 TRPV4 channel activity in astrocytes and neurovascular communication

TRPV4 channels are expressed at astrocytic end-feet (Bayliss, 1902; Berrout, Jin, & O'Neil, 2012; Dunn et al., 2013), suggesting that the channels may play a role in neurovascular communication. Astrocytic end-feet are critical microdomains for coupling neuronal activity with changes in cerebral blood flow (Earley, Waldron, & Brayden, 2004). Findings by Straub et al. demonstrate optically recorded Ca^{2+} signals occurring at astrocytic end-feet (Straub et al., 2006). Neuronal activity induced a rapid increase in end feet Ca^{2+}, which preceded the dilation of adjacent arteries. The authors concluded that spatially restricted IP_3 receptor Ca^{2+} signals at the end-feet induce vasodilation. More recently, Dunn et al. provided the first evidence that TRPV4-mediated Ca^{2+} signals increase the strength of neurovascular coupling by augmenting neuronal activation-induced end-foot Ca^{2+} and arterial dilation. Additionally, amplification of TRPV4 Ca^{2+} signals by Ca^{2+}-induced Ca^{2+} release via IP_3 receptors further enhances neurovascular coupling (Dunn et al., 2013). Interestingly, Kim et al. demonstrated that astrocytic TRPV4 channels mediate pressure-induced constriction of parenchymal arteries, which allows modulation of vascular reactivity to match the neuronal activity (Kim et al., 2015). Thus, TRPV4 channel activity appears to be central to the physiological function of multiple cell types involved in neurovascular communication.

3.3 TRPV4 channel activity in capillary endothelium

Capillaries are the primary location for gas (O_2 and CO_2) and nutrient exchange (glucose, amino acids, etc.). Capillaries are characterized by the absence of SMCs, and therefore, MEPs. Thus, the signaling microdomains that enable TRPV4 channel activity in arterial endothelium could be absent from the capillary endothelium. However, studies show that $TRPV4_{EC}$ channels play an important role in regulating capillary barrier integrity.

3.3.1 $TRPV4_{EC}$ channel activity in cerebral capillary endothelium

Data from cerebral capillary ECs have offered a great amount of evidence for the cellular mechanisms underlying $TRPV4_{EC}$ activity in capillaries. Capillary ECs express functional TRPV4, IP_3R and inwardly rectifying potassium ($K_{IR}2.1$) channels (Dunn et al., 2013; Garcia & Longden, 2020; Longden et al., 2021). In arterial ECs, coupling of $TRPV4_{EC}$ channels with IK/SK channels or eNOS, occurring at MEPs to SMCs, results in SMC relaxation and vasodilation (Marziano et al., 2017; Ottolini & Sonkusare, 2021; Ottolini et al., 2020b; Sonkusare et al., 2012). In the capillaries, however, there are no SMCs surrounding the EC. Additionally, cerebral capillary ECs do not express functional IK/SK channels (Longden et al., 2017). These structural and mechanistic differences suggest a distinct role of $TRPV4_{EC}$ channels in the capillary endothelium, possibly via NO formation. Harraz et al. recently reported a heterogenous manner of $TRPV4_{EC}$ channel regulation in the brain capillary (Harraz et al., 2018). More precisely, cerebral capillary $TRPV4_{EC}$ channels were tonically inhibited by PIP_2, implying that in the brain capillary $TRPV4_{EC}$ channels have a low open probability (Fig. 4). However, G_qPCR activation significantly stimulates $TRPV4_{EC}$ channel activity through PLC activation and hydrolysis of PIP_2 to IP_3 and DAG, thereby, uninhibiting $TRPV4_{EC}$ channel activity (Fig. 4). Activation of $TRPV4_{EC}$ channels in the brain capillaries leads to influx of Na^+ and Ca^{2+} ions, triggering membrane depolarization. Therefore, Garcia and Longden suggested that $TRPV4_{EC}$ channel activity may be useful for EC membrane repolarization after hyperpolarizing electrical signaling, or to prevent passage of electrical signals, which would dampen blood flow through the capillaries (Garcia & Longden, 2020). Recent in vivo studies in mice expressing genetically encoded Ca^{2+} sensor, GCaMP8 in ECs (Chen et al., 2021), demonstrated that neuronal activity-induced IP_3R-dependent endothelial Ca^{2+} signals in cerebral capillaries are reduced by $TRPV4_{EC}$ inhibition (Longden et al., 2021) (Fig. 4). Briefly, this signaling mechanism contributes to the formation of NO in cerebral capillaries and leads to relaxation of contractile pericytes. Contractile pericytes typically surround the capillary ECs, and relaxation of pericytes increases

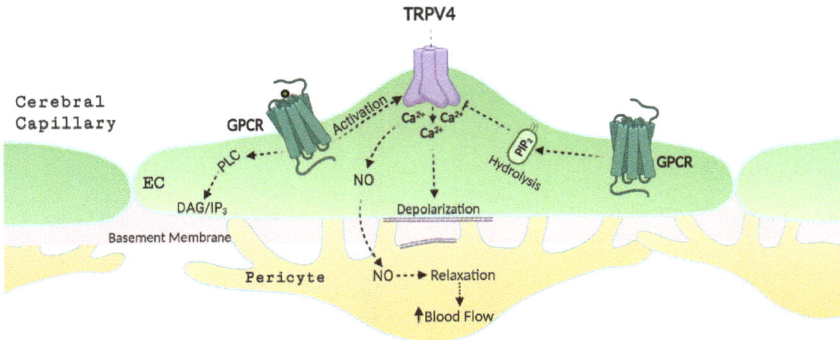

Fig. 4 Schematic depiction of TRPV4 channel activity in cerebral capillaries. Cerebral microvascular endothelial TRPV4 ($TRPV4_{EC}$) channels are tonically inhibited by phosphatidylinositol 4,5-bisphosphate (PIP_2). Endothelial G_q protein-coupled receptor (G_qPCR) stimulation leads to phospholipase C (PLC) activation, hydrolysis of PIP_2 to inositol triphosphate (IP_3) and diacylglycerol (DAG), thereby, uninhibiting $TRPV4_{EC}$ channel activity. Activation of $TRPV4_{EC}$ channels in cerebral capillaries leads to influx of Na^+ and Ca^{2+} ions, triggering membrane depolarization. $TRPV4_{EC}$ channels contribute to the formation of nitric oxide (NO) in cerebral capillaries, which leads to relaxation of contractile pericytes. Relaxation of pericytes elevates blood flow to the capillaries.

blood flow to the capillaries. The PIP_2—$TRPV4_{EC}$ interaction in brain capillaries is reviewed in depth elsewhere (Garcia & Longden, 2020; Jackson, 2022).

3.3.2 $TRPV4_{EC}$ channel activity in pulmonary capillary endothelium

Treatment of isolated lung preparations from wild-type mice with 4α-PDD, 5,6-EET, or 14,15-EET elevated microvascular permeability (Alvarez et al., 2006). This effect was attenuated by the non-selective TRPV4 channel blocker ruthenium red and was absent in the lungs from global $TRPV4^{-/-}$ mice (Alvarez et al., 2006). Furthermore, elevated intravascular pressure-induced, $TRPV4_{EC}$ channel-dependent, Ca^{2+} influx increased lung capillary permeability, but also activated a negative-feedback mechanism involving cGMP-dependent attenuation of $TRPV4_{EC}$ channel activity (Yin et al., 2008) (Fig. 5).

Mitochondrial regulation of Ca^{2+} signals is a crucial controller of EC permeability in the pulmonary circulation. In this regard, Suresh et al. demonstrated that microvascular $TRPV4_{EC}$ channels play a significant role in modulation of mitochondrial fragmentation by mitochondrial reactive oxygen species (mtROS) (Suresh et al., 2019). Mitochondria serve as a "local sink" for rising Ca^{2+}, but increasing intracellular Ca^{2+} also affects

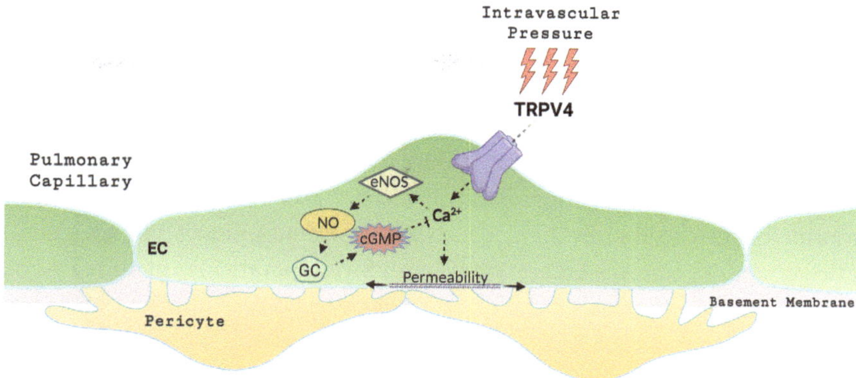

Fig. 5 Schematic depiction of TRPV4 channel activity in pulmonary arteries. Elevated intravascular pressure induces endothelial TRPV4 ($TRPV4_{EC}$) channel-dependent Ca^{2+} influx that increases vascular permeability, but concomitantly activates a negative-feedback mechanism involving NO-GC-cyclic guanosine monophosphate (cGMP)-dependent attenuation of the TRPV4 channel activity.

mitochondrial function and morphology (Jeyaraju, Cisbani, & Pellegrini, 2009; Szabadkai et al., 2006). Furthermore, the authors proposed that mtROS boosts the sensitivity of $TRPV4_{EC}$ to specific agonists, since the observed agonist-induced Ca^{2+} influx was greater in microvascular ECs with high levels of mtROS (Suresh et al., 2019). Reducing the levels of mtROS greatly reduced TRPV4 agonist-induced response in microvascular endothelial cells (MVECs). Additional studies aimed at understanding the mechanistic underpinnings of mtROS production in control and hypoxia-exposed pulmonary capillary ECs will be helpful in elucidating the interplay between $TRPV4_{EC}$ channels and mitochondrial ROS.

3.4 TRPV4 channel activity in SMCs

SMCs are present in arteries and veins throughout the mammalian body. Of all vascular beds, only resistance-sized arteries regulate blood flow and determine blood pressure in both systemic and pulmonary circulatory systems. A hallmark of resistance-sized arteries is pressure-induced myogenic tone, described as pressure-induced contraction of SMCs (myogenic tone). Myogenic tone maintains a resting arterial internal diameter, so that arteries can constrict or dilate in response to changing blood flow and intravascular pressure. SMC-mediated reactivity is controlled by changes in cytosolic Ca^{2+} concentrations. The intracellular concentration of Ca^{2+} is maintained low (~100 nM) by the presence of Ca^{2+}-binding proteins and intracellular organelles that act as storages of Ca^{2+} (Clapham, 2007). A negatively charged intracellular environment, coupled with high extracellular Ca^{2+} concentration (~1–2 mM), accounts for an electrochemical gradient favorable for Ca^{2+} influx into the SMCs. Opening of Ca^{2+} permeable ion channels on the SMC membranes allows extracellular Ca^{2+} to move into the cytosol along the electrochemical gradient. The Ca^{2+} entry mechanisms on SMC membranes can be broadly divided into voltage-gated and non-voltage-gated. The voltage-gated Ca^{2+} entry pathways on SMC membranes include L-type and T-type Ca^{2+} channels and the non-voltage-gated Ca^{2+} entry mechanisms include TRP channels, PIEZO channels, store-operated Ca^{2+} entry (SOCE), purinergic receptors, and Na^+/Ca^{2+} exchangers (NCXs) (Ottolini & Sonkusare, 2021). This section will review the TRPV4-mediated mechanisms that control SMC contraction.

TRP channels participate in the regulation of SMC contractility and proliferation. These functions are achieved either by promoting global or localized increases in intracellular Ca^{2+} or by inducing the activation of ion channels that cause membrane depolarization. Recent studies have highlighted the importance of $TRPV4_{SMC}$ channels in causing both vasodilation and vasoconstriction based on various signaling targets they activate and the vascular beds being examined. TRPV4 channels are expressed in SMCs from rat cerebral arteries, human and rat lung extra-alveolar arteries, rat intralobular PAs, rat MAs, rat and human aorta, and mouse resistance-sized PAs (Alvarez et al., 2006; Gao & Wang, 2010; Marrelli et al., 2007; Martin et al., 2012; Watanabe et al., 2008; Yang et al., 2012). $TRPV4_{SMC}$ channels have been proposed to cause dilation or constriction depending on the signaling targets they activate and the vascular bed under consideration.

3.4.1 $TRPV4_{SMC}$ channel activity in systemic arteries

Ca^{2+} influx through $TRPV4_{SMC}$ channels in cerebral arteries has been reported to promote vasodilation. Ca^{2+} influx through TRPV4 channels increased the activity of ryanodine receptor (RyR)-Ca^{2+}-sensitive K^+ channels (BK) channel signaling, hyperpolarizing SMCs by ~10 mV and causing vasodilation (Earley, 2011). Studies conducted in cerebral pial arteries showed that the endogenous agonist 11,12-EET-activates $TRPV4_{SMC}$ channel currents. The Ca^{2+} influx via $TRPV4_{SMC}$ channels signals through RyR-BK channel to trigger SMC hyperpolarization and vasodilation (Earley et al., 2005) (Fig. 2). Elementary Ca^{2+} release signals (Ca^{2+} sparks) through RyRs located on the sarcoplasmic reticular (SR) membrane activate BK channels (Wellman et al., 2002). Interestingly, reduction of TRPV4 channel expression with antisense oligonucleotides prevented 11,12-EET-induced SMC hyperpolarization and dilation of cerebral pial arteries from male Sprague–Dawley rats (Fig. 2). Furthermore, $TRPV4_{SMC}$ channels appear to play a prominent role in counteracting angiotensin II (Ang II)-induced vasoconstriction in cerebral arteries. Mercado et al. (Mercado et al., 2014) reported that Ang II enhances $TRPV4_{SMC}$ channel activity in cerebral arteries via AKAP150 anchoring of PKCα close to $TRPV4_{SMC}$ channels and subsequent channel phosphorylation (Fig. 2). Indeed, Ang II signaling was found to increase the proximity between AKAP150 and $TRPV4_{SMC}$ channels. Superresolution nanoscopic experiments presented convincing data that $TRPV4_{SMC}$ channel activation by Ang II decreases exponentially with the increasing distance between AKAP150 and $TRPV4_{SMC}$ channels. $TRPV4_{SMC}$ channel activity could not be detected when the distance between AKAP150 and $TRPV4_{SMC}$ channel was higher than 200 nm apart. Interestingly, the distance between AKAP150 and $TRPV4_{SMC}$ channel, and the role of TRPV4 channel in counteracting Ang II-dependent vasoconstriction varied in different vascular beds (Tajada et al., 2017). Furthermore, studies from rat cerebral arteries revealed that the endothelium-derived EETs diffuse to nearby SMCs, activating $TRPV4_{SMC}$ channels (Earley et al., 2005). This activation caused Ca^{2+} entry to the SMC cytosol, which stimulated Ca^{2+} release from RyR, thus, causing an increased frequency of Ca^{2+} sparks (Earley et al., 2005). Ca^{2+} sparks then activated BK channels to hyperpolarize vascular SMCs, leading to vascular relaxation.

In SMCs from small MAs, TRPV4-like currents, activated by 11,12 EET- and 4α-PDD, have been detected. Genetic ablation of TRPV4 channels rendered this current absent (Earley et al., 2009). EETs, via their action of $TRPV4_{SMC}$ channels, were also found to induce smooth muscle hyperpolarization and vascular relaxation in mouse MAs, the effect of which was absent in global $TRPV4^{-/-}$ mice (Earley et al., 2009). Endothelial denudation contributed to only ~50% reduction in 11,12-EET-induced smooth muscle. Thus, the authors suggested that the remaining 50% were endothelium-independent and could be attributed to direct EET action of $TRPV4_{SMC}$ channel activity. In congruence, inhibition of BK channels in SMCs reduced the EET-induced responses by ~50%, confirming that the endothelium-independent component was ~50%. These findings offered a functional link between EETs, $TRPV4_{SMC}$ and BK channels in MAs.

Interestingly, recent work by Chen and coworkers presented a novel understanding of the $TRPV4_{SMC}$ channel signaling domains in resistance-sized mesenteric SMCs. The authors used inducible $TRPV4_{SMC}$ knockout ($TRPV4_{SMC}^{-/-}$) mice to show for the first time that resting mean arterial pressure was lower in $TRPV4_{SMC}^{-/-}$ mice in comparison to control mice (Chen et al., 2022). Chen and colleagues further showed that α1AR stimulation-induced constriction was blunted in small MAs from $TRPV4_{SMC}^{-/-}$ mice, whereas myogenic constriction was increased (Chen et al., 2022). Both α1AR stimulation and increased intraluminal pressure activated $TRPV4_{SMC}$ sparklet activity in MAs from control mice. These data supported the concept that $TRPV4_{SMC}$ channels contribute to α1AR-induced vasoconstriction but oppose myogenic vasoconstriction. Inhibition of BK channels abolished the effect of $TRPV4_{SMC}$ channels on myogenic constriction, revealing a crucial role for $TRPV4_{SMC}$—BK channel signaling in limiting myogenic constriction (Chen et al., 2022). Patch-clamp experiments solidified that a direct activation of $TRPV4_{SMC}$ channels with GSK1016790A (selective activator of TRPV4 channels) increased BK channel currents. α1AR agonism did not appear to increase BK channel currents, implying two functionally opposite pools of $TRPV4_{SMC}$ channels. Chen et al. employed direct stochastic optical reconstruction microscopy (dSTORM) superresolution imaging to elegantly demonstrate α1AR:$TRPV4_{SMC}$ and $TRPV4_{SMC}$:BK channel co-localization, and almost no co-localization of α1ARs and BK channels. Together, these data supported a novel concept of spatially separated and functionally opposite nanodomains of $TRPV4_{SMC}$ channels (Chen et al., 2022). These exciting data also raised the possibility that an imbalance between these two nanodomains could contribute to elevated blood pressure in vascular disorders, such as hypertension.

3.4.2 $TRPV4_{SMC}$ channel activity in pulmonary arteries

TRPV4 channels are expressed in SMCs from PAs (Martin et al., 2012), where they have been suggested to promote constriction (Song et al., 2014a). In studies of chronic hypoxia (CH) animal models, TRPV4 channels appear to be upregulated in PAs, resulting in higher contractility and PAP (Xia et al., 2013; Yang et al., 2012). Therefore, lack of a PAP phenotype in global $TRPV4^{-/-}$ mice could be due to the activation of compensatory mechanisms in SMCs and/or ECs. Recent studies by Daneva and coworkers presented the first measurements of right ventricular systolic pressure (RVSP, an indirect indicator of PAP) in inducible $TRPV4_{SMC}^{-/-}$

mice (Daneva et al., 2021a), demonstrating that $TRPV4_{SMC}$ channels are not involved in the maintenance of resting PAP. In contrast, $TRPV4_{EC}$ channels appear to be crucial for promoting low PAP (Daneva et al., 2021a). In congruence with these findings, the group further showed that the $TRPV4_{EC}$ channels are activated by a $Panx1_{EC}$—$P2Y2R_{EC}$ signaling axis, a signaling complex that is specific for the endothelium of small PAs and is directly responsible for maintaining low PAP (Fig. 3). Baseline RVSP was not altered in smooth muscle Panx1 knockout ($Panx1_{SMC}{}^{-/-}$) mice, indicating a lack of regulation of resting PAP by $Panx1_{SMC}$. It remains largely unexplored what the effect of $Panx1_{SMC}$ is to the vasoreactivity of PAs. Daneva et al. (Daneva et al., 2021b) demonstrated that release of the physiological TRPV4 activator, ATP, is decreased in PAs from inducible $Panx1_{SMC}$ mice, suggesting that there could be a possible interaction between $Panx1_{SMC}$ and $TRPV4_{SMC}$ channels. More studies remain to be conducted to elucidate this possibility and other TRPV4 channel-related signaling complexes in SMCs from PAs.

3.5 TRPV4 channel activity in fibroblasts

Fibroblasts are a type of mesenchymal cells located in the interstitial space of all organs. They contribute to tissue structural framework by producing extracellular matrix (ECM). When inactive, fibroblasts are morphologically and functionally different from myofibroblasts, as they do not express αSMA and lack contractive microfilamentous apparatus, characteristic of myofibroblasts. Markers used to study fibroblasts include thymocyte differentiation antigen 1 (Thy-1, also called CD90). Lineage tracing studies revealed that the origin of fibroblasts was from resident fibroblasts in pressure overload mice during development and fibrosis (Moore-Morris et al., 2014). Shortly after tissue injury, fibroblasts get activated and differentiate into hyper-secretory and hyper-contractile myofibroblasts. Soluble factors (transforming growth factor (TGF), Ang II, platelet-derived growth factor (PDGF) or Endothelin 1) and mechanical forces (stretch and/or ECM rigidity) play an important role in the activation and differentiation of fibroblasts (Adapala et al., 2013, 2020; Frangogiannis, 2019; Pakshir et al., 2020; Rahaman et al., 2014; Stewart & Turner, 2021; Thodeti et al., 2009).

In the past decade, a growing body of literature has identified a strong association between fibrosis and TRPV4 channel signaling (Adapala et al., 2013; Fernandez-Fernandez et al., 2008; Song et al., 2014b) (Fig. 6). Fernandez-Fernandez and colleagues indicated that TRPV4 channel activation regulates the cystic fibrosis transmembrane conductance regulator

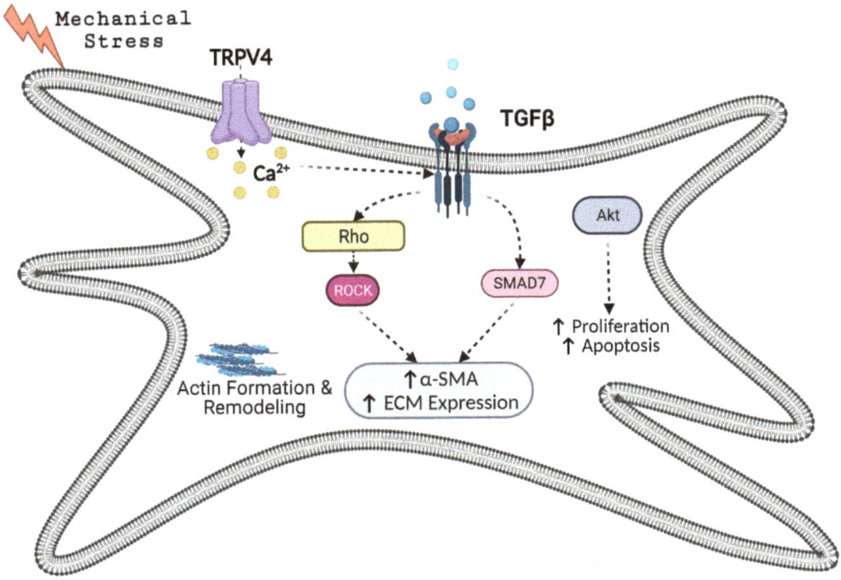

Fig. 6 Schematic depiction of TRPV4 channel activity in fibroblasts. Fibroblast TRPV4 channels act as mechanosensors of extracellular matrix (ECM) stiffness and elicit fibroblast differentiation by increasing α-SMA expression and recruitment of stress fibers in a Rho/ROCK-dependent manner. TRPV4 channels can activate AKT pathways to regulate autophagy-dependent cell proliferation and apoptosis. TRPV4 channels mediate TGF-β 1-induced fibroblast differentiation in a SMAD-dependent manner.

(CFTR) protein, a product of the cystic fibrosis gene (Fernandez-Fernandez et al., 2008). Recent research data point to a positive correlation between the activation of TRPV4 channels and fibroblast differentiation into myofibroblasts in rat heart (Adapala et al., 2013). The most prominent TRPV4 channel signaling mechanism in the differentiation of fibroblasts to myofibroblasts in myocardial, pulmonary and hepatic fibrosis is through TGF-β1 signaling (Adapala et al., 2013; Rahaman et al., 2014; Song et al., 2014b). More precisely, Thodeti et al. proposed that TRPV4 channels act as mechanosensors of ECM stiffness and elicit fibroblast differentiation by increasing α-SMA expression and recruitment into stress fibers (Thodeti, Paruchuri, & Meszaros, 2013) (Fig. 6). The group further suggested that this process may be Rho/ROCK-dependent. Accordingly, increased cell contraction and tensional forces activated additional integrins and triggered the activation of TGF-β1, which established a positive mechanical feedback loop (Thodeti et al., 2013) (Fig. 6). Furthermore, TRPV4 channels can be targeted by microRNA-203, are a vital AKT pathway-dependent component in hepatic fibrosis and regulate autophagy-dependent hepatic stellate cell proliferation and apoptosis (Song et al., 2014b; Zhan et al., 2015) (Fig. 6). In pulmonary fibrosis, TRPV4-mediated TGF-β1-induced fibroblasts differentiation was shown to be SMAD (a family of structurally similar proteins that are the main signal transducers for receptors of the transforming growth factor beta (TGF-β) superfamily)-dependent (Rahaman et al., 2014). The relationship between TRPV4 channels and fibroblast differentiation remains largely unexplored and is mainly studied in vitro. As cell-specific animal models and gene therapies evolve, further mechanistic details into the role of TRPV4 dysfunction in fibrosis will be uncovered.

4. TRPV4 channels in vascular disease conditions

While the role of abnormal Ca^{2+} signaling mechanisms appears to be well established in vascular disorders, the role of TRPV4 channels is still undergoing investigations. Both ECs and SMCs show impaired cellular signaling pathways and contribute to an array of vascular disorders. Recent studies in animal models and patients demonstrate a vital role for aberrant TRPV4 channel activity in the pathogenesis of cardiovascular diseases. Depending on the disease condition and vascular bed under consideration, the activity of $TRPV4_{EC}$ channels is increased or decreased in vascular disorders (Chen & Sonkusare, 2020). Here, we will briefly discuss several vasculopathies of interest vis-à-vis vascular TRPV4 channel dysfunction.

4.1 TRPV4 channel activity in pulmonary hypertension (pH)

Pulmonary hypertension is a serious and debilitating disease, characterized by elevated pulmonary arterial pressure (>25 mm Hg), pulmonary vasoconstriction, and pulmonary vascular remodeling (Chen & Sonkusare, 2020; Daneva et al., 2021a). Recently, abnormal $TRPV4_{EC}$ channel signaling was associated with the pathogenesis of PH (Daneva et al., 2021a, 2021b). Marziano et al. reported that TRPV4 channels are a crucial pathway for endothelial Ca^{2+} entry and vasodilation in small PAs that control pulmonary vascular resistance (Marziano et al., 2017). In congruence with these findings, Daneva and coworkers recently demonstrated, using inducible cell-specific knockout mice, that impairment of the $Cav\text{-}1_{EC}$–$TRPV4_{EC}$ channel signaling axis in resistance-sized PAs contributes to elevated PAP in a mouse model of PH (Daneva et al., 2021a). The involvement of impaired $Cav\text{-}1_{EC}$-dependent signaling to elevated PAP in PH is not surprising. Earlier research established that global $Cav\text{-}1^{-/-}$ mice show elevated PAP, and Cav-1_{EC}-dependent signaling was altered in PH (Bakhshi et al., 2013; Maniatis et al., 2008; Nickel et al., 2015; Oliveira et al., 2019; Zhao et al., 2002). However, using cell-specific knockout mouse models, Daneva et al. provided the first direct evidence for the effect of Cav-1_{EC} on TRPV4_{EC} channel activity under physiological conditions and in PH. Importantly, data from pulmonary arterial hypertension (PAH) patients provided the first evidence that clinical PAH is also associated with the lowering of TRPV4_{EC} channel activity in small PAs (Daneva et al., 2021a). Through a series of detailed mechanistic studies, the impaired Cav-1_{EC}-TRPV4_{EC} channel signaling axis was shown to be due to upregulation of reactive oxygen species accumulation in close proximity to Cav-1_{EC} and TRPV4_{EC} channels. More specifically, Daneva and colleagues demonstrated that NOX1 and iNOS interacted to form PN, an oxidant molecule known to be elevated in the lungs of PH patients, and to have deleterious effects on EC function (Bowers et al., 2004; Radi et al., 1991). Decreasing PN levels or NOX1/iNOS activity rescued Cav-1_{EC}–TRPV4_{EC} signaling, improved vasodilation, and lowered PAP in PH (Daneva et al., 2021a). More recently, Daneva et al. identified a critical role of Panx1_{EC} and P2Y2R_{EC} in stimulating the activity of TRPV4_{EC} channels to reduce PA contractility and maintain low PAP. The group demonstrated critical roles for several key pathways showing that: Panx1_{EC} increases extracellular ATP (eATP) levels in small PAs, which enhances Ca^{2+} influx through TRPV4_{EC} channels, thereby dilating PAs and lowering PAP. Daneva et al. also showed that eATP acts through purinergic P2Y2R_{EC}–PKCα signaling to activate TRPV4_{EC} channels and that Cav-1_{EC} provides a signaling scaffold to ensure spatial proximity among the elements of the Panx1_{EC}–P2Y2R_{EC}–PKCα–TRPV4_{EC} pathway (Daneva et al., 2021b). Overall, these findings reveal a novel signaling axis that can be engaged by physiological stimuli to lower PAP and could also be therapeutically targeted in pulmonary vascular disorders, such as, but not limited to PH.

Endothelial Ca^{2+} has distinct effects on capillary and arterial function (Alvarez et al., 2006; Marziano et al., 2017; Ottolini et al., 2020b; Thorneloe et al., 2012). TRPV4_{EC} channel activity in capillary ECs has mostly been associated with negative effects in pulmonary pathologies. High capillary TRPV4_{EC} channel activity increases capillary endothelial permeability (Thorneloe et al., 2012; Yin et al., 2008) and contributes to lung injury (Alvarez et al., 2006) and pulmonary edema (Thorneloe et al., 2012; Yin et al., 2008). A study by Thorneloe et al. (Thorneloe et al., 2012) showed that chronic treatment with TRPV4 channel inhibitor restores the integrity of the capillary endothelial barrier in heart failure-induced pulmonary edema. Studies by Suresh and colleagues, using a rat model of PAH, showed that Ca^{2+} influx through TRPV4 channels is increased in cultured MVECs in PAH (Suresh et al., 2018). However, arterial $TRPV4_{EC}$ channel activity remains a relatively unexplored area in pulmonary vascular disorders. Daneva et al. showed that acute treatment with TRPV4 channel inhibitor elevates RVSP, and this effect is lost in $TRPV4_{EC}{}^{-/-}$ mice (Daneva et al., 2021a). Capillary and arterial ECs are structurally and functionally heterogeneous (Longden et al., 2017; Sonkusare et al., 2012; Stevens, 2011). Thus, pathological conditions may affect capillary and arterial $TRPV4_{EC}$ channels differently.

EC and SMC components in the TRPV4 channel control of PAP have been difficult to define. Global $TRPV4^{-/-}$ mice do not show any changes in resting RVSP (Xia et al., 2013). Administration of TRPV4 channel activator was shown to reduce RVSP in rats (Pankey et al., 2014), although the exact cell types involved in this effect were not clear. TRPV4 channels were also detected at mRNA and protein levels in pulmonary SMCs (Martin et al., 2012). $TRPV4_{SMC}$ channels promote PA contractility (Song et al., 2014a) and are upregulated following CH (Song et al., 2014a; Xia et al., 2013; Yang et al., 2012). Thus, EC and SMC TRPV4 channels appear to have contrasting effects on contractility, similar to intracellular Ca^{2+} in ECs and SMCs (Ottolini et al., 2019). In other words, it appears that $TRPV4_{EC}$ channels are dilatory, whereas $TRPV4_{SMC}$ channels are contractile in PAs. These findings provide further example of the unique cellular functions of TRPV4 channels in ECs or SMCs, which would be difficult to decipher using global knockout mice. Daneva et al. provided the first evidence that resting RVSP is not altered in $TRPV4_{SMC}{}^{-/-}$ mice, suggesting that $TRPV4_{SMC}$ channels may not play a role in regulating the resting PAP. The lack of a resting PAP phenotype in global $TRPV4^{-/-}$ mice (Xia et al., 2013) but elevated PAP in $TRPV4_{EC}{}^{-/-}$ mice is consistent with previously published findings of unaltered resting systemic blood pressure in $TRPV4^{-/-}$ mice (Hong et al., 2018) and elevated systemic blood pressure in $TRPV4_{EC}{}^{-/-}$ mice (Ottolini et al., 2020a). This discrepancy in results between global and cell-specific knockout mice could be explained by potential developmental compensations in global knockout mice to ensure physiological homeostasis, masking individual cellular function.

4.2 TRPV4 channel activity in systemic hypertension

Hypertension is a disease characterized by elevated systolic/diastolic blood pressure (>130/80 mm Hg). Endothelial dysfunction and increased contractility of resistance-sized arteries are hallmarks of hypertension. Earlier studies focusing on the role of TRPV4 channels in vascular disease revealed that NOS inhibitor-induced hypertension was augmented in global TRPV4$^{-/-}$ mice (Earley et al., 2009). Moreover, TRPV4$_{EC}$ channels activity and activation by acetylcholine were impaired specifically at MEPs in a mouse model of Ang II-induced hypertension (Sonkusare et al., 2014). This abnormality in TRPV4$_{EC}$ channel activity was attributed to reduced levels of the scaffolding protein AKAP150 in hypertension, leading to a loss of coupled openings of TRPV4 channel and acetylcholine—PKC activation of the channel, which lowered Ca^{2+} influx at MEPs. Similarly, Ottolini and Hong reported a reduced TRPV4 channel activity at MEPs in a high-fat diet-induced mouse model of obesity (Kwangseok Hong et al., 2016; Ottolini et al., 2020a). More recently, Chen et al. used superresolution imaging to demonstrate that SMCs possess two functionally opposite pools of TRPV4$_{SMC}$ channels (Chen et al., 2022). Namely, TRPV4$_{SMC}$ channels co-localize either with α1AR (constrictor pool) or BK channels (dilator pool). Logically, the group went on to investigate if there existed an imbalance between TRPV4$_{SMC}$ channels that are co-localized with α1ARs and BK channels (Chen et al., 2022). Chen and colleagues showed that in skeletal muscle arteries from hypertensive patients, α1AR activation of TRPV4$_{SMC}$ channel currents was enhanced, while TRPV4$_{SMC}$ activation of BK channel currents was impaired. Similarly, MAs from Ang II-infused hypertensive mice showed increased activity of constrictor α1AR–TRPV4$_{SMC}$ signaling and impaired dilatory TRPV4$_{SMC}$–BK channel signaling. Collectively, these data identify novel smooth muscle Ca^{2+} signaling nanodomains that regulate blood pressure and demonstrate impaired balance between them in hypertension.

In salt-induced hypertension in mice, hypercontractility of aorta was associated with TRPV4 channel-mediated elevation in endothelial Ca^{2+} and subsequent activation of PLA_2, causing an increase in the formation of prostaglandin $F_{2\alpha}$ ($PGF_{2\alpha}$) by the endothelial cyclo-oxygenase-2 (COX_2) (Zhang et al., 2018). However, in a rat model of salt-induced hypertension, there was enhanced TRPV4 channel expression in mesenteric arteries and higher TRPV4 channel-mediated vasodilation (Gao et al., 2009). These results suggested that a compensatory TRPV4 channel activation may prevent impaired vasodilation in salt-induced hypertension.

4.3 Fibrosis

Fibrosis is a chronic disease characterized by an excessive accumulation of extracellular matrix (ECM) components in tissues and organs

(Zhan & Li, 2018). TRPV4 channels have recently been identified as a crucial regulator in the progression of fibrosis including myocardial, cystic, pulmonary, hepatic and pancreatic fibrosis. The involvement of TRPV4 channels in the pathogenic development of fibrosis implies that the channel may present a novel therapeutic target. However, the mechanisms by which TRPV4 channels regulate fibrosis remain largely unexplored. In this section, we will review established and proposed mechanisms by which TRPV4 channel contributes to vasculature-relevant fibrosis subtypes (cardiac and pulmonary).

4.3.1 TRPV4 channel activity in myocardial fibrosis (MF)

Myocardial fibrosis is characterized by abnormal accumulation of α-SMA, collagens and ECM in the myocardium (Kong, Christia, & Frangogiannis, 2014). Overactive ECM production ultimately leads to cardiac remodeling and cardiac dysfunction. TRPV4 channels are abundantly expressed in rat cardiac fibroblasts (Hatano, Itoh, & Muraki, 2009; Thodeti et al., 2013; Yue, Xie, & Nattel, 2011). TRPV4 channels were activated by 4α-PDD treatment in cardiac fibroblasts, and ruthenium red and siRNA successfully inhibited TRPV4 channel currents (Thodeti et al., 2013; Yue et al., 2011). Adapala and colleagues reported that the expression of TRPV4 channels is significantly amplified by TGF-β1 treatment in cardiac fibroblasts (Adapala et al., 2013). The group also showed that inhibition of TRPV4 channel activity modulates TGF-β1-induced cardiac fibroblast proliferation and differentiation into myofibroblasts. Additionally, a recent study revealed that global TRPV4$^{-/-}$ mice demonstrate reduced fibrotic phenotype in myocardial infarction (Wang et al., 2015). TRPV4 channels present a potential target for the treatment of myocardial fibrosis, but much work remains to be done on understanding the TRPV4-dependent mechanisms in this pathology.

4.3.2 TRPV4 channel activity in idiopathic pulmonary fibrosis (IPF)

IPF is a chronic, progressive and debilitating disorder, characterized by excessive deposition of ECM, which leads to extensive lung remodeling (Sgalla et al., 2018). Rahaman and colleagues reported that TRPV4 expression in pulmonary tissues and cells was upregulated in IPF (Rahaman et al., 2014). In congruence with this observation, genetic and pharmacological inhibition of TRPV4 expression reduced myofibroblast differentiation, an effect that proved to be reversible. Similar to myocardial fibrosis, TRPV4-induced Ca^{2+} influx was required for TGF-β1-mediated lung myofibroblast differentiation. This mechanism appeared to be SMAD-independent. Furthermore, the group suggested that activating TRPV4 channels could potentiate TGF-β1-induced actomyosin remodeling in pulmonary fibrosis (Rahaman et al., 2014). Scheraga and coworkers found that signaling through TRPV4 channels, triggered by changes in ECM stiffness, interacts with LPS-induced signals to mediate macrophage phagocytic function and lung injury resolution. The authors considered this mechanism to be important in regulating macrophage function in the context of pulmonary infection and fibrosis (Scheraga et al., 2016). Gombedza et al. recently demonstrated that TGF-β1 treatment enhanced TRPV4 activation in a PI_3K-dependent manner in normal human lung fibroblasts in vitro (Gombedza et al., 2017). In contrast to this involvement of PI_3K in pulmonary fibrosis, Grove et al. showed that TRPV4 channels promote *trans*-differentiation of human and mouse lung myofibroblasts through interaction with PI_3K, forming intracellular TRPV4–PI_3K complexes (Grove et al., 2019). Furthermore, Al-Azzam and colleagues reported a key involvement of NOX4 and reactive oxygen ROS in TRPV4-mediated TGF-β1-induced lung fibroblast differentiation and *D. farina*-induced airway remodeling (Al-Azzam et al., 2020). Mechanistically, TRPV4 modulated matrix remodeling in a Rho/myocardin-related transcription factor-A manner in the lung in two distinct but dependent ways: one enhances matrix deposition by fibrotic gene activation, whereas the other slows down matrix degradation by increased plasminogen activator inhibitor 1 (Gombedza et al., 2017). Clearly, the path to elucidating the role of TRPV4 channels in the pathogenesis of PF is long, but the importance of the channel in relation to this pathology is obvious. Fibroblast differentiation can be regulated by NO. A recent report demonstrated that NOX4 and generated hydrogen peroxide (H_2O_2) crosstalk with NO pathway increased myofibroblast differentiation. TRPV4 channels have been shown to promote NO production via eNOS activation in ECs, however, their ability to promote NO in fibroblasts has not been explored (Daneva et al., 2021a; Marziano et al., 2017). Based on these findings, it appears that TRPV4 regulates TGF-1/mechanical force-induced fibroblasts differentiation through a Rho–NOX4 signaling complex. Further investigation of TRPV4-mediated cellular effects in pulmonary fibroblasts are necessary to establish if the channels represent a good target for IPF treatment.

4.3.3 TRPV4$_{EC}$ channel activity in angiogenesis

Angiogenesis is the process of forming new blood vessels from existing ones. It is a physiological process that is crucial for maintaining hemostasis in mammalian tissues. There are two categories of angiogenesis: (1) Sprouting (slow)—blood vessels, comprised of a single layer of endothelium, sprout from the pre-existing vessel through changes in inter-endothelial junctions and the subendothelial basement membrane (Eilken & Adams, 2010); and (2) Non-sprouting (fast)—an intravascular process in which pre-existing vessels internally divide while maintaining an intact basement membrane (Mentzer & Konerding, 2014). Both types of angiogenesis are highly synchronized and tightly regulated by biochemical/physiological Carmeliet & Jain, 2011; D'Amore & Smith, 1993; Kevil et al., 1998; Moss, 2013; Przybylski, 2009; Saharinen, Bry, & Alitalo, 2010; Semenza, 2007) and mechanical factors (Carmeliet, 2000; Carmeliet & Jain, 2011; Ingber et al., 1995; Semenza, 2007). The final stage of angiogenesis is "vessel remodeling," where newly formed tubes and sacs undergo remodeling and pruning to form a mature, structured network of vessels (Carmeliet, 2000; Risau, 1997). Not all newly formed vessels become stable and functional as they undergo remodeling. Importantly, blood flow plays a vital role when it comes to determining the fate of each vessel. In pathological angiogenesis, blood vessels often grow into destabilized structures, making them disorganized and leaky due to a lack of pericytes, ECM and properly developed inter-endothelial junctions. In cancer and tumor growth, the endothelium of the affected area requires extra stimulation with pro-angiogenic factors in order to meet the oxygen and nutrient demands. This need causes an elevation of interstitial pressure, blood flow, disproportionate branching, and irregular vessel wall density (Baluk, Hashizume, & McDonald, 2005; Boucher, Baxter, & Jain, 1990). Overall, aberrant angiogenic regulation leads to structurally and functionally abnormal vasculature. Current pharmaceutical therapies that try to treat pathological angiogenesis focus on vascular endothelial growth factor (VEGF) regulation, the most well-studied angiogenesis-related biochemical factor. Therefore, there is a need for a better understanding of the mechanical factors affecting angiogenesis.

In this regard, TRPV4 channels have become an invaluable example for mechanosensor-mediated endothelial regulation. Thus, it is of no surprise that changes in $TRPV4_{EC}$ channel activity may affect angiogenesis. Data on TRPV4-mediated angiogenesis are still rare, but several research groups have delved into elucidating the role of TRPV4 channels in pathological angiogenesis. TRPV4 appears to be important for cytoskeletal remodeling and cell reorientation in response to mechanical stimulation. Thodeti et al. demonstrated that in bovine capillary ECs subjected to cyclic strain, there was elevated TRPV4-mediated Ca^{2+} influx (Thodeti et al., 2009). This further led to activation of PI_3K, and subsequent binding of additional β1 integrins, potentially mediating Rho/Rac signaling necessary for remodeling of the cytoskeleton and EC reorientation (Thodeti et al., 2009). In a follow-up study by Matthews, it was demonstrated that direct forces applied to β1 integrins can activate TRPV4 channels via interaction with integrin-associated transmembrane CD98 protein, found in focal adhesions (Matthews et al., 2010). These in vitro studies strongly suggested that TPRV4 is a mechanosensitive ion channel in ECs and could contribute to angiogenesis.

Most TRPV4 channel–angiogenesis reports draw from cancer studies and focus on pathological phenotypes in tumor-derived endothelial cells (TECs). TECs are characterized by abnormal reorientation after cyclin stress, increased cell spreading, increased stiffness and elevated angiogenesis in vitro. All these characteristics could explain why TECs readily form dilated and non-uniform vessels in in vitro assays. Initially, it was thought that this abnormal mechanosensitivity and angiogenesis was due to high basal Rho activity (Ghosh et al., 2008). However, recent studies indicated that TECs express very low levels of TRPV4 channels (Adapala et al., 2016). Further studies utilized ECs from global TRPV4$^{-/-}$ mice to show that ablation of the ion channels causes significant proliferation, migration, basal Rho activity and abnormal angiogenesis (Adapala et al., 2016; Thoppil et al., 2016). Targeting TRPV4 channels via overexpression or pharmacological activation was shown to restore aberrant mechanosensitivity and angiogenesis as well as abnormal migration and high Rho activity. When wildtype and TRPV4$^{-/-}$ mice were challenged with tumors, the knockout mice had significantly larger tumors and vasculature that appeared to be malformed and dense, with low pericyte coverage. Congruently, treatment with TRPV4 agonists significantly reduced the tumor volumes, when applied in conjunction with cancer chemotherapy agent (Adapala et al., 2016). When targeted with a TRPV4 activator, TEC proliferation significantly decreased (Thoppil et al., 2015).

With the advance of cell-specific knockout animal models, the angiogenesis field has been enriched with more precise and elaborate studies on TRPV4 channels. Kanuluga et al. (Kanugula et al., 2021) used $TRPV4_{EC}^{-/-}$ mice to study the involvement of $TRPV4_{EC}$ channels in tumor angiogenesis. The group found that the knockout mice exhibited enhanced tumor growth and increased expression of p-ERK. Their results reiterated previous findings from global TRPV4$^{-/-}$ mice (Adapala et al., 2016) and confirmed the endothelial-specific role of TRPV4 channels in tumor angiogenesis and growth (Kanugula et al., 2021). Using multiphoton microscopy, the team identified large and convoluted vessels in the $TRPV4_{EC}^{-/-}$ tumors compared to the wildtype tumors, indicating abnormal vessel formation. Furthermore, inactivation of TRPV4 channels resulted in leaky vessels, which contributed to enhanced metastasis.

From mechanistic point of view, $TRPV4_{EC}$ channel activity suppresses tumor angiogenesis, tumor growth, vascular disruption, and metastasis, most likely by controlling VEGF/VEGF receptor 2 ($VEGFR_2$) localization and activity (Kanugula et al., 2021). These results indicate that angiogenesis is regulated by a tight coordination between mechanical and soluble (growth factor) signaling. Previous work by Mammoto et al. demonstrated that ECM stiffness governs $VEGFR_2$ gene promoter activity and expression via a mechano-transcriptional mechanism to stimulate angiogenesis (Mammoto et al., 2009). However, the identity of the proximal mechanosensor(s) involved remains unknown. In this regard, Cappelli and colleagues recently reported that TRPV4 channels regulate angiogenesis in an ECM stiffness-dependent manner in vivo (Cappelli et al., 2019). Furthermore, ablation of $TRPV4_{EC}$ channel activity-induced membrane translocation and phosphorylation of $VEGFR_2$ through the Rho/YAP pathway (Kanugula et al., 2019). Taken together, these findings suggest that mechanical stiffness regulates angiogenesis through a mechano-transcriptional mechanism through both $VEGFR_2$ expression and/or its activation, which is regulated by $TRPV4_{EC}$ channels (Kanugula et al., 2021). In conclusion, targeting $TRPV4_{EC}$ channels could offer a potential novel strategy for vascular normalization in cancer.

5. Conclusion

Vascular TRPV4 channels have a broad array of functions in different cell types in the vasculature. TRPV4 channels exert differential functions, depending on the proteins they co-localize and interact with. TRPV4 channels in the vascular endothelium promote dilation of resistance-sized arteries. TRPV4 channels in SMCs can contribute to vasodilation or vasoconstriction, depending on the downstream signaling pathway they activate. In capillaries, $TRPV4_{EC}$ channels play an important role in maintaining healthy permeability to accommodate oxygen exchange. Fibroblast TRPV4 channels facilitate tissue connectivity and maintenance of structural framework for vascular beds. Strong evidence supports a role of TRPV4 channel-mediated maintenance of healthy vascular function and vascular homeostasis. Therefore, it is not surprising that impaired

TRPV4 channel signaling is associated with multiple vascular pathologies. Recent advancements in scientific technology and genetic animal models have improved our understanding of TRPV4-dependent signaling pathways and their roles in controlling vascular function. Future investigations may result in successful targeting of TRPV4 channels for therapeutic benefit in multiple vascular diseases.

References

Adapala, R. K., et al. (2013). TRPV4 channels mediate cardiac fibroblast differentiation by integrating mechanical and soluble signals. *Journal of Molecular and Cellular Cardiology*, *54*, 45–52.

Adapala, R. K., et al. (2016). Activation of mechanosensitive ion channel TRPV4 normalizes tumor vasculature and improves cancer therapy. *Oncogene*, *35*(3), 314–322.

Adapala, R. K., et al. (2020). TRPV4 deletion protects heart from myocardial infarction-induced adverse remodeling via modulation of cardiac fibroblast differentiation. *Basic Research in Cardiology*, *115*(2), 14.

Al-Azzam, N., et al. (2020). Transient receptor potential vanilloid channel regulates fibroblast differentiation and airway remodeling by modulating redox signals through NADPH oxidase 4. *Scientific Reports*, *10*(1), 9827.

Alvarez, D. F., et al. (2006). Transient receptor potential vanilloid 4-mediated disruption of the alveolar septal barrier: A novel mechanism of acute lung injury. *Circulation Research*, *99*(9), 988–995.

Bagher, P., et al. (2012). Low intravascular pressure activates endothelial cell TRPV4 channels, local Ca2+ events, and IKCa channels, reducing arteriolar tone. *Proceedings of the National Academy of Sciences of the United States of America*, *109*(44), 18174–18179.

Bakhshi, F. R., et al. (2013). Nitrosation-dependent caveolin 1 phosphorylation, ubiquitination, and degradation and its association with idiopathic pulmonary arterial hypertension. *Pulmonary Circulation*, *3*(4), 816–830.

Baluk, P., Hashizume, H., & McDonald, D. M. (2005). Cellular abnormalities of blood vessels as targets in cancer. *Current Opinion in Genetics & Development*, *15*(1), 102–111.

Bayliss, W. M. (1902). On the local reactions of the arterial wall to changes of internal pressure. *The Journal of Physiology*, *28*(3), 220–231.

Benfenati, V., et al. (2007). Expression and functional characterization of transient receptor potential vanilloid-related channel 4 (TRPV4) in rat cortical astrocytes. *Neuroscience*, *148*(4), 876–892.

Bernatchez, P. N., et al. (2005). Dissecting the molecular control of endothelial NO synthase by caveolin-1 using cell-permeable peptides. *Proceedings of the National Academy of Sciences of the United States of America*, *102*(3), 761–766.

Berridge, M. J., Bootman, M. D., & Roderick, H. L. (2003). Calcium signalling: Dynamics, homeostasis and remodelling. *Nature Reviews. Molecular Cell Biology*, *4*(7), 517–529.

Berrout, J., Jin, M., & O'Neil, R. G. (2012). Critical role of TRPP2 and TRPC1 channels in stretch-induced injury of blood-brain barrier endothelial cells. *Brain Research*, *1436*, 1–12.

Boucher, Y., Baxter, L. T., & Jain, R. K. (1990). Interstitial pressure gradients in tissue-isolated and subcutaneous tumors: Implications for therapy. *Cancer Research*, *50*(15), 4478–4484.

Bowers, R., et al. (2004). Oxidative stress in severe pulmonary hypertension. *American Journal of Respiratory and Critical Care Medicine*, *169*(6), 764–769.

Cao, S., et al. (2018). Transient receptor potential vanilloid 4 (TRPV4) activation by arachidonic acid requires protein kinase A-mediated phosphorylation. *The Journal of Biological Chemistry*, *293*(14), 5307–5322.

Cappelli, H. C., et al. (2019). Mechanosensitive TRPV4 channels stabilize VE-cadherin junctions to regulate tumor vascular integrity and metastasis. *Cancer Letters*, *442*, 15–20.

Carmeliet, P. (2000). Mechanisms of angiogenesis and arteriogenesis. *Nature Medicine*, *6*(4), 389–395.

Carmeliet, P., & Jain, R. K. (2011). Molecular mechanisms and clinical applications of angiogenesis. *Nature*, *473*(7347), 298–307.

Chen, Y. L., & Sonkusare, S. K. (2020). Endothelial TRPV4 channels and vasodilator reactivity. *Current Topics in Membranes*, *85*, 89–117.

Chen, Y. L., et al. (2021). Calcium signal profiles in vascular endothelium from Cdh5-GCaMP8 and Cx40-GCaMP2 mice. *Journal of Vascular Research*, *58*(3), 159–171.

Chen, Y. L., et al. (2022). Novel smooth muscle ca(2+)-signaling nanodomains in blood pressure regulation. *Circulation* (p. 101161CIRCULATIONAHA121058607).

Clapham, D. E. (2007). Calcium signaling. *Cell*, *131*(6), 1047–1058.

Clapham, D. E., et al. (2003). International Union of Pharmacology. XLIII. Compendium of voltage-gated ion channels: Transient receptor potential channels. *Pharmacological Reviews*, *55*(4), 591–596.

D'Amore, P. A., & Smith, S. R. (1993). Growth factor effects on cells of the vascular wall: A survey. *Growth Factors*, *8*(1), 61–75.

Daneva, Z., et al. (2021a). Caveolar peroxynitrite formation impairs endothelial TRPV4 channels and elevates pulmonary arterial pressure in pulmonary hypertension. *Proceedings of the National Academy of Sciences of the United States of America*, *118*(17).

Daneva, Z., et al. (2021b). Endothelial pannexin 1-TRPV4 channel signaling lowers pulmonary arterial pressure in mice. *eLife*, *10*.

Daneva, Z., et al. (2022). Endothelial Pannexin1-purinergic signaling pulmonary hypertension. *The FASEB Journal*, *36*(S1).

Darby, W. G., et al. (2018). Shear stress sensitizes TRPV4 in endothelium-dependent vasodilatation. *Pharmacological Research*, *133*, 152–159.

Davis, M. J., Hill, M. A., & Kuo, L. (2008). Local regulation of microvascular perfusion. In R. F. Tuma, W. N. Duran, & K. Ley (Eds.), *Handbook of physiology: The cardiovascular system, microcirculation (Section 2)* (*2nd ed.*, pp. 161–284). San Diego, CA: Academic Press.

Deng, Z., et al. (2018). Cryo-EM and X-ray structures of TRPV4 reveal insight into ion permeation and gating mechanisms. *Nature Structural & Molecular Biology*, *25*(3), 252–260.

D'Hoedt, D., et al. (2008). Stimulus-specific modulation of the cation channel TRPV4 by PACSIN 3. *The Journal of Biological Chemistry*, *283*(10), 6272–6280.

Di Carlo, S. E., & Peduto, L. (2018). The perivascular origin of pathological fibroblasts. *The Journal of Clinical Investigation*, *128*(1), 54–63.

Du, J., et al. (2014). TRPV4, TRPC1, and TRPP2 assemble to form a flow-sensitive heteromeric channel. *The FASEB Journal*, *28*(11), 4677–4685.

Dunn, K. M., et al. (2013). TRPV4 channels stimulate Ca2+−induced Ca2+ release in astrocytic endfeet and amplify neurovascular coupling responses. *Proceedings of the National Academy of Sciences of the United States of America*, *110*(15), 6157–6162.

Earley, S. (2011). Endothelium-dependent cerebral artery dilation mediated by transient receptor potential and Ca2+−activated K+ channels. *Journal of Cardiovascular Pharmacology*, *57*(2), 148–153.

Earley, S., & Brayden, J. E. (2015). Transient receptor potential channels in the vasculature. *Physiological Reviews*, *95*(2), 645–690.

Earley, S., Waldron, B. J., & Brayden, J. E. (2004). Critical role for transient receptor potential channel TRPM4 in myogenic constriction of cerebral arteries. *Circulation Research*, *95*(9), 922–929.

Earley, S., et al. (2005). TRPV4 forms a novel Ca2+ signaling complex with ryanodine receptors and BKCa channels. *Circulation Research*, *97*(12), 1270–1279.
Earley, S., et al. (2009). TRPV4-dependent dilation of peripheral resistance arteries influences arterial pressure. *American Journal of Physiology. Heart and Circulatory Physiology*, *297*(3), H1096–H1102.
Eilken, H. M., & Adams, R. H. (2010). Dynamics of endothelial cell behavior in sprouting angiogenesis. *Current Opinion in Cell Biology*, *22*(5), 617–625.
Ellinsworth, D. C., et al. (2016). Endothelium-derived hyperpolarization and coronary vasodilation: Diverse and integrated roles of Epoxyeicosatrienoic acids, hydrogen peroxide, and gap junctions. *Microcirculation*, *23*(1), 15–32.
Fan, H. C., Zhang, X., & McNaughton, P. A. (2009). Activation of the TRPV4 ion channel is enhanced by phosphorylation. *The Journal of Biological Chemistry*, *284*(41), 27884–27891.
Fernandez-Fernandez, J. M., et al. (2008). Functional coupling of TRPV4 cationic channel and large conductance, calcium-dependent potassium channel in human bronchial epithelial cell lines. *Pflügers Archiv*, *457*(1), 149–159.
Ferrandiz-Huertas, C., et al. (2014). Trafficking of ThermoTRP channels. *Membranes (Basel)*, *4*(3), 525–564.
Filosa, J. A., et al. (2006). Local potassium signaling couples neuronal activity to vasodilation in the brain. *Nature Neuroscience*, *9*(11), 1397–1403.
Frangogiannis, N. G. (2019). The extracellular matrix in ischemic and nonischemic heart failure. *Circulation Research*, *125*(1), 117–146.
Gao, F., & Wang, D. H. (2010). Hypotension induced by activation of the transient receptor potential vanilloid 4 channels: Role of Ca2+—activated K+ channels and sensory nerves. *Journal of Hypertension*, *28*(1), 102–110.
Gao, F., et al. (2009). Salt intake augments hypotensive effects of transient receptor potential vanilloid 4: Functional significance and implication. *Hypertension*, *53*(2), 228–235.
Garcia, D. C. G., & Longden, T. A. (2020). Ion channels in capillary endothelium. *Current Topics in Membranes*, *85*, 261–300.
Garcia-Elias, A., et al. (2008). IP3 receptor binds to and sensitizes TRPV4 channel to osmotic stimuli via a calmodulin-binding site. *The Journal of Biological Chemistry*, *283*(46), 31284–31288.
Garcia-Elias, A., et al. (2013). Phosphatidylinositol-4,5-biphosphate-dependent rearrangement of TRPV4 cytosolic tails enables channel activation by physiological stimuli. *Proceedings of the National Academy of Sciences of the United States of America*, *110*(23), 9553–9558.
Ghosh, K., et al. (2008). Tumor-derived endothelial cells exhibit aberrant rho-mediated mechanosensing and abnormal angiogenesis in vitro. *Proceedings of the National Academy of Sciences of the United States of America*, *105*(32), 11305–11310.
Gombedza, F., et al. (2017). Mechanosensitive transient receptor potential vanilloid 4 regulates Dermatophagoides farinae-induced airway remodeling via 2 distinct pathways modulating matrix synthesis and degradation. *The FASEB Journal*, *31*(4), 1556–1570.
Grove, L. M., et al. (2019). Translocation of TRPV4-PI3Kgamma complexes to the plasma membrane drives myofibroblast transdifferentiation. *Science Signaling*, *12*(607).
Harraz, O. F., et al. (2018). PIP2 depletion promotes TRPV4 channel activity in mouse brain capillary endothelial cells. *eLife*, *7*.
Hartmannsgruber, V., et al. (2007). Arterial response to shear stress critically depends on endothelial TRPV4 expression. *PLoS One*, *2*(9), e827.
Hatano, N., Itoh, Y., & Muraki, K. (2009). Cardiac fibroblasts have functional TRPV4 activated by 4alpha-phorbol 12,13-didecanoate. *Life Sciences*, *85*(23–26), 808–814.
Heathcote, H. R., et al. (2019). Endothelial TRPV4 channels modulate vascular tone by ca(2+) -induced ca(2+) release at inositol 1,4,5-trisphosphate receptors. *British Journal of Pharmacology*, *176*(17), 3297–3317.

Hennigs, J. K., et al. (2019). The P2-receptor-mediated ca(2+) signalosome of the human pulmonary endothelium - implications for pulmonary arterial hypertension. *Purinergic Signal, 15*(3), 299–311.

Hong, K., et al. (2018). TRPV4 (transient receptor potential vanilloid 4) channel-dependent negative feedback mechanism regulates Gq protein-coupled receptor-induced vasoconstriction. *Arteriosclerosis, Thrombosis, and Vascular Biology, 38*(3), 542–554.

Ingber, D. E., et al. (1995). Cell shape, cytoskeletal mechanics, and cell cycle control in angiogenesis. *Journal of Biomechanics, 28*(12), 1471–1484.

Jackson, W. F. (2022). Endothelial ion channels and cell-cell communication in the microcirculation. *Frontiers in Physiology, 13*, 805149.

Jeyaraju, D. V., Cisbani, G., & Pellegrini, L. (2009). Calcium regulation of mitochondria motility and morphology. *Biochimica et Biophysica Acta, 1787*(11), 1363–1373.

Kanugula, A. K., et al. (2019). Novel noncanonical regulation of soluble VEGF/VEGFR2 signaling by mechanosensitive ion channel TRPV4. *The FASEB Journal, 33*(1), 195–203.

Kanugula, A. K., et al. (2021). Endothelial TRPV4 channels prevent tumor growth and metastasis via modulation of tumor angiogenesis and vascular integrity. *Angiogenesis, 24*(3), 647–656.

Kevil, C. G., et al. (1998). Vascular permeability factor/vascular endothelial cell growth factor-mediated permeability occurs through disorganization of endothelial junctional proteins. *The Journal of Biological Chemistry, 273*(24), 15099–15103.

Kim, K. J., et al. (2015). Astrocyte contributions to flow/pressure-evoked parenchymal arteriole vasoconstriction. *The Journal of Neuroscience, 35*(21), 8245–8257.

Kisler, K., et al. (2017). Cerebral blood flow regulation and neurovascular dysfunction in Alzheimer disease. *Nature Reviews. Neuroscience, 18*(7), 419–434.

Kohler, R., et al. (2006). Evidence for a functional role of endothelial transient receptor potential V4 in shear stress-induced vasodilatation. *Arteriosclerosis, Thrombosis, and Vascular Biology, 26*(7), 1495–1502.

Konduri, G. G., et al. (2004). P2Y purine receptor responses and expression in the pulmonary circulation of juvenile rabbits. *American Journal of Physiology. Heart and Circulatory Physiology, 287*(1), H157–H164.

Kong, P., Christia, P., & Frangogiannis, N. G. (2014). The pathogenesis of cardiac fibrosis. *Cellular and Molecular Life Sciences, 71*(4), 549–574.

Kwangseok Hong, E. L. C., Marziano, C., & Sonkusare, S. K. (2016). Loss of local Ca2+ signaling networks in the endothelium in diet induced obesity. *The FASEB Journal, 30*(S1) (p. Ib607).

Lamande, S. R., et al. (2011). Mutations in TRPV4 cause an inherited arthropathy of hands and feet. *Nature Genetics, 43*(11), 1142–1146.

Li, J., Mahajan, A., & Tsai, M. D. (2006). Ankyrin repeat: A unique motif mediating protein-protein interactions. *Biochemistry, 45*(51), 15168–15178.

Longden, T. A., et al. (2017). Capillary K(+)-sensing initiates retrograde hyperpolarization to increase local cerebral blood flow. *Nature Neuroscience, 20*(5), 717–726.

Longden, T. A., et al. (2021). Local IP3 receptor-mediated ca(2+) signals compound to direct blood flow in brain capillaries. *Science Advances*, 7(30).

Ma, X., et al. (2010). Functional role of vanilloid transient receptor potential 4-canonical transient receptor potential 1 complex in flow-induced Ca2+ influx. *Arteriosclerosis, Thrombosis, and Vascular Biology, 30*(4), 851–858.

Mammoto, A., et al. (2009). A mechanosensitive transcriptional mechanism that controls angiogenesis. *Nature, 457*(7233), 1103–1108.

Maniatis, N. A., et al. (2008). Increased pulmonary vascular resistance and defective pulmonary artery filling in caveolin-1 −/− mice. *American Journal of Physiology. Lung Cellular and Molecular Physiology, 294*(5), L865–L873.

Marrelli, S. P., et al. (2007). PLA2 and TRPV4 channels regulate endothelial calcium in cerebral arteries. *American Journal of Physiology. Heart and Circulatory Physiology, 292*(3), H1390–H1397.

Martin, E., et al. (2012). Involvement of TRPV1 and TRPV4 channels in migration of rat pulmonary arterial smooth muscle cells. *Pflügers Archiv, 464*(3), 261–272.

Marziano, C., et al. (2017). Nitric oxide-dependent feedback loop regulates transient receptor potential vanilloid 4 (TRPV4) channel cooperativity and endothelial function in small pulmonary arteries. *Journal of the American Heart Association, 6*(12).

Matthews, B. D., et al. (2010). Ultra-rapid activation of TRPV4 ion channels by mechanical forces applied to cell surface beta1 integrins. *Integrative Biology (Cambridge), 2*(9), 435–442.

Mendoza, S. A., et al. (2010). TRPV4-mediated endothelial Ca2+ influx and vasodilation in response to shear stress. *American Journal of Physiology. Heart and Circulatory Physiology, 298*(2), H466–H476.

Mentzer, S. J., & Konerding, M. A. (2014). Intussusceptive angiogenesis: Expansion and remodeling of microvascular networks. *Angiogenesis, 17*(3), 499–509.

Mercado, J., et al. (2014). Local control of TRPV4 channels by AKAP150-targeted PKC in arterial smooth muscle. *The Journal of General Physiology, 143*(5), 559–575.

Minshall, R. D., et al. (2003). Caveolin regulation of endothelial function. *American Journal of Physiology. Lung Cellular and Molecular Physiology, 285*(6), L1179–L1183.

Moore-Morris, T., et al. (2014). Resident fibroblast lineages mediate pressure overload-induced cardiac fibrosis. *The Journal of Clinical Investigation, 124*(7), 2921–2934.

Moss, A. (2013). The angiopoietin:Tie 2 interaction: A potential target for future therapies in human vascular disease. *Cytokine & Growth Factor Reviews, 24*(6), 579–592.

Nickel, N. P., et al. (2015). Elafin reverses pulmonary hypertension via Caveolin-1-dependent bone morphogenetic protein signaling. *American Journal of Respiratory and Critical Care Medicine, 191*(11), 1273–1286.

Nilius, B., & Flockerzi, V. (2014). Mammalian transient receptor potential (TRP) cation channels. Preface. *Handbook of Experimental Pharmacology, 223*, v–vi.

Nilius, B., Watanabe, H., & Vriens, J. (2003). The TRPV4 channel: Structure-function relationship and promiscuous gating behaviour. *Pflügers Archiv, 446*(3), 298–303.

Oliveira, S. D. S., et al. (2019). Injury-induced shedding of extracellular vesicles depletes endothelial cells of Cav-1 (Caveolin-1) and enables TGF-beta (transforming growth factor-beta)-dependent pulmonary arterial hypertension. *Arteriosclerosis, Thrombosis, and Vascular Biology, 39*(6), 1191–1202.

Ottolini, M., Hong, K., & Sonkusare, S. K. (2019). Calcium signals that determine vascular resistance. *Wiley Interdisciplinary Reviews. Systems Biology and Medicine, 11*(5), e1448.

Ottolini, M., & Sonkusare, S. K. (2021). The calcium signaling mechanisms in arterial smooth muscle and endothelial cells. *Comprehensive Physiology, 11*(2), 1831–1869.

Ottolini, M., et al. (2020a). Local Peroxynitrite impairs endothelial transient receptor potential vanilloid 4 channels and elevates blood pressure in obesity. *Circulation, 141*(16), 1318–1333.

Ottolini, M., et al. (2020b). Mechanisms underlying selective coupling of endothelial ca(2+) signals with eNOS vs. IK/SK channels in systemic and pulmonary arteries. *The Journal of Physiology, 598*(17), 3577–3596.

Pakshir, P., et al. (2020). The myofibroblast at a glance. *Journal of Cell Science, 133*(13).

Pankey, E. A., et al. (2014). Analysis of responses to the TRPV4 agonist GSK1016790A in the pulmonary vascular bed of the intact-chest rat. *American Journal of Physiology. Heart and Circulatory Physiology, 306*(1), H33–H40.

Phelps, C. B., et al. (2010). Differential regulation of TRPV1, TRPV3, and TRPV4 sensitivity through a conserved binding site on the ankyrin repeat domain. *The Journal of Biological Chemistry, 285*(1), 731–740.

Plant, T. D., & Strotmann, R. (2007). TRPV4: A multifunctional nonselective Cation Channel with complex regulation. In W. B. Liedtke, & S. Heller (Eds.), *TRP Ion Channel function in sensory transduction and cellular signaling cascades*. Boca Raton (FL): CRC Press/Taylor & Francis.

Przybylski, M. (2009). A review of the current research on the role of bFGF and VEGF in angiogenesis. *Journal of Wound Care, 18*(12), 516–519.

Radi, R., et al. (1991). Peroxynitrite oxidation of sulfhydryls. The cytotoxic potential of superoxide and nitric oxide. *The Journal of Biological Chemistry, 266*(7), 4244–4250.

Rahaman, S. O., et al. (2014). TRPV4 mediates myofibroblast differentiation and pulmonary fibrosis in mice. *The Journal of Clinical Investigation, 124*(12), 5225–5238.

Risau, W. (1997). Mechanisms of angiogenesis. *Nature, 386*(6626), 671–674.

Saharinen, P., Bry, M., & Alitalo, K. (2010). How do angiopoietins tie in with vascular endothelial growth factors? *Current Opinion in Hematology, 17*(3), 198–205.

Saliez, J., et al. (2008). Role of caveolar compartmentation in endothelium-derived hyperpolarizing factor-mediated relaxation: Ca2+ signals and gap junction function are regulated by caveolin in endothelial cells. *Circulation, 117*(8), 1065–1074.

Scheraga, R. G., et al. (2016). TRPV4 mechanosensitive Ion Channel regulates lipopolysaccharide-stimulated macrophage phagocytosis. *Journal of Immunology, 196*(1), 428–436.

Semenza, G. L. (2007). Vasculogenesis, angiogenesis, and arteriogenesis: Mechanisms of blood vessel formation and remodeling. *Journal of Cellular Biochemistry, 102*(4), 840–847.

Senadheera, S., et al. (2013). Pregnancy-induced remodelling and enhanced endothelium-derived hyperpolarization-type vasodilator activity in rat uterine radial artery: Transient receptor potential vanilloid type 4 channels, caveolae and myoendothelial gap junctions. *Journal of Anatomy, 223*(6), 677–686.

Sgalla, G., et al. (2018). Idiopathic pulmonary fibrosis: Pathogenesis and management. *Respiratory Research, 19*(1), 32.

Song, S., et al. (2014a). Flow shear stress enhances intracellular Ca2+ signaling in pulmonary artery smooth muscle cells from patients with pulmonary arterial hypertension. *American Journal of Physiology. Cell Physiology, 307*(4), C373–C383.

Song, Y., et al. (2014b). TRPV4 channel inhibits TGF-beta1-induced proliferation of hepatic stellate cells. *PLoS One, 9*(7), e101179.

Sonkusare, S. K., et al. (2012). Elementary Ca2+ signals through endothelial TRPV4 channels regulate vascular function. *Science, 336*(6081), 597–601.

Sonkusare, S. K., et al. (2014). AKAP150-dependent cooperative TRPV4 channel gating is central to endothelium-dependent vasodilation and is disrupted in hypertension. *Science Signaling*, 7(333) (p. ra66).

Stevens, T. (2011). Functional and molecular heterogeneity of pulmonary endothelial cells. *Proceedings of the American Thoracic Society, 8*(6), 453–457.

Stewart, L., & Turner, N. A. (2021). Channelling the force to reprogram the matrix: Mechanosensitive ion channels in cardiac fibroblasts. *Cell, 10*(5).

Straub, S. V., et al. (2006). Dynamic inositol trisphosphate-mediated calcium signals within astrocytic endfeet underlie vasodilation of cerebral arterioles. *The Journal of General Physiology, 128*(6), 659–669.

Strotmann, R., Schultz, G., & Plant, T. D. (2003). Ca2+−dependent potentiation of the nonselective cation channel TRPV4 is mediated by a C-terminal calmodulin binding site. *The Journal of Biological Chemistry, 278*(29), 26541–26549.

Strotmann, R., et al. (2000). OTRPC4, a nonselective cation channel that confers sensitivity to extracellular osmolarity. *Nature Cell Biology, 2*(10), 695–702.

Strotmann, R., et al. (2010). Interdomain interactions control Ca2+−dependent potentiation in the cation channel TRPV4. *PLoS One, 5*(5), e10580.

Sullivan, M. N., et al. (2012). Optical recording reveals novel properties of GSK1016790A-induced vanilloid transient receptor potential channel TRPV4 activity in primary human endothelial cells. *Molecular Pharmacology*, *82*(3), 464–472.
Suresh, K., et al. (2018). Reactive oxygen species induced ca(2+) influx via TRPV4 and microvascular endothelial dysfunction in the SU5416/hypoxia model of pulmonary arterial hypertension. *American Journal of Physiology. Lung Cellular and Molecular Physiology*, *314*(5), L893–L907.
Suresh, K., et al. (2019). Regulation of mitochondrial fragmentation in microvascular endothelial cells isolated from the SU5416/hypoxia model of pulmonary arterial hypertension. *American Journal of Physiology. Lung Cellular and Molecular Physiology*, *317*(5), L639–L652.
Syed, N. I., et al. (2010). Characterisation of P2X receptors expressed in rat pulmonary arteries. *European Journal of Pharmacology*, *649*(1–3), 342–348.
Szabadkai, G., et al. (2006). Mitochondrial dynamics and Ca2+ signaling. *Biochimica et Biophysica Acta*, *1763*(5–6), 442–449.
Tajada, S., et al. (2017). Distance constraints on activation of TRPV4 channels by AKAP150-bound PKCalpha in arterial myocytes. *The Journal of General Physiology*, *149*(6), 639–659.
Takahashi, N., et al. (2014). TRPV4 channel activity is modulated by direct interaction of the ankyrin domain to PI(4,5)P(2). *Nature Communications*, *5*, 4994.
Thodeti, C. K., Paruchuri, S., & Meszaros, J. G. (2013). A TRP to cardiac fibroblast differentiation. *Channels (Austin, Tex.)*, 7(3), 211–214.
Thodeti, C. K., et al. (2009). TRPV4 channels mediate cyclic strain-induced endothelial cell reorientation through integrin-to-integrin signaling. *Circulation Research*, *104*(9), 1123–1130.
Thoppil, R. J., et al. (2015). TRPV4 channel activation selectively inhibits tumor endothelial cell proliferation. *Scientific Reports*, *5*, 14257.
Thoppil, R. J., et al. (2016). TRPV4 channels regulate tumor angiogenesis via modulation of rho/rho kinase pathway. *Oncotarget*, 7(18), 25849–25861.
Thorneloe, K. S., et al. (2012). An orally active TRPV4 channel blocker prevents and resolves pulmonary edema induced by heart failure. *Science Translational Medicine*, *4*(159) (p. 159ra148).
Tomilin, V., et al. (2016). Role of renal TRP channels in physiology and pathology. *Seminars in Immunopathology*, *38*(3), 371–383.
Vriens, J., et al. (2005). Modulation of the Ca2 permeable cation channel TRPV4 by cytochrome P450 epoxygenases in vascular endothelium. *Circulation Research*, *97*(9), 908–915.
Wang, M., et al. (2015). Screening genes associated with myocardial infarction and transverse aortic constriction using a combined analysis of miRNA and mRNA microarray. *Gene*, *571*(2), 245–248.
Watanabe, H., et al. (2002a). Heat-evoked activation of TRPV4 channels in a HEK293 cell expression system and in native mouse aorta endothelial cells. *The Journal of Biological Chemistry*, *277*(49), 47044–47051.
Watanabe, H., et al. (2002b). Activation of TRPV4 channels (hVRL-2/mTRP12) by phorbol derivatives. *The Journal of Biological Chemistry*, *277*(16), 13569–13577.
Watanabe, H., et al. (2003). Anandamide and arachidonic acid use epoxyeicosatrienoic acids to activate TRPV4 channels. *Nature*, *424*(6947), 434–438.
Watanabe, H., et al. (2008). TRP channel and cardiovascular disease. *Pharmacology & Therapeutics*, *118*(3), 337–351.
Wellman, G. C., et al. (2002). Ca2+ sparks and their function in human cerebral arteries. *Stroke*, *33*(3), 802–808.

White, C. R., & Frangos, J. A. (2007). The shear stress of it all: The cell membrane and mechanochemical transduction. *Philosophical Transactions of the Royal Society of London. Series B, Biological Sciences*, *362*(1484), 1459–1467.

Wu, S., et al. (2009). Ca2+ entry via alpha1G and TRPV4 channels differentially regulates surface expression of P-selectin and barrier integrity in pulmonary capillary endothelium. *American Journal of Physiology. Lung Cellular and Molecular Physiology*, *297*(4), L650–L657.

Xia, Y., et al. (2013). TRPV4 channel contributes to serotonin-induced pulmonary vasoconstriction and the enhanced vascular reactivity in chronic hypoxic pulmonary hypertension. *American Journal of Physiology. Cell Physiology*, *305*(7), C704–C715.

Yang, X. R., et al. (2012). Upregulation of osmo-mechanosensitive TRPV4 channel facilitates chronic hypoxia-induced myogenic tone and pulmonary hypertension. *American Journal of Physiology. Lung Cellular and Molecular Physiology*, *302*(6), L555–L568.

Yin, J., et al. (2008). Negative-feedback loop attenuates hydrostatic lung edema via a cGMP-dependent regulation of transient receptor potential vanilloid 4. *Circulation Research*, *102*(8), 966–974.

Yue, L., Xie, J., & Nattel, S. (2011). Molecular determinants of cardiac fibroblast electrical function and therapeutic implications for atrial fibrillation. *Cardiovascular Research*, *89*(4), 744–753.

Zhan, L., & Li, J. (2018). The role of TRPV4 in fibrosis. *Gene*, *642*, 1–8.

Zhan, L., et al. (2015). Transient receptor potential vanilloid 4 inhibits rat HSC-T6 apoptosis through induction of autophagy. *Molecular and Cellular Biochemistry*, *402*(1–2), 9–22.

Zhang, D. X., et al. (2009). Transient receptor potential vanilloid type 4-deficient mice exhibit impaired endothelium-dependent relaxation induced by acetylcholine in vitro and in vivo. *Hypertension*, *53*(3), 532–538.

Zhang, P., et al. (2018). TRPV4 (transient receptor potential vanilloid 4) mediates endothelium-dependent contractions in the aortas of hypertensive mice. *Hypertension*, *71*(1), 134–142.

Zhao, Y. Y., et al. (2002). Defects in caveolin-1 cause dilated cardiomyopathy and pulmonary hypertension in knockout mice. *Proceedings of the National Academy of Sciences of the United States of America*, *99*(17), 11375–11380.

Zhao, L., et al. (2014). Calcineurin/nuclear factor of activated T cells-coupled vanilliod transient receptor potential channel 4 ca2+ sparklets stimulate airway smooth muscle cell proliferation. *American Journal of Respiratory Cell and Molecular Biology*, *50*(6), 1064–1075.

Zheng, X., et al. (2013). Arachidonic acid-induced dilation in human coronary arterioles: Convergence of signaling mechanisms on endothelial TRPV4-mediated Ca2+ entry. *Journal of the American Heart Association*, *2*(3), e000080.

CHAPTER TWO

Endothelial TRPV4 channels in lung edema and injury

Swapnil K. Sonkusare[a,b,*] and Victor E. Laubach[c]
[a]Robert M. Berne Cardiovascular Research Center, University of Virginia, Charlottesville, VA, United States
[b]Department of Molecular Physiology and Biological Physics, University of Virginia, Charlottesville, VA, United States
[c]Department of Surgery, University of Virginia, Charlottesville, VA, United States
*Corresponding author: e-mail address: swapnil.sonkusare@virginia.edu

Contents

Abstract

The alveolo-capillary barrier is relatively impermeable, and facilitates gas exchange *via* the large alveolar surface in the lung. Disruption of alveolo-capillary barrier leads to accumulation of edema fluid in lung injury. Studies in animal models of various forms of lung injury provide evidence that TRPV4 channels play a critical role in disruption of the alveolo-capillary barrier and pathogenesis of lung injury. TRPV4 channels from capillary endothelial cells, alveolar epithelial cells, and immune cells have been implicated in the pathogenesis of lung injury. Recent studies in endothelium-specific TRPV4 knockout mice point to a central role for endothelial TRPV4 channels in lung injury. In this chapter, we review the findings on the pathological roles of endothelial TRPV4 channels in different forms of lung injury and future directions for further investigation.

Current Topics in Membranes, Volume 89
ISSN 1063-5823
https://doi.org/10.1016/bs.ctm.2022.07.001

1. Introduction

In 1969, Cosens and Manning reported a Drosophila mutant that was defective in light sensing and exhibited only transient light-induced receptor potentials (TRPs) instead of the normal sustained response (Cosens & Manning, 1969). This finding was explained by a defect in an ion channel, which later led to the discovery of the large family of TRP channels. The mammalian TRP channel superfamily consists of 28 members divided into six subfamilies based on their amino acid sequence homology: TRPC (Canonical), TRPM (Melastatin), TRPML (Mucolipins), TRPV (Vanilloid), TRPP (Polycystic), and TRPA (Ankyrin-rich protein). TRP channels exhibit diverse permeation and gating properties and regulate a plethora of signaling mechanisms that control physiological function in multiple cell types. TRP channels can be activated by polymodal stimuli, including physical (voltage, temperature, and mechanical stimuli) and chemical (neurohumoral mediators and second messengers) stimuli (Jara-Oseguera & Islas, 2013; Nieto-Posadas, Jara-Oseguera, & Rosenbaum, 2011). The TRPV subfamily can be further subdivided into six isoforms: TRPV1–TRPV6. Within the vasculature, TRPV channels are expressed in smooth muscle cells, endothelial cells, fibroblasts, and perivascular nerves (Filosa, Yao, & Rath, 2013; Ottolini & Sonkusare, 2021). Studies on TRPV channels at the plasma membrane demonstrate crucial roles for these channels in Ca^{2+} homeostasis in the vasculature. TRPV1 and TRPV4 are the most studied members of the TRPV family in the systemic and pulmonary vasculature.

In the lungs, TRPV4 channels are expressed in multiple cell types, including epithelial cells, immune cells, and cells in the vascular wall. TRPV4 channels have been associated with many physiological functions and pathological conditions in the lung. Studies in different models suggest a critical role for TRPV4 channels in alveolo-capillary barrier in the lung (Rajan et al., 2021), resulting in lung edema and injury. The alveolo–capillary barrier plays an essential role in gas exchange, permeability regulation, fluid clearance, and host defense. This barrier is formed by alveolar epithelium, capillary endothelium, and a basement membrane that separates alveolar epithelium and capillary endothelium. Disrupted alveolo-capillary barrier, which results in accumulation of edema fluid in alveolar spaces, is an important pathological feature of lung edema and injury. This chapter focuses on the role of TRPV4 channels in disrupting pulmonary alveolo-capillary barrier, thereby causing lung edema and injury. The role of TRPV4 channels in the airway and immune cells has been discussed in detail elsewhere (Belvisi & Birrell, 2017; Rajan et al., 2021).

2. Structure and properties

Recently elucidated single-particle cryo-electron microscopy (cryo-EM) structure of *Xenopus tropicalis* TRPV4 channel under near-atomic resolution revealed a symmetric tetramer formed by four subunits, each with six transmembrane segments (S1–S6) and linking loops (Deng et al., 2018). Among the six transmembrane segments, S5, S6, and the inter-connecting loop form the central cation-permeable pore, and S3 and S4 appear to form an agonist-binding pocket that influences channel gating (recently reviewed in (Chen & Sonkusare, 2020; Rosenbaum et al., 2020; Toft-Bertelsen & MacAulay, 2021). Moreover, the inner helices formed an intracellular gate in the ion-conduction pore but lacked an extra-cellular gate in the selectivity filter that was seen in other TRPV family members, possibly explaining the less selective nature of the channel compared to other TRPV channels. Anomalous X-ray diffraction analyses also identified a single ion-binding site in an unusually wide selectivity filter (Deng et al., 2018). Thus, unique structural features of TRPV4 channels may underlie their distinct gating mechanisms and selectivity features when compared to other ion channels of TRP family.

The intracellular N- and C-termini of TRPV4 channels include multiple regions for protein–protein interactions that regulate the assembly and activity of the channel and its interaction with the cytoskeleton (Harteneck & Schultz, 2007; Nilius & Voets, 2013; White et al., 2016). The N-terminus contains ankyrin repeat domains, a feature found in all channels of TRPC and TRPV subfamilies, and proline-rich domain (Everaerts, Nilius, & Owsianik, 2010; Inada, Procko, Sotomayor, & Gaudet, 2012; Phelps, Huang, Lishko, Wang, & Gaudet, 2008). The C-terminal tail contains a TRP box that holds binding sites for calmodulin, actin, and tubulin and an amino acid stretch (Asp-Ala-Pro-Leu) similar to a PDZ-binding like motif.

TRPV4 is a cation channel characterized by a moderately high Ca^{2+} permeability ratio (P_{Ca}/P_{Na} of ~6) and a single-channel conductance of ~90 pS (Clapham, Montell, Schultz, Julius, & International Union of Pharmacology, 2003). Residues Asp672 and Asp682 in the ionic pathway generated by the pore-forming loop between the S5 and S6 domains modulate the ion selectivity: alanine substitution of either or both asparagines causes a reduction in the relative permeability for divalent cations (Voets et al., 2002). Intracellular Ca^{2+} can inhibit or potentiate TRPV4 channels in a concentration-dependent manner (Sonkusare et al., 2012, 2014; Strotmann, Harteneck, Nunnenmacher, Schultz, & Plant, 2000), a mechanism that has been attributed to residue F707 (Watanabe et al., 2003). The biogenesis of TRPV4 channels involves N-glycosylation on residue N651 (Toft-Bertelsen & MacAulay, 2021) and oligomerization in the ER, followed by a transfer to the Golgi apparatus for subsequent maturation. The channel usually organizes as a homotetrameric structure (Arniges, Fernandez-Fernandez, Albrecht, Schaefer, & Valverde, 2006) requiring both the N- and the C-termini (Ryskamp et al., 2014), while heterotetramers can occur with TRPC1 (Ma et al., 2011), TRPP2 (Stewart, Smith, Sandford, & Edwardson, 2010), and TRPC1–TRPP2 channels (Du et al., 2014). Thus, studies of the biophysical properties of TRPV4 channels and TRPV4 heteromultimer may provide crucial insights into the mechanisms for altered channel function in lung injury.

3. TRPV4 channels in the lung

TRPV4 channels are expressed in numerous cells of the respiratory tract (airway smooth muscle cells, bronchial epithelial cells, alveolar type 1 (AT1) and type 2 (AT2) epithelial cells), pulmonary vascular cells (endothelial cells, smooth muscle cells and fibroblasts), and inflammatory cells (macrophages, neutrophils and T-cells) (Baxter et al., 2014; Daneva, Laubach, & Sonkusare, 2019; Groot-Kormelink, Fawcett, Wright, Gosling, & Kent, 2012; Hamanaka et al., 2010; Jia et al., 2004; Majhi et al., 2015; Marziano et al., 2017; Rahaman et al., 2014; Xia et al., 2013; Yang et al., 2012). While long and flat AT1 cells maintain the integrity of alveolar epithelial barrier, cubic AT2 cells produce surfactants to reduce surface tension and enhance gas exchange across the epithelium. Accordingly, epithelial barrier function was impaired and epithelial surfactant levels were lower in global $TRPV4^{-/-}$ mice (Weber et al., 2020). TRPV4 agonist GSK1016790A (GSK101) increased endothelial permeability in pulmonary capillaries (Rajan et al., 2021; Villalta & Townsley, 2013; Willette et al., 2008). Thus, TRPV4 channel activity appears to be essential for the regulation of epithelial and endothelial barrier function, and therefore, may play a role in defense against invading pathogens and toxicants in the lung.

TRPV4 channel dysfunction is observed in many diseases of the lung, including pulmonary hypertension, asthma, acute respiratory distress syndrome, chronic obstructive pulmonary disorder (COPD), cystic fibrosis (*CF*), lung fibrosis, pulmonary edema, and lung injury. TRPV4 channels in vascular smooth muscle and endothelial cells from small pulmonary arteries have distinct effects on pulmonary arterial pressure. Endothelium-specific TRPV4 knockout mice exhibited an elevated pulmonary arterial pressure, whereas smooth muscle-specific TRPV4 knockout mice showed unaltered pulmonary arterial pressure (Daneva, Marziano, et al., 2021; Daneva, Ottolini, et al., 2021). In pulmonary hypertension, the activity of smooth muscle TRPV4 channels is increased whereas the activity of endothelial TRPV4 channels is reduced (Daneva, Marziano, et al., 2021; Goldenberg et al., 2015). Airway smooth muscle TRPV4 channels increase the contractile response of the airway in asthma and promote transforming growth factor beta (TGFβ)-dependent remodeling of the airway (Bonvini et al., 2020). TRPV4 channels from multiple cell types in the lung contribute to the pathogenesis of acute lung injury (ALI) and acute respiratory distress syndrome (ARDS). In alveolar epithelial and endothelial cells, TRPV4 channel overactivation impairs epithelial and endothelial septal barrier, leading to edema and lung injury (Villalta & Townsley, 2013). TRPV4 channels from neutrophils increase ROS production (Scheraga, Southern, Grove, & Olman, 2017) and facilitate neutrophil activation in lung injury (Yin et al., 2016). In contrast, some studies show essential roles for TRPV4 channels in maintaining the integrity of alveolar barrier and protecting against edema formation (Weber et al., 2020). TRPV4 channels are potential therapeutic targets for lung disorders. Many specific TRPV4 channel activators and inhibitors have been developed in the past two decades (Chen & Sonkusare, 2020). Indeed, one TRPV4 inhibitor—GSK2798745—has been tested clinically and was found to be well-tolerated (Goyal et al., 2019). Identification of GSK2798745 as a clinically safe TRPV4 inhibitor may lead to additional clinical trials in individuals suffering from lung edema and injury.

4. TRPV4 channels in the pulmonary endothelium

Studies over the past two decades establish the functional expression of TRPV4 channels in arterial, capillary, and septal endothelium in the lung. Most studies have focused on the contribution of capillary endothelial TRPV4 channels to pulmonary pathologies, including lung edema and lung injury. However, the physiological roles of capillary endothelial TRPV4 channels have remained unclear. Recent studies show that the activation of endothelial TRPV4 channels dilates small, resistance-sized pulmonary arteries. In small pulmonary arteries, Ca^{2+} influx events through TRPV4 channels (TRPV4 sparklets) selectively activate endothelial nitric oxide synthase (eNOS) signaling and NO release (Daneva et al., 2019; Marziano et al., 2017; Ottolini, Daneva, et al., 2020). In resistance-sized systemic arteries, however, Ca^{2+} influx events through TRPV4 channels activate endothelial Ca^{2+}-activated K^{+} (IK and SK) channels (Ottolini, Daneva, et al., 2020; Ottolini, Hong, et al., 2020; Ottolini, Hong, & Sonkusare, 2019; Ottolini & Sonkusare, 2021; Sonkusare et al., 2012, 2014). Studies in endothelium-specific TRPV4 channel knockout mice suggested that endothelial TRPV4 channels lower pulmonary artery constriction and resting pulmonary arterial pressure (Daneva, Marziano, et al., 2021; Daneva, Ottolini, et al., 2021). Moreover, in mouse models of pulmonary hypertension and pulmonary arterial hypertension patients, the activity of endothelial TRPV4 channels was decreased in small pulmonary arteries, contributing to reduced vasodilation and elevated pulmonary arterial pressure (Daneva, Marziano, et al., 2021). On the contrary, smooth muscle-specific TRPV4 channel knockout did not result in alteration of basal pulmonary arterial pressure (Daneva, Marziano, et al., 2021; Daneva, Ottolini, et al., 2021), indicating that smooth muscle TRPV4 channels may not play a role in the regulation of pulmonary arterial pressure under normal conditions. Other studies linked smooth muscle TRPV4 channels with hypoxia- and G-protein coupled receptor signaling-induced constriction of pulmonary arteries and elevation of pulmonary arterial pressure (Goldenberg et al., 2015; Xia et al., 2013; Yang et al., 2012). Thus, endothelial and smooth muscle TRPV4 channels appear to differentially regulate the pulmonary arterial pressure under normal conditions and in pulmonary hypertension. Recent studies also showed that the efflux of adenosine triphosphate (ATP) through endothelial Pannexin 1, an ATP efflux pathway at the cell membrane, activates endothelial TRPV4 channels *via* P2 purinergic receptor signaling and lowers pulmonary arterial pressure (Daneva, Ottolini, et al., 2021). Moreover, this pathway was shown to be activated by increases in flow/shear stress. Since pulmonary arteries are a high-flow, low-resistance circulation, shear stress-activation of endothelial pannexin 1–P2Y2 receptor–TRPV4 signaling pathway needs to be investigated as a new mechanism for flow-induced dilation of pulmonary arteries.

Pulmonary capillary endothelium is structurally and functionally different from arterial endothelium (Stevens, 2011) (Fig. 1). TRPV4 channel

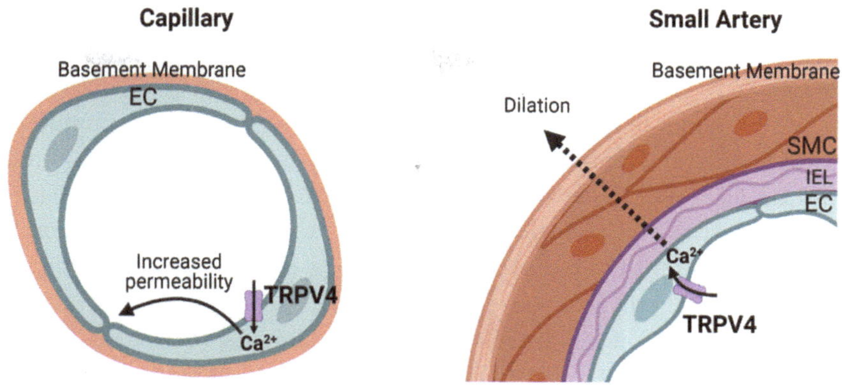

Fig. 1 Distinct functions of endothelial TRPV4 channels in capillaries and arteries. *Left*, TRPV4 channels in capillary endothelial cells increase endothelial permeability; *right*, TRPV4 channels in arterial endothelial cells relax smooth muscle cells and cause vasodilation.

activity in capillary endothelial cells has mostly been associated with increased permeability and detrimental effects in pulmonary pathologies (Villalta & Townsley, 2013). Endothelial Ca^{2+} signaling mechanisms involving TRPV4 channels have distinct effects on capillary and arterial function (Fig. 1) (Alvarez et al., 2006; Marziano et al., 2017; Ottolini, Daneva, et al., 2020; Willette et al., 2008). TRPV4 channels couple to disparate targets and regulatory mechanisms in capillary and arterial endothelium (Harraz, Longden, Eubanks, & Nelson, 2018; Longden et al., 2017; Sonkusare et al., 2012; Sonkusare, Dalsgaard, Bonev, & Nelson, 2016). Capillary endothelium regulates capillary permeability but, unlike pulmonary arterial endothelium, does not play an active role in the regulation of pulmonary arterial pressure. Capillary endothelial TRPV4 channel activity has been associated with endothelial hyperpermeability, pulmonary edema, and lung injury. Current evidence in the literature suggests that pathological conditions affect capillary and arterial TRPV4 channels differently (Daneva, Marziano, et al., 2021; Suresh et al., 2018). Studies in primary microvascular endothelial cells, which likely represent capillary endothelium, suggest that the activity of endothelial TRPV4 channels is elevated in pulmonary hypertension, contributing to endothelial cell migration and proliferation (Suresh et al., 2018, 2015). This is in contrast with pulmonary artery endothelium, where the activity of TRPV4 channels is reduced in pulmonary hypertension (Daneva, Marziano, et al., 2021). Since capillary and alveolar septal endothelium is directly involved in lung edema and injury, the following sections will review the findings on TRPV4 channels in capillary or alveolar septal endothelium.

5. TRPV4 channels in lung edema and injury

The endothelium of the pulmonary circulation is an important barrier to protect the lung tissue from toxic substances and pathogens circulating in the blood. However, in response to bacterial infections of the lung tissue, endothelial cell permeability increases to facilitate the migration of immune cells from the blood to the alveoli. As a result, protein-rich fluid accumulates in the alveolar space, causing life-threatening pulmonary edema (Rajan et al., 2021). Pulmonary edema can be cardiogenic (heart failure) or non-cardiogenic (ALI or ARDS) (Komiya, Akaba, Kozaki, Kadota, & Rubin, 2017). Overwhelming evidence in the literature suggests that TRPV4 is a key ion channel for increasing pulmonary endothelial permeability in both cardiogenic and non-cardiogenic pulmonary edema. TRPV4 channel activation can induce lung endothelial/epithelial barrier disruption, a critical feature of lung injury, suggesting that TRPV4 channel inhibition could be of therapeutic value (Fig. 2). Interestingly, a recent study showed that macrophage TRPV4 channel protects the lung from injury upon

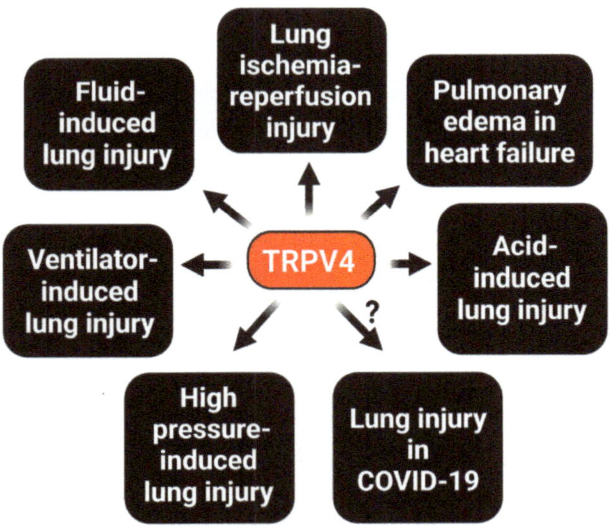

Fig. 2 Role of TRPV4 channels in different forms of pulmonary edema and lung injury. TRPV4 channels play an important role lung ischemia/reperfusion injury, heart failure-induced pulmonary edema, fluid-, acid-, ventilator-, and high pressure-induced lung injury. A potential role for TRPV4 channels has also been proposed in pulmonary edema and lung injury in COVID-19.

intratracheal *Pseudomonas aeruginosa* exposure in mice (Scheraga, Abraham, et al., 2020). TRPV4 channels enhanced macrophage bacterial clearance and reduced the levels of proinflammatory cytokines. These studies indicate that TRPV4 channels in different pulmonary cell types can play distinct roles in lung injury, which can lead to different pathologies depending on whether infection-driven (e.g., *Pseudomonas* infection) or sterile-driven (e.g., ischemia-reperfusion) inflammatory processes are involved. Further studies using cell-specific knockout models will be necessary for understanding the cell type-specific roles of TRPV4 channels in lung injury.

5.1 Heart failure-induced pulmonary edema

Heart failure-induced pulmonary edema resulting from high pulmonary venous pressure (PVP) is a major cause of morbidity and mortality in heart failure patients. High PVP results in increased capillary permeability, ultimately leading to pulmonary edema. Current treatment options do not address high PVP-induced pulmonary edema in heart failure. Recent evidence from rodent lungs suggests that heart failure-induced pulmonary edema is driven by activation of pulmonary capillary endothelial TRPV4 channels (Thorneloe et al., 2012). Furthermore, in both acute and chronic heart failure models, pretreatment with TRPV4 antagonist, GSK2193874, inhibited the formation of pulmonary edema and enhanced arterial oxygenation. GSK2193874 treatment also resolved pulmonary edema in a mouse model of myocardial infarction. These findings identified a crucial role for TRPV4 channels in the pathogenesis of heart failure-induced pulmonary edema and suggest that TRPV4 inhibition is a potential therapeutic strategy for pulmonary edema in heart failure patients (Thorneloe et al., 2012).

5.2 Lung ischemia-reperfusion injury

Lung ischemia-reperfusion injury (IRI), involving severe inflammation, edema, alveolar damage, and hypoxemia, is a major cause of primary graft failure after transplant (de Perrot, Liu, Waddell, & Keshavjee, 2003; Haywood et al., 2021). IRI is characterized by robust inflammation, alveolar damage, and increased vascular permeability. TRPV4 channel activation has been linked to lung injury through endothelial barrier disruption, edema, and inflammatory cell recruitment and activation. Therefore, TRPV4 channels may be a promising therapeutic target to mitigate lung IRI and decrease the incidence of primary graft dysfunction after transplant.

In this regard, Haywood and colleagues demonstrated that endothelial TRPV4 channel plays a key role in the development of inflammation and pulmonary edema after IR (Haywood et al., 2022). Using clinically relevant measurements of lung function (PaO_2 and compliance), the authors found that endothelium-specific TRPV4 knockout or pharmacologic inhibition of TRPV4 channels preserves lung function and reduces inflammation and edema after IR. These findings supported previous reports demonstrating that TRPV4 channel activation can lead to increased endothelial permeability, edema, and immune cell infiltration and activation (Dutta et al., 2020; Rayees et al., 2019; Scheraga, Southern, Grove, & Olman, 2020; Villalta & Townsley, 2013). Studies in endothelium-specific TRPV4 knockout mice suggested that the primary protective effect of TRPV4 inhibition in lung IRI is localized to the pulmonary endothelium (Haywood et al., 2022). Therefore, pharmacological inhibition of TRPV4 channels may be an effective strategy to treat lung IRI and prevent primary graft dysfunction after transplantation.

Studies in pulmonary arteries have shown that ATP efflux through endothelial pannexin 1 can activate TRPV4 channels through P2 purinergic receptor signaling. In this regard, Sharma et al. used pannexin 1 endothelial knockout mice to demonstrate that endothelial pannexin 1 plays a key role in mediating vascular permeability, inflammation, edema, leukocyte infiltration, and lung dysfunction after IR (Sharma et al., 2018). Since endothelium-specific deletion of TRPV4 or pannexin 1 appears to be protective against lung IRI, it is plausible that an endothelial pannexin 1–TRPV4 interaction is involved in the pathogenies of lung IRI. A recent study by Daneva and colleagues showed an interaction between pannexin 1 and TRPV4 channels in the intact endothelium from small pulmonary arteries (Daneva, Ottolini, et al., 2021). Pannexin 1–TRPV4 interaction in capillary endothelial cells and its role in lung IRI has not yet been verified.

5.3 Fluid-induced acute lung injury

Rapid administration of fluids, or fluid resuscitation, is a common clinical practice. Notably, administration of bolus intravenous fluid has been associated with respiratory dysfunction and increased mortality (Maitland et al., 2011). Administration of intravenous saline solution resulted in pulmonary edema in rats and mice (Bihari et al., 2017). Global knockout of TRPV4 channels and administration of TRPV4 inhibitor prevented the development of intravenous fluid-induced pulmonary edema, suggesting that the activation of TRPV4 channels is a central mechanism in fluid-induced acute lung injury (Bihari et al., 2017). However, the exact cell-types contributing to the effect of TRPV4 channels in fluid-induced lung injury remain unknown.

5.4 Ventilator-induced lung injury

Mechanical ventilation is a crucial tool for supporting critically ill patients but can cause lung injury. Ventilator-induced lung injury is associated with disrupted pulmonary endothelial barrier and release of pro-inflammatory cytokines. TRPV4 channels play an essential role in the activation of macrophages in ventilator-induced lung injury (Hamanaka et al., 2010). Hamanaka et al. showed that high pressure ventilation induces pulmonary edema in mice, and this effect is absent in TRPV4$^{-/-}$ mice (Hamanaka et al., 2007). In a follow-up study, this group reported that alveolar macrophages from wild-type mice restored the susceptibility of TRPV4$^{-/-}$ mice to ventilator-induced lung injury, implying that TRPV4 channel activation in alveolar macrophages is critical to ventilator-induced lung injury. Michalik and colleagues demonstrated that pharmacological inhibition of TRPV4 channels and serum glucocorticoid-regulated kinase 1 (SGK1) reduces endothelial Ca^{2+} influx and vascular leakage in an isolated lung model of ventilation-induced lung injury (Michalick et al., 2017). Further studies demonstrated that lung ventilation promotes endothelial Ca^{2+} influx and barrier disruption through the activation of TRPV4 channels, presumably through phosphorylation at serine 824 residue by SGK1. Thus, TRPV4 and SGK1 were proposed as promising new targets for prevention or treatment of ventilator-induced lung injury (Michalick et al., 2017). A decrease in cytokine release and endothelial permeability following TRPV4 inhibition suggested that TRPV4 inhibitors may be utilized as a prophylactic treatment to improve the pathological response of lung cells during ventilation and potentially support patients receiving mechanical ventilation (Pairet et al., 2018). Intriguingly, exosomes derived from human adipocytes protected mice against ventilator-induced lung injury *via* inhibition of Ca^{2+} influx through TRPV4 channels (Yu et al., 2020). Thus, new approaches to lower TRPV4 channel activity may be effective in clinical treatment of lung edema and injury.

5.5 High intravascular pressure-induced lung injury

Abnormally high intravascular pressure can increase endothelial permeability at the alveolar-septal barrier. Disruption of the alveolar-septal barrier leads to acute lung injury, patchy alveolar flooding, and hypoxemia. Multiple studies implicate TRPV4 channels in disruption of the alveolar-septal barrier and pathogenesis of acute lung injury (Alvarez et al., 2006). Jian et al. showed that high intravascular pressure-induced disruption of alveolar-septal barrier was reduced in the lungs from global TRPV4$^{-/-}$ mice. Further, the authors suggested a critical role for epoxyeicosatrienoic acid-dependent activation of TRPV4 channels in disruption of alveolar-septal barrier following high intravascular pressure (Jian, King, Al-Mehdi, Liedtke, & Townsley, 2008). In a follow-up study, the group showed that increased lung endothelial permeability in response to 14,15-epoxyeicosatrienoic acid (14,15-EET) in rat lungs requires Ca^{2+} entry *via* TRPV4 channels and functional coupling of TRPV4 channels with Ca^{2+}-sensitive, intermediate, and small conductance (IK and SK) K^+ channels (Lin et al., 2015). Such "functional coupling" suggests that endothelial TRPV4 channels likely form signaling microdomains with IK and SK channels and dictate the extent of lung endothelial injury caused by 14,15-EET.

5.6 Acid exposure-induced lung injury

The treatment of acute lung injury caused by exposure to reactive chemicals remains challenging because of the lack of mechanism-based therapeutic approaches. In a study by Balakrishna and colleagues, the effects of two new TRPV4 inhibitors (GSK2220691, GSK2337429A) were examined in mice exposed to hydrochloric acid, mimicking acid exposure and acid aspiration injury, and to chlorine gas, a severe chemical threat with frequent exposures in domestic and occupational environments and transportation accidents. The authors demonstrated that TRPV4 inhibitors are potent and efficacious countermeasures against severe chemical exposures, acting against exaggerated inflammatory responses, and protecting tissue barriers and cardiovascular function (Balakrishna et al., 2014). These findings pointed to TRPV4 channel-dependent mechanisms of chemical-induced lung injury and provided proof-of-principle for TRPV4 channel inhibition as a treatment strategy.

6. Potential role of TRPV4 channels in lung edema and injury in COVID-19 patients

Disruption of the alveolo-capillary barrier as a direct consequence of severe acute respiratory syndrome coronavirus type 2 (SARS-CoV2) infection is a critical driver of lethality in the recent coronavirus disease 2019 (COVID-19) pandemic. Therefore, there is an urgent need for alveolo-capillary barrier-stabilizing drugs in the COVID-19 pandemic. TRPV4 channel inhibition is protective in various animal models of lung edema as described above, and overactivation of TRPV4 channels damages the alveolo-capillary barrier of the lungs (Rajan et al., 2021; Villalta & Townsley, 2013). Moreover, TRPV4 channels are expressed in all the cell types of the alveolo-capillary unit, including AT1 and AT2 cells and capillary endothelial cells. TRPV4 channels also regulate the activation of alveolar macrophages and neutrophils, which contribute to alveolo-capillary barrier disruption *via* the release of proteases, cytokines, and reactive oxygen species (Balakrishna et al., 2014; Dutta et al., 2020; Hamanaka et al., 2010; Scheraga, Southern, et al., 2020; Yin et al., 2016). Since SARS-CoV2 also infects endothelial cells (Liu et al., 2021), a recent article by Kuebler and colleagues (Kuebler, Jordt, & Liedtke, 2020) proposed that TRPV4 channel inhibition could be a promising approach to protect the alveolo-capillary barrier of the lungs in COVID-19 pandemic. The authors suggested that GSK2798745 (TRPV4 channel inhibitor) may be a powerful therapeutic option in COVID-19 patients. Indeed, *in vitro* screening has identified TRPV4 channel as a target for endothelial barrier stabilization in COVID-19 (Michalick et al., 2022). In the first clinical trial, GSK2798745 did not affect lipopolysaccharide (LPS)-induced elevation of total protein and neutrophils in the bronchoalveolar lavage despite blood and lung exposures that were predicted to be efficacious (Mole et al., 2020). However, the central role of TRPV4 channels in mediating vascular inflammation and endothelial barrier dysfunction provides rationale supporting investigation of TRPV4-targeting therapies against COVID-19 pandemic.

7. Conclusions and future directions

The current evidence in the literature points to a critical role for TRPV4 channels from endothelial cells, immune cells, and alveolar epithelial cells in the pathogenesis of lung injury. Modulating TRPV4 channel activity is a potentially useful therapeutic approach for lung edema and injury, but clinically successful drugs or genetic strategies need to be established. It should be noted that most studies in TRPV4 knockout mice have used global knockout mice, making it difficult to decipher the contribution of TRPV4 channels from individual cell types in the lung. Recent studies in endothelium-specific TRPV4 channel knockout mice indicate a central role for endothelial TRPV4 channels in lung ischemia-reperfusion injury. Similar studies using neutrophil/macrophage/alveolar epithelial cell-specific knockout mice are needed for confirming the contribution of TRPV4 channels from these cell types to lung injury. The interactions amongst immune cell, endothelial cells, and alveolar epithelial cells, and potential roles of TRPV4 channels in regulating these interactions will be an interesting area for future investigations on lung injury. For example, it is plausible that the release of cytokines from immune cells alters the activity of endothelial TRPV4 channels in lung injury. Previous studies in other vascular beds have shown that inflammatory cytokines can alter the activity of endothelial TRPV4 channels (Maier-Begandt et al., 2021).

TRPV4 channels are expressed and are functional in endothelial cells, immune cells, alveolar epithelial cells, and vascular and airway smooth muscle cells in the lung. Studies in cell-specific knockout mice will provide much needed information on the physiological roles of TRPV4 channels from different cell types in the lung. Although numerous studies in the literature point to a critical role of endothelial TRPV4 channels in lung injury, a direct recording of ionic currents through endothelial TRPV4 channels (Chen et al., 2022; Ottolini, Daneva, et al., 2020; Ottolini, Hong, et al., 2020) in a model of lung injury has not yet been reported. Furthermore, individual Ca^{2+} influx events through TRPV4 channels (TRPV4 sparklets) (Chen et al., 2021; Sonkusare et al., 2012) have not been recorded in endothelial cells in a lung injury model. Since intracellular Ca^{2+} is a major contributor to the effects of TRPV4 channels on endothelial permeability, studies of TRPV4 sparklets will provide a more in-depth understanding of alterations in TRPV4 channel activity in lung injury. Direct studies of TRPV4 channel activity– TRPV4 channel currents and TRPV4 sparklets–will enable a more thorough examination of TRPV4 channels as a therapeutic target in lung injury.

TRPV4 channels can interact with multiple other ion channels. Recent studies demonstrate direct or indirect interactions of TRPV4 channels with IK/SK channels, large conductance, Ca^{2+}-activated BK channels and inward rectifier K^+ channels (Chen et al., 2022; Earley, Heppner, Nelson, & Brayden, 2005; Sonkusare et al., 2012, 2016) in vascular endothelial and smooth muscle cells. Such interactions could contribute to the deleterious effect of TRPV4 channel activity in lung injury and may provide new targets to alter the harmful effects of TRPV4 channels. It should also be noted that multiple scaffolding proteins, including caveolin-1 (Cav-1) (Daneva, Marziano, et al., 2021; Daneva, Ottolini, et al., 2021) and A-kinase anchoring protein 150 (AKAP150) (Mercado et al., 2014; Ottolini, Hong, et al., 2020; Sonkusare et al., 2014) promote the activity of vascular TRPV4 channels. Whether alterations in these scaffolding proteins contribute to abnormal activity of TRPV4 channels in lung injury has not been investigated. Finally, nitric oxide is a major signaling molecule in the pulmonary circulation. Studies have shown that nitric oxide can limit the coupling amongst endothelial TRPV4 channels, thereby reducing their activity in pulmonary endothelial cells (Marziano et al., 2017). Thus, nitric oxide–TRPV4 channel interaction could be explored as a pathological mechanism for lung injury and potential therapeutic target for restoring TRPV4 channel activity. Future studies are expected to solidify our understanding of TRPV4 channel as a therapeutic target in lung injury and identify new proteins that can be targeted to lower the harmful effects of TRPV4 channels in lung injury.

Funding

This work was in part funded by awards HL146914, HL142808, and HL157407 from the National Institutes of Health.

References

Alvarez, D. F., King, J. A., Weber, D., Addison, E., Liedtke, W., & Townsley, M. I. (2006). Transient receptor potential vanilloid 4-mediated disruption of the alveolar septal barrier: A novel mechanism of acute lung injury. *Circulation Research, 99*, 988–995.

Arniges, M., Fernandez-Fernandez, J. M., Albrecht, N., Schaefer, M., & Valverde, M. A. (2006). Human TRPV4 channel splice variants revealed a key role of ankyrin domains in multimerization and trafficking. *The Journal of Biological Chemistry, 281*, 1580–1586.

Balakrishna, S., Song, W., Achanta, S., Doran, S. F., Liu, B., Kaelberer, M. M., et al. (2014). TRPV4 inhibition counteracts edema and inflammation and improves pulmonary function and oxygen saturation in chemically induced acute lung injury. *American Journal of Physiology. Lung Cellular and Molecular Physiology, 307*, L158–L172.

Baxter, M., Eltom, S., Dekkak, B., Yew-Booth, L., Dubuis, E. D., Maher, S. A., et al. (2014). Role of transient receptor potential and pannexin channels in cigarette smoke-triggered ATP release in the lung. *Thorax, 69*, 1080–1089.

Belvisi, M. G., & Birrell, M. A. (2017). The emerging role of transient receptor potential channels in chronic lung disease. *The European Respiratory Journal, 50*, 1601357.

Bihari, S., Dixon, D. L., Lawrence, M. D., De Bellis, D., Bonder, C. S., Dimasi, D. P., et al. (2017). Fluid-induced lung injury-role of TRPV4 channels. *Pflügers Archiv, 469*, 1121–1134.

Bonvini, S. J., Birrell, M. A., Dubuis, E., Adcock, J. J., Wortley, M. A., Flajolet, P., et al. (2020). Novel airway smooth muscle-mast cell interactions and a role for the TRPV4-ATP axis in non-atopic asthma. *The European Respiratory Journal, 56*, 1901458.

Chen, Y. L., Baker, T. M., Lee, F., Shui, B., Lee, J. C., Tvrdik, P., et al. (2021). Calcium signal profiles in vascular endothelium from Cdh5-GCaMP8 and Cx40-GCaMP2 mice. *Journal of Vascular Research, 58*, 159–171.

Chen, Y. L., Daneva, Z., Kuppusamy, M., Ottolini, M., Baker, T. M., Klimentova, E., et al. (2022). Novel smooth muscle ca(2+)-signaling nanodomains in blood pressure regulation. *Circulation*. 101161circulationaha121058607.

Chen, Y. L., & Sonkusare, S. K. (2020). Endothelial TRPV4 channels and vasodilator reactivity. *Current Topics in Membranes, 85*, 89–117.

Clapham, D. E., Montell, C., Schultz, G., Julius, D., & International Union of Pharmacology. (2003). International union of pharmacology. XLIII. Compendium of voltage-gated ion channels: Transient receptor potential channels. *Pharmacological Reviews, 55*, 591–596.

Cosens, D. J., & Manning, A. (1969). Abnormal electroretinogram from a drosophila mutant. *Nature, 224*, 285–287.

Daneva, Z., Laubach, V. E., & Sonkusare, S. K. (2019). Novel regulators and targets of redox signaling in pulmonary vasculature. *Current Opinion in Physiology, 9*, 87–93.

Daneva, Z., Marziano, C., Ottolini, M., Chen, Y. L., Baker, T. M., Kuppusamy, M., et al. (2021). Caveolar peroxynitrite formation impairs endothelial TRPV4 channels and elevates pulmonary arterial pressure in pulmonary hypertension. *Proceedings of the National Academy of Sciences of the United States of America, 118*, e2023130118.

Daneva, Z., Ottolini, M., Chen, Y. L., Klimentova, E., Kuppusamy, M., Shah, S. A., et al. (2021). Endothelial pannexin 1-TRPV4 channel signaling lowers pulmonary arterial pressure in mice. *eLife, 10*, e67777.

de Perrot, M., Liu, M., Waddell, T. K., & Keshavjee, S. (2003). Ischemia-reperfusion-induced lung injury. *American Journal of Respiratory and Critical Care Medicine, 167*, 490–511.

Deng, Z., Paknejad, N., Maksaev, G., Sala-Rabanal, M., Nichols, C. G., Hite, R. K., et al. (2018). Cryo-EM and X-ray structures of TRPV4 reveal insight into ion permeation and gating mechanisms. *Nature Structural & Molecular Biology, 25*, 252–260.

Du, J., Ma, X., Shen, B., Huang, Y., Birnbaumer, L., & Yao, X. (2014). TRPV4, TRPC1, and TRPP2 assemble to form a flow-sensitive heteromeric channel. *The FASEB Journal, 28*, 4677–4685.

Dutta, B., Arya, R. K., Goswami, R., Alharbi, M. O., Sharma, S., & Rahaman, S. O. (2020). Role of macrophage TRPV4 in inflammation. *Laboratory Investigation, 100*, 178–185.

Earley, S., Heppner, T. J., Nelson, M. T., & Brayden, J. E. (2005). TRPV4 forms a novel Ca2+ signaling complex with ryanodine receptors and BKCa channels. *Circulation Research, 97*, 1270–1279.

Everaerts, W., Nilius, B., & Owsianik, G. (2010). The vanilloid transient receptor potential channel TRPV4: From structure to disease. *Progress in Biophysics and Molecular Biology, 103*, 2–17.

Filosa, J. A., Yao, X., & Rath, G. (2013). TRPV4 and the regulation of vascular tone. *Journal of Cardiovascular Pharmacology, 61*, 113–119.

Goldenberg, N. M., Wang, L., Ranke, H., Liedtke, W., Tabuchi, A., & Kuebler, W. M. (2015). TRPV4 is required for hypoxic pulmonary vasoconstriction. *Anesthesiology, 122*, 1338–1348.

Goyal, N., Skrdla, P., Schroyer, R., Kumar, S., Fernando, D., Oughton, A., et al. (2019). Clinical pharmacokinetics, safety, and tolerability of a novel, first-in-class TRPV4 Ion Channel inhibitor, GSK2798745, in healthy and heart failure subjects. *American Journal of Cardiovascular Drugs*, *19*, 335–342.

Groot-Kormelink, P. J., Fawcett, L., Wright, P. D., Gosling, M., & Kent, T. C. (2012). Quantitative GPCR and ion channel transcriptomics in primary alveolar macrophages and macrophage surrogates. *BMC Immunology*, *13*, 57.

Hamanaka, K., Jian, M. Y., Townsley, M. I., King, J. A., Liedtke, W., Weber, D. S., et al. (2010). TRPV4 channels augment macrophage activation and ventilator-induced lung injury. *American Journal of Physiology. Lung Cellular and Molecular Physiology*, *299*, L353–L362.

Hamanaka, K., Jian, M. Y., Weber, D. S., Alvarez, D. F., Townsley, M. I., Al-Mehdi, A. B., et al. (2007). TRPV4 initiates the acute calcium-dependent permeability increase during ventilator-induced lung injury in isolated mouse lungs. *American Journal of Physiology. Lung Cellular and Molecular Physiology*, *293*, L923–L932.

Harraz, O. F., Longden, T. A., Eubanks, D. H., & Nelson, M. T. (2018). PIP2depletion promotes TRPV4 channel activity in mouse brain capillary endothelial cells. *eLife*, *7*, 1–24.

Harteneck, C., & Schultz, G. (2007). TRPV4 and TRPM3 as volume-regulated cation channels. In W. Liedtke, & S. Heller (Eds.), *TRP ion channel function in sensory transduction and cellular signaling cascades*. Boca Raton, FL: CRC Press/Taylor & Francis.

Haywood, N., Ta, H. Q., Rotar, E., Daneva, Z., Sonkusare, S. K., & Laubach, V. E. (2021). Role of the purinergic signaling network in lung ischemia-reperfusion injury. *Current Opinion in Organ Transplantation*, *26*, 250–257.

Haywood, N., Ta, H. Q., Zhang, A., Charles, E. J., Rotar, E., St, N., et al. (2022). Endothelial transient receptor potential vanilloid 4 channels mediate lung ischemia-reperfusion injury. *The Annals of Thoracic Surgery*, *113*, 1256–1264.

Inada, H., Procko, E., Sotomayor, M., & Gaudet, R. (2012). Structural and biochemical consequences of disease-causing mutations in the ankyrin repeat domain of the human TRPV4 channel. *Biochemistry*, *51*, 6195–6206.

Jara-Oseguera, A., & Islas, L. D. (2013). The role of allosteric coupling on thermal activation of thermo-TRP channels. *Biophysical Journal*, *104*, 2160–2169.

Jia, Y., Wang, X., Varty, L., Rizzo, C. A., Yang, R., Correll, C. C., et al. (2004). Functional TRPV4 channels are expressed in human airway smooth muscle cells. *American Journal of Physiology. Lung Cellular and Molecular Physiology*, *287*, L272–L278.

Jian, M. Y., King, J. A., Al-Mehdi, A. B., Liedtke, W., & Townsley, M. I. (2008). High vascular pressure-induced lung injury requires P450 epoxygenase-dependent activation of TRPV4. *American Journal of Respiratory Cell and Molecular Biology*, *38*, 386–392.

Komiya, K., Akaba, T., Kozaki, Y., Kadota, J. I., & Rubin, B. K. (2017). A systematic review of diagnostic methods to differentiate acute lung injury/acute respiratory distress syndrome from cardiogenic pulmonary edema. *Critical Care*, *21*, 228.

Kuebler, W. M., Jordt, S. E., & Liedtke, W. B. (2020). Urgent reconsideration of lung edema as a preventable outcome in COVID-19: Inhibition of TRPV4 represents a promising and feasible approach. *American Journal of Physiology. Lung Cellular and Molecular Physiology*, *318*, L1239–L1243.

Lin, M. T., Jian, M. Y., Taylor, M. S., Cioffi, D. L., Yap, F. C., Liedtke, W., et al. (2015). Functional coupling of TRPV4, IK, and SK channels contributes to ca(2+)-dependent endothelial injury in rodent lung. *Pulmonary Circulation*, *5*, 279–290.

Liu, F., Han, K., Blair, R., Kenst, K., Qin, Z., Upcin, B., et al. (2021). SARS-CoV-2 infects endothelial cells in vivo and in vitro. *Frontiers in Cellular and Infection Microbiology*, *11*, 701278.

Longden, T. A., Dabertrand, F., Koide, M., Gonzales, A. L., Tykocki, N. R., Brayden, J. E., et al. (2017). Capillary K(+)-sensing initiates retrograde hyperpolarization to increase local cerebral blood flow. *Nature Neuroscience*, *20*, 717–726.

Ma, X., Cheng, K. T., Wong, C. O., O'Neil, R. G., Birnbaumer, L., Ambudkar, I. S., et al. (2011). Heteromeric TRPV4-C1 channels contribute to store-operated ca(2+) entry in vascular endothelial cells. *Cell Calcium*, *50*, 502–509.

Maier-Begandt, D., Comstra, H. S., Molina, S. A., Kruger, N., Ruddiman, C. A., Chen, Y. L., et al. (2021). A venous-specific purinergic signaling cascade initiated by pannexin 1 regulates TNFalpha-induced increases in endothelial permeability. *Science Signaling*, *14*, eaba2940.

Maitland, K., Kiguli, S., Opoka, R. O., Engoru, C., Olupot-Olupot, P., Akech, S. O., et al. (2011). Mortality after fluid bolus in African children with severe infection. *The New England Journal of Medicine*, *364*, 2483–2495.

Majhi, R. K., Sahoo, S. S., Yadav, M., Pratheek, B. M., Chattopadhyay, S., & Goswami, C. (2015). Functional expression of TRPV channels in T cells and their implications in immune regulation. *The FEBS Journal*, *282*, 2661–2681.

Marziano, C., Hong, K., Cope, E. L., Kotlikoff, M. I., Isakson, B. E., & Sonkusare, S. K. (2017). Nitric oxide-dependent feedback loop regulates transient receptor potential vanilloid 4 (TRPV4) channel cooperativity and endothelial function in small pulmonary arteries. *Journal of the American Heart Association*, *6*, e007157.

Mercado, J., Baylie, R., Navedo, M. F., Yuan, C., Scott, J. D., Nelson, M. T., et al. (2014). Local control of TRPV4 channels by AKAP150-targeted PKC in arterial smooth muscle. *The Journal of General Physiology*, *143*, 559–575.

Michalick, L., Erfinanda, L., Weichelt, U., van der Giet, M., Liedtke, W., & Kuebler, W. M. (2017). Transient receptor potential vanilloid 4 and serum glucocorticoid-regulated kinase 1 are critical mediators of lung injury in overventilated mice in vivo. *Anesthesiology*, *126*, 300–311.

Michalick, L., Mandzimba-Maloko, B., Hamedi, S., Dohmen, M., Brack, M. C., Schulz, S., et al. (2022). In vitro screening identifies TRPV4 and PAR1 as targets for endothelial barrier stabilization in COVID-19. *The FASEB Journal*, *36*.

Mole, S., Harry, A., Fowler, A., Hotee, S., Warburton, J., Waite, S., et al. (2020). Investigating the effect of TRPV4 inhibition on pulmonary-vascular barrier permeability following segmental endotoxin challenge. *Pulmonary Pharmacology & Therapeutics*, *64*, 101977.

Nieto-Posadas, A., Jara-Oseguera, A., & Rosenbaum, T. (2011). TRP channel gating physiology. *Current Topics in Medicinal Chemistry*, *11*, 2131–2150.

Nilius, B., & Voets, T. (2013). The puzzle of TRPV4 channelopathies. *EMBO Reports*, *14*, 152–163.

Ottolini, M., Daneva, Z., Chen, Y. L., Cope, E. L., Kasetti, R. B., Zode, G. S., et al. (2020). Mechanisms underlying selective coupling of endothelial Ca2+ signals with eNOS vs. IK/SK channels in systemic and pulmonary arteries. *Journal of Physiology*, *598*, 3577–3596.

Ottolini, M., Hong, K., Cope, E. L., Daneva, Z., DeLalio, L. J., Sokolowski, J. D., et al. (2020). Local peroxynitrite impairs endothelial transient receptor potential vanilloid 4 channels and elevates blood pressure in obesity. *Circulation*, *141*, 1318–1333.

Ottolini, M., Hong, K., & Sonkusare, S. K. (2019). Calcium signals that determine vascular resistance. *Wiley Interdisciplinary Reviews. Systems Biology and Medicine, e1448*.

Ottolini, M., & Sonkusare, S. K. (2021). The calcium signaling mechanisms in arterial smooth muscle and endothelial cells. *Comprehensive Physiology*, *11*, 1831–1869.

Pairet, N., Mang, S., Fois, G., Keck, M., Kuhnbach, M., Gindele, J., et al. (2018). TRPV4 inhibition attenuates stretch-induced inflammatory cellular responses and lung barrier dysfunction during mechanical ventilation. *PLoS One*, *13*, e0196055.

Phelps, C. B., Huang, R. J., Lishko, P. V., Wang, R. R., & Gaudet, R. (2008). Structural analyses of the ankyrin repeat domain of TRPV6 and related TRPV ion channels. *Biochemistry*, *47*, 2476–2484.

Rahaman, S. O., Grove, L. M., Paruchuri, S., Southern, B. D., Abraham, S., Niese, K. A., et al. (2014). TRPV4 mediates myofibroblast differentiation and pulmonary fibrosis in mice. *The Journal of Clinical Investigation*, *124*, 5225–5238.

Rajan, S., Schremmer, C., Weber, J., Alt, P., Geiger, F., & Dietrich, A. (2021). Ca(2 +) signaling by TRPV4 channels in respiratory function and disease. *Cells*, *10*, 822.

Rayees, S., Joshi, J. C., Tauseef, M., Anwar, M., Baweja, S., Rochford, I., et al. (2019). PAR2-mediated cAMP generation suppresses TRPV4-dependent ca(2+) signaling in alveolar macrophages to resolve TLR4-induced inflammation. *Cell Reports*, *27*, 793–805.

Rosenbaum, T., Benitez-Angeles, M., Sanchez-Hernandez, R., Morales-Lazaro, S. L., Hiriart, M., Morales-Buenrostro, L. E., et al. (2020). TRPV4: A physio and Pathophysiologically significant Ion Channel. *International Journal of Molecular Sciences*, *21*, 3837.

Ryskamp, D. A., Jo, A. O., Frye, A. M., Vazquez-Chona, F., MacAulay, N., Thoreson, W. B., et al. (2014). Swelling and eicosanoid metabolites differentially gate TRPV4 channels in retinal neurons and glia. *The Journal of Neuroscience*, *34*, 15689–15700.

Scheraga, R. G., Abraham, S., Grove, L. M., Southern, B. D., Crish, J. F., Perelas, A., et al. (2020). TRPV4 protects the lung from bacterial pneumonia via MAPK molecular pathway switching. *Journal of Immunology*, *204*, 1310–1321.

Scheraga, R. G., Southern, B. D., Grove, L. M., & Olman, M. A. (2017). The role of transient receptor potential vanilloid 4 in pulmonary inflammatory diseases. *Frontiers in Immunology*, *8*, 503.

Scheraga, R. G., Southern, B. D., Grove, L. M., & Olman, M. A. (2020). The role of TRPV4 in regulating innate immune cell function in lung inflammation. *Frontiers in Immunology*, *11*, 1211.

Sharma, A. K., Charles, E. J., Zhao, Y., Narahari, A. K., Baderdinni, P. K., Good, M. E., et al. (2018). Pannexin-1 channels on endothelial cells mediate vascular inflammation during lung ischemia-reperfusion injury. *American Journal of Physiology. Lung Cellular and Molecular Physiology*, *315*, L301–L312.

Sonkusare, S. K., Bonev, A. D., Ledoux, J., Liedtke, W., Kotlikoff, M. I., Heppner, T. J., et al. (2012). Elementary Ca2+ signals through endothelial TRPV4 channels regulate vascular function. *Science*, *336*, 597–601.

Sonkusare, S. K., Dalsgaard, T., Bonev, A. D., Hill-Eubanks, D. C., Kotlikoff, M. I., Scott, J. D., et al. (2014). AKAP150-dependent cooperative TRPV4 channel gating is central to endothelium-dependent vasodilation and is disrupted in hypertension. *Science Signaling*, *7*, ra66.

Sonkusare, S. K., Dalsgaard, T., Bonev, A. D., & Nelson, M. T. (2016). Inward rectifier potassium (Kir2.1) channels as end-stage boosters of endothelium-dependent vasodilators. *The Journal of Physiology*, *594*, 3271–3285.

Stevens, T. (2011). Functional and molecular heterogeneity of pulmonary endothelial cells. *Proceedings of the American Thoracic Society*, *8*, 453–457.

Stewart, A. P., Smith, G. D., Sandford, R. N., & Edwardson, J. M. (2010). Atomic force microscopy reveals the alternating subunit arrangement of the TRPP2-TRPV4 heterotetramer. *Biophysical Journal*, *99*, 790–797.

Strotmann, R., Harteneck, C., Nunnenmacher, K., Schultz, G., & Plant, T. D. (2000). OTRPC4, a nonselective cation channel that confers sensitivity to extracellular osmolarity. *Nature Cell Biology*, *2*, 695–702.

Suresh, K., Servinsky, L., Jiang, H., Bigham, Z., Yun, X., Kliment, C., et al. (2018). Reactive oxygen species induced ca(2+) influx via TRPV4 and microvascular endothelial dysfunction in the SU5416/hypoxia model of pulmonary arterial hypertension. *American Journal of Physiology. Lung Cellular and Molecular Physiology*, *314*, L893–L907.

Suresh, K., Servinsky, L., Reyes, J., Baksh, S., Undem, C., Caterina, M., et al. (2015). Hydrogen peroxide-induced calcium influx in lung microvascular endothelial cells involves TRPV4. *American Journal of Physiology. Lung Cellular and Molecular Physiology*, *309*, L1467–L1477.

Thorneloe, K. S., Cheung, M., Bao, W., Alsaid, H., Lenhard, S., Jian, M. Y., et al. (2012). An orally active TRPV4 channel blocker prevents and resolves pulmonary edema induced by heart failure. *Science Translational Medicine*, *4* (159ra148).

Toft-Bertelsen, T. L., & MacAulay, N. (2021). TRPing to the point of clarity: Understanding the function of the complex TRPV4 Ion Channel. *Cells*, *10*, 165.

Villalta, P. C., & Townsley, M. I. (2013). Transient receptor potential channels and regulation of lung endothelial permeability. *Pulmonary Circulation*, *3*, 802–815.

Voets, T., Prenen, J., Vriens, J., Watanabe, H., Janssens, A., Wissenbach, U., et al. (2002). Molecular determinants of permeation through the cation channel TRPV4. *The Journal of Biological Chemistry*, *277*, 33704–33710.

Watanabe, H., Vriens, J., Janssens, A., Wondergem, R., Droogmans, G., & Nilius, B. (2003). Modulation of TRPV4 gating by intra- and extracellular Ca2+. *Cell Calcium*, *33*, 489–495.

Weber, J., Rajan, S., Schremmer, C., Chao, Y. K., Krasteva-Christ, G., Kannler, M., et al. (2020). TRPV4 channels are essential for alveolar epithelial barrier function as protection from lung edema. *JCI Insight*, *5*, e134464.

White, J. P., Cibelli, M., Urban, L., Nilius, B., McGeown, J. G., & Nagy, I. (2016). TRPV4: Molecular conductor of a diverse orchestra. *Physiological Reviews*, *96*, 911–973.

Willette, R. N., Bao, W., Nerurkar, S., Yue, T. L., Doe, C. P., Stankus, G., et al. (2008). Systemic activation of the transient receptor potential vanilloid subtype 4 channel causes endothelial failure and circulatory collapse: Part 2. *The Journal of Pharmacology and Experimental Therapeutics*, *326*, 443–452.

Xia, Y., Fu, Z., Hu, J., Huang, C., Paudel, O., Cai, S., et al. (2013). TRPV4 channel contributes to serotonin-induced pulmonary vasoconstriction and the enhanced vascular reactivity in chronic hypoxic pulmonary hypertension. *American Journal of Physiology. Cell Physiology*, *305*, C704–C715.

Yang, X. R., Lin, A. H., Hughes, J. M., Flavahan, N. A., Cao, Y. N., Liedtke, W., et al. (2012). Upregulation of osmo-mechanosensitive TRPV4 channel facilitates chronic hypoxia-induced myogenic tone and pulmonary hypertension. *American Journal of Physiology. Lung Cellular and Molecular Physiology*, *302*, L555–L568.

Yin, J., Michalick, L., Tang, C., Tabuchi, A., Goldenberg, N., Dan, Q., et al. (2016). Role of transient receptor potential vanilloid 4 in neutrophil activation and acute lung injury. *American Journal of Respiratory Cell and Molecular Biology*, *54*, 370–383.

Yu, Q., Wang, D., Wen, X., Tang, X., Qi, D., He, J., et al. (2020). Adipose-derived exosomes protect the pulmonary endothelial barrier in ventilator-induced lung injury by inhibiting the TRPV4/ca(2+) signaling pathway. *American Journal of Physiology. Lung Cellular and Molecular Physiology*, *318*, L723–L741.

CHAPTER THREE

Cardiac TRPV4 channels

Vivian C. Onyali and Timothy L. Domeier*
Department of Medical Pharmacology and Physiology, University of Missouri School of Medicine, Columbia, MO, United States
*Corresponding author: e-mail address: domeiert@health.missouri.edu

Contents

Abstract

Transient Receptor Potential Vanilloid 4 (TRPV4) is expressed in numerous cell types within the heart, yet the expression levels, subcellular localization, and functional relevance of TRPV4 in cardiac myocytes is under-appreciated. Recent data indicate a critical role of TRPV4 in both atrial and ventricular myocyte biology, with expression levels and channel function increasing following pathological scenarios including ischemia, myocardial infarction, mechanical stress, and inflammation. Excessive activation of TRPV4 at the cellular level contributes to enhanced Ca^{2+} entry which predisposes the cardiac myocyte to pro-arrhythmic Ca^{2+} overload and electrophysiological abnormalities. At the organ level, excessive TRPV4 activity associates with cardiac hypercontractility, cardiac damage, ventricular arrhythmia, and atrial fibrillation. This manuscript chapter describes the emerging literature on TRPV4 in cardiac myocytes in physiology and disease.

1. Introduction

Calcium is a ubiquitous, universal and multifaceted intracellular mediator in both excitable and non-excitable cells that regulates/modulates a diverse array of physiological functions. Cellular Ca^{2+} homeostasis is tightly controlled in cardiac contractile cells (myocytes) and Ca^{2+} dysregulation

Current Topics in Membranes, Volume 89
ISSN 1063-5823
https://doi.org/10.1016/bs.ctm.2022.06.004

leads to abnormalities such as altered contractile function, cellular Ca^{2+} overload, hyperexcitability, cell death, and activation of pathological remodeling pathways (Inoue, Kurahara, & Hiraishi, 2019). The TRP channels are a superfamily of channels found in many organs including the heart (Nilius, Owsianik, Voets, & Peters, 2007; Nilius, Voets, & Peters, 2005; White et al., 2016). Most TRP channels act as Ca^{2+} entry pathways in the plasma membrane, contributing to the cytosolic free Ca^{2+} concentration. Altered TRP channel function can lead to dysfunction in Ca^{2+} signaling and significant dysregulation of homeostatic processes (Watanabe, Murakami, Ohba, Takahashi, & Ito, 2008). In cardiac muscle cells an expanding literature indicates a key role of TRPV channels in cardiac physiology/pathophysiology, particularly in pathological states where the channels predispose the heart to contractile abnormalities, electrophysiological changes, and arrhythmia (Miller, Koch, Veteto, Domeier, & Rubinstein, 2021).

2. Cardiac myocyte excitation-contraction coupling

Cardiac excitation-contraction coupling (ECC) occurs via the Ca^{2+}-induced Ca^{2+} release (CICR) process (Bers, 2001). The action potential upstroke (excitation) is due to opening of fast voltage-gated Na^{+} channels and Ca^{2+} channels, and the resulting Ca^{2+} influx through the L-type Ca^{2+} channel (LTCC) triggers additional Ca^{2+} release through Ryanodine Receptors (RyR) of the SR Ca^{2+} store. A salient feature of the ventricular myocardium is the presence of a well-structured t-tubule network (Setterberg, Le, Frisk, Li, & Louch, 2021). T-tubules are surface sarcolemma invaginations that penetrate into the myocyte interior and afford close communication between the sarcolemmal LTCC and the RyR of the SR membrane. The ventricular myocyte CICR process leads to a rapid increase in intracellular Ca^{2+}, uniform myocyte contraction, and well-coordinated ventricular systole. A notable difference between atrial and ventricular myocyte cellular architecture is the lesser developed (or absent) t-tubule network (Setterberg et al., 2021), with the functional consequence of slower Ca^{2+} signal propagation from the surface sarcolemma inward to cell interior via CICR between neighboring RyR release clusters (Kockskamper et al., 2001). Therefore, atrial muscle ECC tends to be slower and less-coordinated than that of the ventricular muscle. For diastolic cardiac relaxation to occur Ca^{2+} must be removed from the cytosol, which occurs primarily via sarcolemmal Na^{+}/Ca^{2+} exchange and SR Ca^{2+} re-sequestration via the sarco/endoplasmic reticulum Ca^{2+} ATPase (SERCA). Over the entirety of the cardiac cycle the cardiac myocyte remains in Ca^{2+} flux balance when Ca^{2+} influx into the cytosol via LTCC is counterbalanced by NCX Ca^{2+} extrusion, and when RyR SR Ca^{2+} release is counterbalanced by SERCA-mediated SR Ca^{2+} sequestration (Eisner, 2018). Thus, additional Ca^{2+} influx via sarcolemmal TRP channels shifts Ca^{2+} flux balance toward cellular and SR Ca^{2+} accumulation. While such changes can augment Ca^{2+} transients during ECC and lead to increased cardiac contractile strength, under pathological scenarios excessive Ca^{2+} influx can be detrimental as it associates with cellular Ca^{2+} overload, electrophysiological abnormalities, and cardiac muscle damage (Nakayama et al., 2007). These cellular findings manifest at the organ level with adverse chamber remodeling, cardiac dysfunction, and arrhythmia.

3. TRPV4 expression in cardiac myocytes

TRPV4 is a high-flux cation channel with selectivity for Ca^{2+} over other cations (Ca^{2+}:Na^{+} permeability ratio of ~6:1) (Owsianik, Talavera, Voets, & Nilius, 2006). Considering ion gradients and fluxes during cardiac muscle ECC, the addition of a sarcolemmal TRPV4 Ca^{2+} influx channel would significantly shift Ca^{2+} flux balance toward myocyte Ca^{2+} accumulation and predispose the myocyte to adverse cellular Ca^{2+} overload. TRPV4 relative expression levels, sarcolemmal versus intracellular localization, and functional activity in cardiac myocytes are therefore of significant interest. Comprehensive knowledge of TRPV4 expression, localization, and function in cardiac muscle is limited due to lesser focus on TRPV4 in cardiac muscle versus other cell types (White et al., 2016). While TRPV4 plays a fundamental role in physiological processes in many cell types, the channel appears to play an adaptive role in muscle biology during pathological processes, similar to other TRP channels (Falcon et al., 2019). The paucity of data in cardiac myocytes likely stems from several factors. Molecular studies frequently examine "cardiac" mRNA or protein expression in whole heart homogenates which include many non-myocyte cell types, and although cardiac muscle cells are the primary contractile cell type in the heart they represent only ~25–35% of the total cells (Pinto et al., 2016). Changes in cardiac myocyte TRPV4 expression are therefore difficult to measure using whole heart homogenates due to high levels of TRPV4 expression in other cell types within the organ (e.g., endothelial cells, smooth muscle cells, and fibroblasts). Isolated, purified cardiac myocyte populations (Jones et al., 2019) or single-cell approaches are required to obtain quantitative assessment of TRPV4 levels in the cardiac myocyte. Cardiac myocytes also typically express low levels of TRPV4 in the absence of pathological stress. Thus, experimental investigations using young, non-pathological models may not reveal detectable cardiac myocyte TRPV4 expression or function. Furthermore, cardiac myocyte TRPV4 activity is also stimulus-dependent, leading to hidden phenotypes which must be unmasked using select stimuli including agonist stimulation, hypo-osmotic stress, mechanical stretch, inflammation, or other pathological insults (White et al., 2016). In sum, these experimental challenges frequently lead to negative experimental results, and at present, an under-appreciation for the important role of TRPV4 in cardiac myocyte biology. In recent years new emphasis has been placed on TRPV4 channel function in cardiac myocytes due to emerging data that TRPV4 expression and/or function increases in cardiac myocytes and contributes to pathological Ca^{2+} handling following hypoxia (Wu et al., 2017), ischemia-reperfusion (Jones et al., 2019; Peana, Polo-Parada, & Domeier, 2021), myocardial infarction (Adapala, Kanugula, Paruchuri, Chilian, & Thodeti, 2020; Wu et al., 2019), sustained stretch (Lu et al., 2017; Veteto, Peana, Lambert, McDonald, & Domeier, 2020), inflammation, and atrial fibrillation (Liao et al., 2020; Yang et al., 2022).

4. TRPV4 localization in cardiac myocytes

One of the initial reports of TRPV4 expression and localization in acutely isolated rat ventricular myocytes suggested complex and dynamic cellular TRPV4 localization, with nuclear/peri-nuclear localization shifting to sarcolemmal distribution following hypoosmotic stress (Zhao et al., 2012). Subsequent studies in ventricular myocytes of aged (24–27 month) mice, which express higher levels of TRPV4 than young (3–6 month) counterparts, revealed similar complexity in channel localization. Using high-resolution confocal imaging and immunocytochemistry in fixed cells (Fig. 1), TRPV4 co-localized with the microtubule perinuclear cage (co-localization with α-tubulin), with cytosolic microtubule networks,

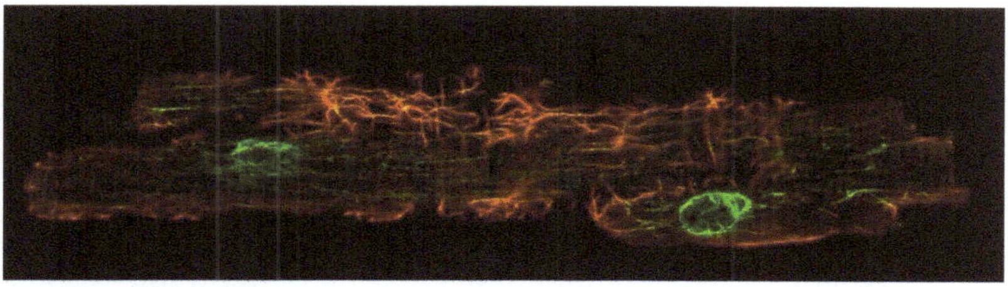

Fig. 1 TRPV4 localization in ventricular myocyte with aging. Immunocytochemistry of TRPV4 (green) and α-tubulin (red) in a ventricular myocyte of an aged mouse, with overlay of TRPV4 and α-tubulin in yellow. Immunocytochemistry Methods described in Jones et al. (2019).

and with the sarcolemma including t-tubule networks (co-localization with Caveolin-3) (Jones et al., 2019). Immunohistochemistry in left-ventricular sections of aged mice and elderly human aortic valve replacement patients also indicated striated t-tubule patterning (Jones et al., 2019). TRPV4 cellular localization was further confirmed in live cardiac myocytes using a transgenic mCherry-TRPV4 fusion protein mouse model, with observed co-localization of the mCherry fusion protein with both microtubule networks (SiR tubulin) and the sarcolemma/t-tubule networks (CellMask Deep Red membrane dye) (Jones et al., 2019). In ventricular myocytes of young mice, immunoprecipitation experiments confirmed functional co-localization between TRPV4 and the L-type Ca^{2+} channel, and immunocytochemistry indicated surface sarcolemmal, but in contrast to data in aged mice, not t-tubule localization (Chaigne et al., 2021). In atrial myocytes TRPV4 expression has also been observed, and localization shifts from perinuclear areas to cytosolic and/or membrane regions following progression of sterile pericarditis. However, these studies did not use high-resolution imaging approaches and the distinct sub-cellular distribution in the atrium remains to be determined. Additional investigations are needed to resolve the complex inter-relationships between TRPV4 expression, physiological/pathological status of the heart/organism, and stimulus-dependent channel localization. Nevertheless, it is critical to highlight that TRPV4 functional activity at the sarcolemmal membrane is consistently observed in both atrial and ventricular myocytes (Veteto et al., 2020). Thus, the channel is present and functional in a cellular domain where it will have significant impact on cell, organ, and *in vivo* physiology.

5. Role of TRPV4 in cardiac myocyte Ca^{2+} handling

The first indication that TRPV4 may alter Ca^{2+} cycling in ventricular tissue was in the LV papillary muscle preparation, where the TRPV4 gating stimulus of hypoosmotic stress induced an increase in tissue force development. Although Ca^{2+} was not monitored directly in this study, the effects were prevented by the non-specific TRP channel inhibitor ruthenium red, pointing toward a sarcolemmal ion flux underlying the increase in contractile strength (Li et al., 2008). Additional data in support of cardiac muscle TRPV4 were obtained in embryonic stem (ES) cell-derived (Qi et al., 2015) and induced pluripotent stem cell-derived (iPSCs) cardiac myocytes (Lu et al., 2017), and in these studies TRPV4 agonists or cellular stretch increased levels of intracellular Ca^{2+} (Qi et al., 2015). Interestingly, iPSCs from patients with dilated cardiomyopathy (DCM) had excessive stretch-sensitivity, which could be prevented by TRPV4 pharmacological inhibition or genetic knock-down (Lu et al., 2017). Similarly, H9C2 cells and neonatal rat ventricular myocytes subjected to hypoxia-reoxygenation challenge *in vitro* exhibited an increase in TRPV4 expression, a TRPV4-dependent increase in Ca^{2+}, and a reduction in cell viability (Wu et al., 2017). It is noteworthy that myocyte cell culture models do not exhibit the same cellular architecture of the adult cardiac myocyte, and therefore do not faithfully recapitulate Ca^{2+} handling mechanisms during ECC. These approaches add unique and significant insight to the field yet associate with limitations in mechanistic insight.

A series of mechanistic investigations in acutely isolated ventricular myocytes of an aging mouse model revealed key aspects of Ca^{2+} handling in response to TRPV4 gating stimuli of hypoosmotic stress (Jones et al., 2019; Peana, Polo-Parada, & Domeier, 2022) or longitudinal myocyte stretch (Veteto et al., 2020). The studies utilized global TRPV4 knockout mice (Jones et al., 2019) or TRPV4 antagonism (Jones et al., 2019; Peana et al., 2022; Veteto et al., 2020) to confirm TRPV4-dependent effects. Consistent with the addition of a sarcolemmal cation flux (Veteto et al., 2020), TRPV4 led to increased SR Ca^{2+} content, enhancement of systolic RyR SR Ca^{2+} release, and initial positive inotropic effects on the Ca^{2+} transient during ECC (Jones et al., 2019; Peana et al., 2022; Veteto et al., 2020). However, with time the inotropic effects devolved into cellular Ca^{2+} overload (Veteto et al., 2020), pro-arrhythmic Ca^{2+} signaling

(Peana et al., 2022), and cell damage (Jones et al., 2019). In the same studies, observations at the cellular level manifest as findings at the organ level in *ex vivo* perfused hearts. The enhanced ventricular myocyte Ca^{2+} transient following hypoosmotic stress in isolated cells associated with a phase of cardiac hypercontractility in perfused hearts following ischemia-reperfusion and associated osmotic stress (Jones et al., 2019). Similarly, the increase in myocyte Ca^{2+} transient amplitude following myocyte stretch associated with a period of cardiac hypercontractility following left-ventricular preload elevation and associated ventricular muscle stretch (Veteto et al., 2020). Much like the positive inotropic cellular effects transitioning into adverse cellular Ca^{2+} overload (Veteto et al., 2020), pro-arrhythmic Ca^{2+} signaling (Peana et al., 2022), and cell damage (Jones et al., 2019), at the organ level the hypercontractile phase transitioned into cardiac pump dysfunction following sustained preload challenge (Veteto et al., 2020), and ventricular arrhythmia (Peana et al., 2022) and cardiac damage (Jones et al., 2019) following ischemia-reperfusion. Consistent with the notion of a "Goldilocks effect" of TRPV4 channel activation (Miller et al., 2021), whereby modest amounts of TRPV4 activity could be beneficial yet excessive activation is catastrophic, application of the potent GSK101 agonist to atrial myocytes isolated from a rat atrial fibrillation model increased cellular Ca^{2+} entry yet predisposed myocytes to cell death (Liao et al., 2020). The GSK101 agonist also increased the amplitude of Ca^{2+} transients in young mice and this effect was absent in mice with global TRPV4 deletion (Chaigne et al., 2021). Interestingly, the mice in this study exhibited several changes in cellular and in vivo ECG parameters (described below), yet did not present with notable pathological findings such as cellular damage or arrhythmia. Proof-of-concept studies in young mice with TRPV4 Overexpression (at levels ~2 fold greater than observed in aged mice) indicate a striking Ca^{2+} handling phenotype (Jones et al., 2019), yet with minimal cardiac pathology following sustained preload elevation (Veteto et al., 2020) or ischemia-reperfusion injury (Domeier laboratory, unpublished data). Such findings are consistent with the "multi-hit" model of cardiac disease (Lorin, Vogeli, & Niggli, 2015), where the additional ionic flux via TRPV4 is only detrimental if the cardiac cell is unable to compensate for the additional influx with its typical cohort of ion channels and transporters. Alternatively, potential pro-arrhythmic Ca^{2+} signaling at the cellular level may not trigger or sustain arrhythmia in the absence of concomitant adverse cardiac chamber remodeling. Nevertheless, the emerging data on cardiac myocyte TRPV4 provide strong support that excessive cardiac muscle TRPV4 activity leads to pathological changes following several gating stimuli, most notably hypoosmotic stress following ischemia-reperfusion injury. Indeed, pharmacological TRPV4 inhibition or genetic TRPV4 deletion have been shown to reduce infarct size following ischemia-reperfusion *in vivo* in several studies, although the cell type/s contributing to the detrimental effect of excessive TRPV4 activation remains an active area of investigation (Adapala et al., 2020; Dong et al., 2017; Wu et al., 2017; Wu et al., 2019).

6. Role of TRPV4 in cardiac myocyte electrophysiology and arrhythmia

TRPV4 is a high-flux cation channel which will conduct inward (i.e., depolarizing) current based on physiological ionic gradients (Owsianik et al., 2006). However, pharmacological TRPV4 activation with the agonist GSK101 did not significantly change resting membrane potential in mouse ventricular myocytes (Chaigne et al., 2021). Agonist stimulation did prolong ventricular action potential duration, and this effect was prevented using pharmacological TRPV4 inhibition and absent in myocytes isolated from global TRPV4 knockout mice. Furthermore, ECG measurements in mice *in vivo* revealed a shorter Q-T interval in TRPV4 knockout mice than wildtype controls. Alternatively stated, the TRPV4-dependent prolongation of the ventricular action potential associated with the expected organ-level prolongation of the Q-T interval. Q-T prolongation is known to associate with ventricular arrhythmia, yet no arrhythmias were observed in this study which is to be expected in young mice in the absence of pathological stress. In contrast, Peana et al. used sharp-electrode electrophysiology in *ex vivo* isolated hearts of aged mice to examine the role of TRPV4 in ventricular electrophysiology following pathological ischemia-reperfusion injury (Peana et al., 2022). After ischemia-reperfusion, a population of depolarized myocytes emerged with slower upstroke kinetics, decreased action potential amplitude, and action potential prolongation. This population of cells was not observed with acute TRPV4 inhibition indicating a role for TRPV4 in pathological electrical activity. Complimentary Ca^{2+} measurements were performed in *ex vivo* hearts of aged cardiac-specific

GCaMP6f transgenic mice and revealed populations of ventricular myocytes with Ca^{2+} overload behavior after ischemia-reperfusion. The pathological effects on myocyte Ca^{2+} handling were also prevented by TRPV4 antagonism. Finally, isolated hearts exhibited ventricular arrhythmia following ischemia-reperfusion, and arrhythmia was prevented with pharmacological TRPV4 inhibition. Taken together, these data indicate that TRPV4 contributes to Ca^{2+} overload, electrophysiological derangements, and ventricular arrhythmia following ischemia-reperfusion injury. A single study has examined atrial myocyte electrophysiology in a rat model of atrial fibrillation (Liao et al., 2020), and similar to findings in ventricular myocytes (Chaigne et al., 2021), the TRPV4 agonist GSK101 increased Ca^{2+} levels and led to action potential prolongation in atrial myocytes. Interestingly, while the increased Ca^{2+} entry via TRPV4 would be expected to predispose atrial myocytes to increased automaticity and triggered activity (Dobrev & Wehrens, 2017), action-potential prolongation is typically anti-arrhythmic in the setting of atrial fibrillation. In this study TRPV4 inhibition *in vivo* prevented adverse atrial remodeling and fibrosis. Therefore, it remains unclear if TRPV4 inhibition prevents atrial fibrillation via anti-arrhythmic effects on atrial myocytes, or anti-fibrotic effects on other cell types within the heart such as fibroblasts. Clearly, additional work is needed to study the mechanistic role of TRPV4 in atrial and ventricular arrhythmogenesis.

7. Conclusions

The cardiac field has only recently focused on TRPV4 in atrial and ventricular myocytes, and there are more unanswered, than answered, questions in the field. Although limited, studies thus far indicate that excessive TRPV4 activity causes severe cardiac pathology including cardiac damage, adverse chamber remodeling, and ventricular arrhythmia following ischemia-reperfusion injury (Fig. 2). Furthermore, the literature on atrial pathophysiology is also emerging with key links between inflammation and atrial fibrillation. Considering TRPV4 inhibitors were well-tolerated in high-risk heart failure patients in clinical trials (Goyal et al., 2019), these drugs hold promise as therapeutic agents to prevent adverse cardiac remodeling and arrhythmias with aging and during progression of cardiovascular disease.

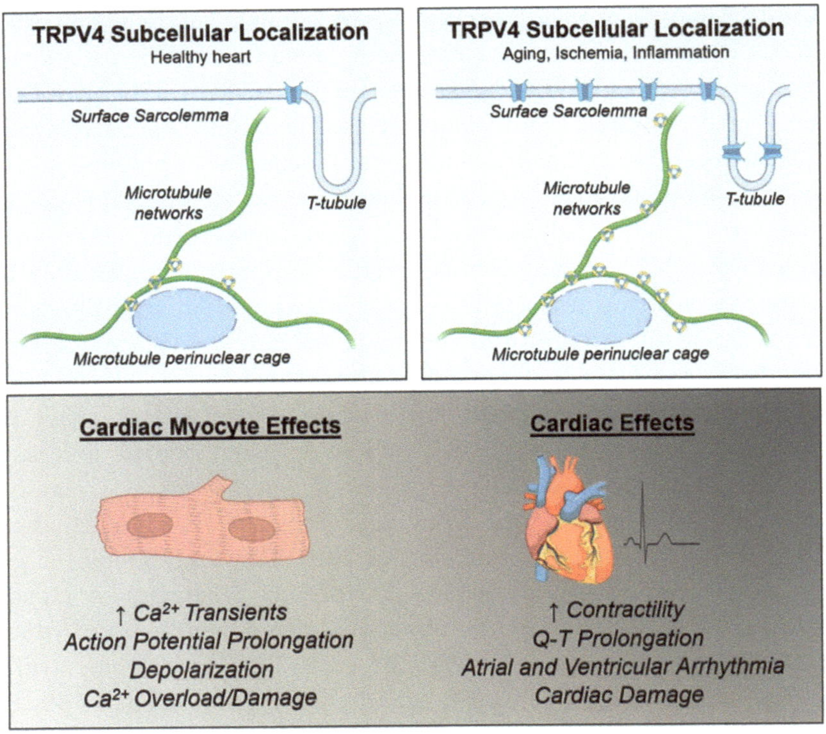

Fig. 2 Working model of TRPV4 localization and function in cardiac myocytes. Under physiological conditions/homeostasis, TRPV4 channels (blue channels) are not abundantly expressed and localized in the perinuclear area. With aging, ischemia, inflammation, or other pathological stressors, TRPV4 exhibits an increase in expression and broad localization at the microtubule perinuclear cage, intracellular microtubule networks, the surface sarcolemma, and t-tubule networks. At the cellular level, TRPV4 increases Ca^{2+} transients during excitation-contraction coupling and lengthens the cardiac action potential. With severe pathological stress, TRPV4 contributes to membrane depolarization, Ca^{2+} overload, and cellular damage. At the organ level such findings manifest as excessive contractility, Q-T interval prolongation, atrial and ventricular arrhythmias, and cardiac damage. Figure created using BioRender.com.

References

Adapala, R. K., Kanugula, A. K., Paruchuri, S., Chilian, W. M., & Thodeti, C. K. (2020). TRPV4 deletion protects heart from myocardial infarction-induced adverse remodeling via modulation of cardiac fibroblast differentiation. *Basic Research in Cardiology*, *115*, 14.

Bers, D. M. (2001). *Excitation-contraction coupling and cardiac contractile force* (2nd ed.). Dordrecht: Kluwer Academic Publishers.

Chaigne, S., Cardouat, G., Louradour, J., Vaillant, F., Charron, S., Sacher, F., et al. (2021). Transient receptor potential vanilloid 4 channel participates in mouse ventricular electrical activity. *American Journal of Physiology. Heart and Circulatory Physiology*, *320*, H1156–H1169.

Dobrev, D., & Wehrens, X. H. T. (2017). Calcium-mediated cellular triggered activity in atrial fibrillation. *The Journal of Physiology, 595*, 4001–4008.

Dong, Q., Li, J., Wu, Q. F., Zhao, N., Qian, C., Ding, D., et al. (2017). Blockage of transient receptor potential vanilloid 4 alleviates myocardial ischemia/reperfusion injury in mice. *Scientific Reports*, 7, 42678.

Eisner, D. A. (2018). Ups and downs of calcium in the heart. *The Journal of Physiology, 596*, 19–30.

Falcon, D., Galeano-Otero, I., Calderon-Sanchez, E., Del Toro, R., Martin-Bornez, M., Rosado, J. A., et al. (2019). TRP channels: Current perspectives in the adverse cardiac remodeling. *Frontiers in Physiology, 10*, 159.

Goyal, N., Skrdla, P., Schroyer, R., Kumar, S., Fernando, D., Oughton, A., et al. (2019). Clinical pharmacokinetics, safety, and tolerability of a novel, first-in-class TRPV4 ion channel inhibitor, GSK2798745, in healthy and heart failure subjects. *American Journal of Cardiovascular Drugs, 19*, 335–342.

Inoue, R., Kurahara, L. H., & Hiraishi, K. (2019). TRP channels in cardiac and intestinal fibrosis. *Seminars in Cell & Developmental Biology, 94*, 40–49.

Jones, J. L., Peana, D., Veteto, A. B., Lambert, M. D., Nourian, Z., Karasseva, N. G., et al. (2019). TRPV4 increases cardiomyocyte calcium cycling and contractility yet contributes to damage in the aged heart following hypoosmotic stress. *Cardiovascular Research, 115*, 46–56.

Kockskamper, J., Sheehan, K. A., Bare, D. J., Lipsius, S. L., Mignery, G. A., & Blatter, L. A. (2001). Activation and propagation of Ca(2+) release during excitation-contraction coupling in atrial myocytes. *Biophysical Journal, 81*, 2590–2605.

Li, J., Wang, M. H., Wang, L., Tian, Y., Duan, Y. Q., Luo, H. Y., et al. (2008). Role of transient receptor potential vanilloid 4 in the effect of osmotic pressure on myocardial contractility in rat. *Sheng Li Xue Bao, 60*, 181–188.

Liao, J., Wu, Q., Qian, C., Zhao, N., Zhao, Z., Lu, K., et al. (2020). TRPV4 blockade suppresses atrial fibrillation in sterile pericarditis rats. *JCI Insight, 5*.

Lorin, C., Vogeli, I., & Niggli, E. (2015). Dystrophic cardiomyopathy: role of TRPV2 channels in stretch-induced cell damage. *Cardiovascular Research, 106*, 153–162.

Lu, J., Lee, Y. K., Ran, X., Lai, W. H., Li, R. A., Keung, W., et al. (2017). An abnormal TRPV4-related cytosolic Ca(2+) rise in response to uniaxial stretch in induced pluripotent stem cells-derived cardiomyocytes from dilated cardiomyopathy patients. *Biochimica et Biophysica Acta - Molecular Basis of Disease, 1863*, 2964–2972.

Miller, M., Koch, S. E., Veteto, A., Domeier, T., & Rubinstein, J. (2021). Role of known transient receptor potential vanilloid channels in modulating cardiac mechanobiology. *Frontiers in Physiology, 12*, 734113.

Nakayama, H., Chen, X., Baines, C. P., Klevitsky, R., Zhang, X., Zhang, H., et al. (2007). Ca2+- and mitochondrial-dependent cardiomyocyte necrosis as a primary mediator of heart failure. *The Journal of Clinical Investigation, 117*, 2431–2444.

Nilius, B., Owsianik, G., Voets, T., & Peters, J. A. (2007). Transient receptor potential cation channels in disease. *Physiological Reviews, 87*, 165–217.

Nilius, B., Voets, T., & Peters, J. (2005). TRP channels in disease. *Science's STKE, 2005*. re8.

Owsianik, G., Talavera, K., Voets, T., & Nilius, B. (2006). Permeation and selectivity of TRP channels. *Annual Review of Physiology, 68*, 685–717.

Peana, D., Polo-Parada, L., & Domeier, T. L. (2021). Arrhythmogenesis in the aged heart following ischaemia-reperfusion: Role of Transient Receptor Potential Vanilloid 4. *Cardiovascular Research*.

Peana, D., Polo-Parada, L., & Domeier, T. L. (2022). Arrhythmogenesis in the aged heart following ischaemia-reperfusion: Role of transient receptor potential vanilloid 4. *Cardiovascular Research, 118*, 1126–1137.

Pinto, A. R., Ilinykh, A., Ivey, M. J., Kuwabara, J. T., D'Antoni, M. L., Debuque, R., et al. (2016). Revisiting cardiac cellular composition. *Circulation Research*, *118*, 400–409.

Qi, Y., Li, Z., Kong, C. W., Tang, N. L., Huang, Y., Li, R. A., et al. (2015). Uniaxial cyclic stretch stimulates TRPV4 to induce realignment of human embryonic stem cell-derived cardiomyocytes. *Journal of Molecular and Cellular Cardiology*, *87*, 65–73.

Setterberg, I. E., Le, C., Frisk, M., Li, J., & Louch, W. E. (2021). The physiology and pathophysiology of T-tubules in the heart. *Frontiers in Physiology*, *12*, 718404.

Veteto, A. B., Peana, D., Lambert, M. D., McDonald, K. S., & Domeier, T. L. (2020). Transient receptor potential vanilloid-4 contributes to stretch-induced hypercontractility and time-dependent dysfunction in the aged heart. *Cardiovascular Research*, *116*, 1887–1896.

Watanabe, H., Murakami, M., Ohba, T., Takahashi, Y., & Ito, H. (2008). TRP channel and cardiovascular disease. *Pharmacology & Therapeutics*, *118*, 337–351.

White, J. P., Cibelli, M., Urban, L., Nilius, B., McGeown, J. G., & Nagy, I. (2016). TRPV4: Molecular conductor of a diverse orchestra. *Physiological Reviews*, *96*, 911–973.

Wu, Q., Lu, K., Zhao, Z., Wang, B., Liu, H., Zhang, S., et al. (2019). Blockade of transient receptor potential vanilloid 4 enhances antioxidation after myocardial ischemia/reperfusion. *Oxidative Medicine and Cellular Longevity*, *2019*, 7283683.

Wu, Q. F., Qian, C., Zhao, N., Dong, Q., Li, J., Wang, B. B., et al. (2017). Activation of transient receptor potential vanilloid 4 involves in hypoxia/reoxygenation injury in cardiomyocytes. *Cell Death & Disease*, *8*, e2828.

Yang, S., Zhao, Z., Zhao, N., Liao, J., Lu, Y., Zhang, S., et al. (2022). Blockage of transient receptor potential vanilloid 4 prevents postoperative atrial fibrillation by inhibiting NLRP3-inflammasome in sterile pericarditis mice. *Cell Calcium*, *104*, 102590.

Zhao, Y., Huang, H., Jiang, Y., Wei, H., Liu, P., Wang, W., et al. (2012). Unusual localization and translocation of TRPV4 protein in cultured ventricular myocytes of the neonatal rat. *European Journal of Histochemistry*, *56*, e32.

CHAPTER FOUR

Evolving concepts of TRPV4 in controlling flow-sensitivity of the renal nephron

Kyrylo Pyrshev, Anna Stavniichuk, Viktor N. Tomilin, Oleg Zaika, and Oleh Pochynyuk*
Department of Integrative Biology and Pharmacology, the University of Texas Health Science Center at Houston, Houston, TX, United States
*Corresponding author: e-mail address: oleh.m.pochynyuk@uth.tmc.edu

Contents

Abstract

Kidneys are central for whole body water and electrolyte balance by first filtering plasma at the glomeruli and then processing the filtrate along the renal nephron until the final urine is produced. Renal nephron epithelial cells mediate transport of water and solutes which is under the control of systemic hormones as well as local mechanical stimuli arising from alterations in fluid flow. TRPV4 is a mechanosensitive Ca^{2+} channel abundantly expressed in different segments of the renal nephron. The accumulated evidence suggests a critical role for TRPV4 in sensing variations in flow rates. In turn, TRPV4 activation triggers numerous downstream cellular responses stimulated by elevated intracellular Ca^{2+} concentrations $[Ca^{2+}]_i$. In this review, we discuss the recent concepts in flow-mediated regulation of solute homeostasis by TRPV4 in different segments of renal nephron. Specifically, we summarize the evidence for TRPV4 involvement in endocytosis-mediated albumin uptake in the proximal tubule, reactive oxygen species (ROS) generation in the ascending loop of Henle, and maintaining K^+ homeostasis in the connecting tubule/collecting duct. Finally, we outline the function and significance of TRPV4 in the setting of polycystic kidney disease.

Current Topics in Membranes, Volume 89
ISSN 1063-5823
https://doi.org/10.1016/bs.ctm.2022.06.001

1. Overview of TRPV4 structure, gating, and mechanosensitivity

TRPV4 is a tetrameric transmembrane protein belonging to the vanilloid subfamily of the transient receptor potential channels (Rosenbaum et al., 2020; Voets et al., 2002). Each TRPV4 subunit has 6 membrane (S1-S6) spanning domains. The subunits are assembled into a square-shaped functional channel to delineate the central conducting pore allowing passage of mono (Na^+, K^+) and divalent (Ca^{2+}) cations (Deng et al., 2018). Single channel TRPV4 conductance exhibits notable outward rectification (~90 pS for outward current and ~50 pS for inward current; Watanabe et al., 2002; Watanabe, Vriens, Janssens, et al., 2003; Watanabe, Vriens, Prenen, et al., 2003), which is also characteristic of other TRP channels (Venkatachalam & Montell, 2007). TRPV4 is roughly 6 times more selective for Ca^{2+} than Na^+ (Voets et al., 2002). This indicates that the channel is better suited to initiate Ca^{2+}-dependent intracellular signaling pathways and to control intracellular Ca^{2+} concentration ($[Ca^{2+}]_i$), rather than to induce a substantial Na^+ movement across the plasma membrane to affect membrane voltage. Since its cloning a little over 20 years ago as a mammalian analog of the osmosensitive protein, OSM-9 in *Caenorhabditis elegans* (Liedtke et al., 2000), TRPV4 has become a widely recognized mechanosensitive channel activated by flow-induced shear stress and osmotically-driven plasma membrane stretch (Rosenbaum et al., 2020). In addition, TRPV4 activity can be augmented by warm temperatures above 25 °C, poly-unsaturated fatty acids, and 4α-phorbol ester derivatives (Everaerts, Nilius, & Owsianik, 2010; Guilak, Leddy, & Liedtke, 2010; O'Neil & Heller, 2005). Earlier studies suggested that TRPV4 is not a "true" mechanosensitive channel due to a relatively slow (second range) activation kinetics in response to mechanical stimuli (Berrout et al., 2012; Everaerts et al., 2010; Mamenko et al., 2013; Nilius & Honore, 2012; Nilius, Vriens, Prenen, Droogmans, & Voets, 2004; Wu, Gao, Brown, Heller, & O'Neil, 2007). It was proposed that the mechanosensitive TRPV4-mediated Ca^{2+} influx is secondary to activation of a plasma membrane associated signaling cascade (s), most likely phospholipase A_2 (PLA_2) (Vriens et al., 2004). However, more recent evidence demonstrated TRPV4 activation within milliseconds by mechanical stimuli applied to cell-substrate contacts in cultured chondrocytes (Servin-Vences,

Moroni, Lewin, & Poole, 2017) and in renal epithelial cells (Gualdani et al., 2020). Such fast activation kinetics is reminiscent of that seen in "true" mechanosensitive channels, such as Piezo1 (Nilius & Honore, 2012; Servin-Vences et al., 2017). In agreement with this notion, the cryo-EM structure of *Xenopus tropicalis* TRPV4 (78% sequence identity with human analog) revealed unique organization of the S3–S4 transmembrane domains which form multiple interactions with the pore region at S5–S6 (Deng et al., 2018). This implies that S3–S4 movement can lead to the pore opening at S5–S6 due to acute distortion of the plasma membrane, thus pointing to a direct mechanism of TRPV4 activation by mechanical stimuli.

TRPV4 has a broad expression pattern across the body. This includes central nervous system, sensory neurons, such as dorsal root ganglia (DRG), circumventricular organs, organum vasculosum laminae terminalis (OVLT); osteoblasts and chondrocytes, vasculature (both endothelium and smooth muscle cells), and a variety of epithelial tissues, including skin, alveoli, trachea, bile ducts, pancreas, sweat glands, and kidney (Liedtke et al., 2000; Liedtke & Friedman, 2003; Masuyama et al., 2008; Muramatsu et al., 2007; Toft-Bertelsen & MacAulay, 2021; Venkatachalam & Montell, 2007). The kidney and more specifically the renal epithelium has probably the highest levels of TRPV4 expression (Liedtke et al., 2000). Somewhat paradoxically, numerous TRPV4 loss-of-function and gain-of-function mutations linked to clinical syndromes, such as Charcot-Marie-Tooth disease type 2C, skeletal dysplasia, osteoarthropathies, familial digital arthropathy-brachydactyly, etc., describe abnormalities of TRPV4 activity in non-renal tissues (reviewed in Nilius & Voets, 2013). It was originally anticipated that osmosensitive TRPV4 channel is well-positioned to play a central role in systemic water balance by controlling renal water handling and urinary concentrating ability. However, this hypothesis was not supported by experimental data. In fact, TRPV4 deletion did not affect urinary production and osmolarity at baseline and after water deprivation test (Janas et al., 2016; Liedtke & Friedman, 2003). Moreover, the total and plasma membrane expression of arginine vasopressin (AVP)-regulated AQP2 water channel in the collecting duct is comparable in WT and $TRPV4^{-/-}$ mice (Janas et al., 2016). It is quite possible that the lack of apparent renal water-wasting phenotype in the knockouts served as a strong discouraging factor for further investigation of the significance of TRPV4 in the renal epithelium. Luckily, the state of the field has been greatly improved after revisiting the central paradigm that the mechanosensitive TRPV4 channel is critical for flow/pressure sensitivity rather than osmosensitivity. The current review summarizes recent experimental advances/concepts of the TRPV4 significance in different parts of the renal nephron with a focus on its relevance for kidney function and how channel dysfunction can contribute to pathophysiology.

2. TRPV4 expression in different segments of the renal nephron

Kidney is central for controlling systemic homeostasis of water and solutes. This is achieved by repeated filtering of blood plasma (up to 40 times or 180L/day) at the glomeruli. The filtered load is processed/reabsorbed by sequential segments of the renal nephron from the proximal tubule to the collecting duct until the final urine with a variable volume and composition (containing waste products and excess of water and solutes) is formed. Each segment of the renal tubule has well-characterized functional roles and unique set of plasma membrane channels and transporters to execute reabsorption/secretion. Regardless of such diversity, the renal epithelial cells in different segments are commonly exposed to high and often quite variable rates of fluid flow. Multiple reports demonstrated acute elevations in $[Ca^{2+}]_i$ in native and cultured renal epithelial cells in response to increased flow over the apical plasma membrane (Berrout et al., 2012; Liu et al., 2003, 2005, 2009; Mamenko, Zaika, Jin, O'Neil, & Pochynyuk, 2011; Satlin, Carattino, Liu, & Kleyman, 2006; Tomilin et al., 2019; Weinbaum, Duan, Satlin, Wang, & Weinstein, 2010; Woda, Leite Jr., Rohatgi, & Satlin, 2002). It is now appreciated that the associated mechanical forces (such as flow-induced shear stress) can serve as important signaling cues in modulating transport of water and solutes (Praetorius & Leipziger, 2010; Sipos, Vargas, & Peti-Peterdi, 2010). It is reasonable to propose that the mechanosensitive nature of TRPV4 activation is well-suited for sensing alterations in the renal tubule flow rates. In line with this, abundant TRPV4 mRNA and protein expression was found in the proximal tubule, the late distal convoluted tubule, and throughout the connecting tubule and collecting duct (Berrout et al., 2012; Janas et al., 2016). Another study also reported TRPV4 mRNA expression in water-impermeable renal tubule segments, specifically in the ascending loop of Henle (Tian et al., 2004). From functional perspective, the accumulated experimental evidence demonstrated that either genetic deletion or pharmacological TRPV4 blockade with

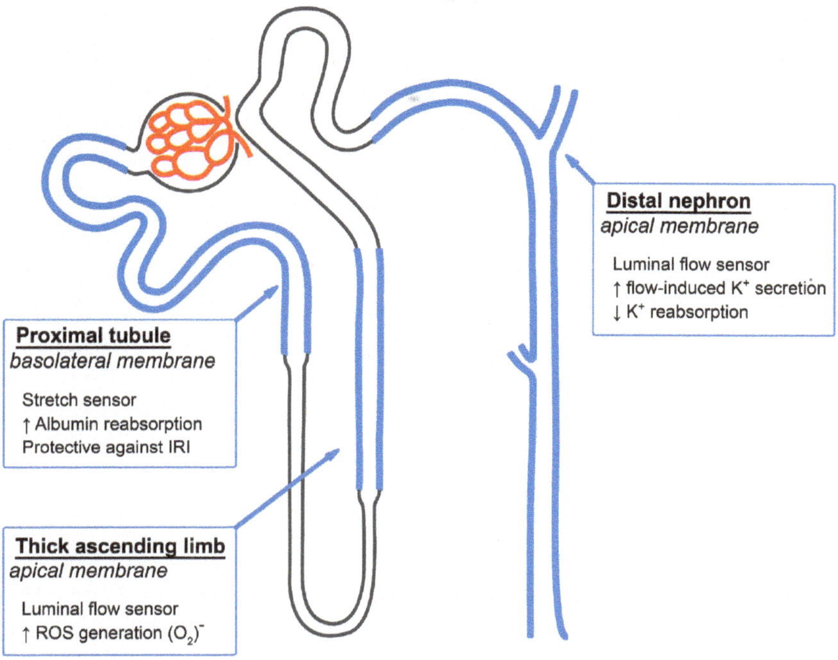

Fig. 1 Physiological roles of TRPV4 in the renal nephron. Schematic representation of the renal nephron showing the sites of TRPV4 expression (highlighted in blue). For each segment, TRPV4 localization (apical/basolateral) is specified and the major established roles are outlined.

selective antagonists abolish $[Ca^{2+}]_i$ elevation to increased flow or pressure-induced plasma membrane stretch in freshly isolated segments of the renal tubule, primary, and immortalized cultures of renal epithelial cells (Berrout et al., 2012; Mamenko et al., 2017; Tomilin, Pyrshev, Khayyat, Zaika, & Pochynyuk, 2021; Tomilin, Reif, Zaika, Wallace, & Pochynyuk, 2018). Fig. 1 shows the sites of TRPV4 expression in the renal nephron and outlines specific roles of the channel, as will be discussed below.

3. TRPV4 functions as a pressure sensor in the proximal tubule

The proximal tubule serves as a powerhouse in reabsorption of the majority (roughly 2/3) of the filtered load (Curthoys & Moe, 2014). In addition, the proximal tubule is the principal site for reabsorption of glucose and phosphate, and also megalin-dependent endocytosis of large size molecules, such as albumin, which escape through the glomerular filtration barrier (Curthoys & Moe, 2014). Using a combination of immunolocalization tools and *LacZ* reporter mice, it was shown that TRPV4 is present predominantly on the basolateral membrane, while a smaller amount can also be found on the apical membrane of proximal tubule cells (Gualdani et al., 2020; Janas et al., 2016). Furthermore, selective TRPV4 agonist, GSK1016790A stimulated whole cell currents in primary cultured proximal tubule cells from WT but not from TRPV4 $^{-/-}$ mice (Gualdani et al., 2020). Interestingly, inhibition of Piezo with GsMTx-4, but not TRPV4 deletion abolished $[Ca^{2+}]_i$ responses by elevated flow over the apical membrane. At the same time, stretching of the elastic silicon support led to TRPV4-dependent Ca^{2+} influx, which is consistent with predominantly basolateral TRPV4 expression (Gualdani et al., 2020). It appears that TRPV4 serves as a flow/pressure sensor in the proximal tubule, and can also be activated by hypotonicity-induced osmotic stretch of the plasma membrane (Gualdani et al., 2020). The proximal tubule is not exposed to notable alterations in osmolarity due to isotonic nature of water-solute reabsorption and leakiness of the tight junctions (Curthoys & Moe, 2014). Instead, increased perfusion (i.e. ultrafiltrate delivery to the proximal tubule) will augment intraluminal volume and impose mechanical stretch of the basolateral plasma membrane, thus leading to TRPV4 activation. It was next demonstrated that mechanosensitive TRPV4-dependent Ca^{2+} influx is instrumental for stimulation of endocytosis-mediated albumin uptake by the proximal tubule (Gualdani et al., 2020; Janas et al., 2016). Thus, Angiotensin II-induced albuminuria was more pronounced in TRPV4 $^{-/-}$ mice than in WT mice (Gualdani et al., 2020). In addition, TRPV4 expression is protective against ischemia-reperfusion injury of AQP1-positive proximal tubule cells (Mannaa et al., 2018). It was shown that hypoxia stimulates Ca^{2+} influx via TRPV4, which is beneficial for the proximal tubule cell survival (Mannaa et al., 2018). The molecular details of this mechanism are currently not elucidated.

4. Flow-sensitive TRPV4 $[Ca^{2+}]_i$ signaling and O_2^- production in the ascending loop of Henle

Water-impermeable ascending loop of Henle plays a central role in maintaining salt-water balance by creating the cortico-medullary osmotic gradient to set up urinary concentrating ability and reabsorbing approximately 25% of the filtered NaCl load (reviewed in Gonzalez-Vicente, Saez, Monzon, Asirwatham, & Garvin, 2019). Activity of the segment-specific Na^+-K^+-Cl^- contransporter (NKCC2, *SLC12A1*) is central for transport in the ascending loop of Henle. Indeed, Bartter's syndrome type 1 due to inactivating NKCC2 mutations is associated with polyuria, hypovolemia, and hypotension (Hebert, 2003). On the contrary, augmented NKCC2 activity has been reported in hypertensive animal models, such as Dahl salt-sensitive rats (Haque, Ares, Caceres, & Ortiz, 2011). It has been further demonstrated that generation of reactive oxygen species (ROS) leads to stimulation of NKCC2, which might contribute to the development of salt-sensitive hypertension (Silva, Ortiz, Hong, & Garvin, 2006; Taylor, Glocka, Liang, & Cowley Jr., 2006). Superoxide (O_2^-) production is stimulated by radial stretch of the ascending loop of Henle due to increased luminal flow (Hong, Silva, & Garvin, 2010). It appears that flow-stimulated $[Ca^{2+}]_i$ influx via TRPV4 is critical for O_2^- generation in perfused ascending loops of Henle (Cabral, Capurro, & Garvin, 2015; Saez, Hong, Cabral, & Garvin, 2020). Thus, pharmacological inhibition with RN1734 and shRNA knockdown of TRPV4 expression significantly decreased $[Ca^{2+}]_i$ responses to luminal flow (Cabral et al., 2015). Furthermore, ablation of TRPV4 activity had a greater effect in decreasing O_2^- generation in perfused ascending loops of Henle of Dahl salt-sensitive than in normotensive rats (Saez et al., 2020). Whether TRPV4 inhibition would be beneficial in promoting natriuresis and reducing blood pressure in the setting of salt-sensitive hypertension has not been investigated.

5. Kaliuretic actions of TRPV4 in the distal nephron

The connecting tubule and the collecting duct are the most heterogeneous segments of the renal nephron containing two completely different cell types: principal and intercalated with distinct morphology and function (Pearce et al., 2015; Roy, Al-bataineh, & Pastor-Soler, 2015). This mosaic architecture is instrumental for adjusting reabsorption/secretion rates of water, Na^+, K^+, Cl^-, H^+, HCO_3^- in a selective manner depending on dynamic changes in their intake in order to maintain systemic homeostasis. It has been well documented that the collecting duct cells are exposed to highest variations in luminal flow rates dependent on physiological conditions (Cai, Xin, Pollock, & Pollock, 2000). Predominantly apical TRPV4 localization has been shown in both principal and intercalated cells of the connecting tubule and the collecting duct of WT but not in $TRPV4^{-/-}$ mice (Berrout et al., 2012). Furthermore, either blockade or genetic deletion of TRPV4 abolished elevations in $[Ca^{2+}]_i$ in both principal and intercalated cells of freshly-isolated split-open distal nephron segments in response to physiologically relevant levels of flow (Berrout et al., 2012; Mamenko et al., 2017; Zaika et al., 2013). Similar results were obtained in immortalized cultures of the collecting duct cells: M1 and $mpkCCD_{c14}$ (Du, Wong, Sun, Huang, & Yao, 2012; Mamenko et al., 2017; Wu et al., 2007). Altogether, the current evidence strongly suggests that TRPV4 serves as a sensor of tubular flow in the distal nephron.

Kidneys play a central role in maintaining K^+ homeostasis by excreting more than 90% of dietary K^+ intake (Palmer, 2015; Youn & McDonough, 2009). Distal segments of renal nephron, the connecting tubule and the collecting duct, are the primary sites of controlled K^+ secretion and thus the major determinants of K^+ levels in the urine (Giebisch & Wang, 2010; Wang & Giebisch, 2009). This is mainly attributed to the activity of renal outer medullary K^+ (ROMK, $K_{ir}1.1$) channel on the apical membrane of principal cells (Hebert, Desir, Giebisch, & Wang, 2005; Wang, Schwab, & Giebisch, 1990). However, it has been well-documented that K^+ secretion is markedly augmented by elevated flow, thus giving rise to the phenomenon of flow-induced K^+ secretion (FIKS) (Liu, Morimoto, Woda, Kleyman, & Satlin, 2007; Satlin et al., 2006; Woda, Bragin, Kleyman, & Satlin, 2001). FIKS is independent of ROMK and is mediated by the activity of the large conductance Ca^{2+}-activated K^+ (BK) channel (Liu et al., 2007; Taniguchi et al., 2007; Taniguchi & Imai, 1998; Woda et al., 2001; Wu et al., 2007). BK is a voltage- and Ca^{2+}-activated channel due to the presence of a Ca^{2+}-binding domain (so called RCK) in the C-terminus of the pore forming α subunit (Rothberg, 2012; Wen, Cornelius, & Sansom, 2014). This implies that flow-induced Ca^{2+} influx via TRPV4 is well-suited for BK channel activation. Consistently, stimulation of TRPV4 with luminal but not basolateral 4α-PDD (TRPV4 channel agonist) increased K^+ secretion in perfused collecting ducts of WT but not TRPV4 $^{-/-}$ mice (Taniguchi et al., 2007). Increased dietary K^+ intake augments BK expression and activity in the distal nephron to facilitate kaliuresis via aldosterone-dependent and -independent mechanisms (Holtzclaw, Grimm, & Sansom, 2011; Rieg et al., 2007). BK-deficient mice exhibit impaired urinary K^+ excretion, hyperkalemia, and hyperaldosteronism when placed on a high K^+ diet (Grimm, Irsik, Settles, Holtzclaw, & Sansom, 2009; Rieg et al., 2007). Importantly, dietary K^+ supplementation also augments TRPV4 expression and increases TRPV4-dependent Ca^{2+} influx in distal nephron cells (Mamenko et al., 2017). Deletion of TRPV4 reduces BK single channel activity, decreases

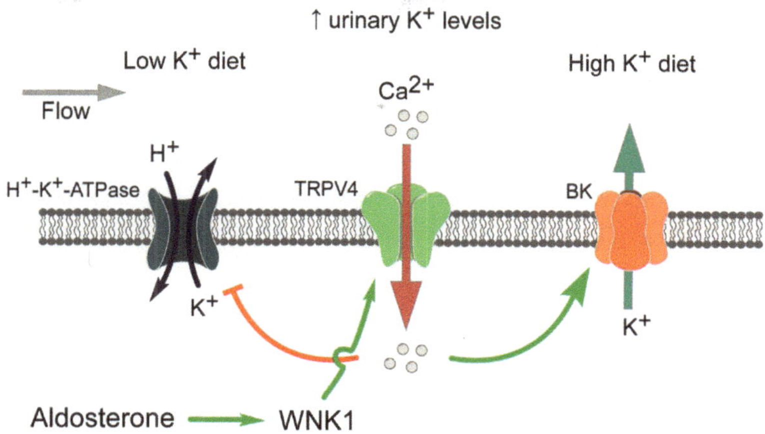

Fig. 2 Regulation of urinary K^+ excretion by TRPV4 in the distal nephron. Increased flow rates during K^+ deficient (low K^+) and high K^+ diet stimulate TRPV4-dependent Ca^{2+} influx to inhibit H^+-K^+ ATPase-mediated K^+ reabsorption and BK-mediated K^+ secretion, respectively, to augment urinary K^+ excretion. Green and red arrows represent stimulatory/inhibitory actions, respectively. BK—large conductance maxi K^+ channel; WNK1—with-no-lysine(K) kinase type 1.

urinary K^+ levels, and leads to hyperkalemia, recapitulating the phenotype of BK deficiency (Grimm et al., 2009; Mamenko et al., 2017; Rieg et al., 2007). These findings demonstrate that the functional coupling between flow-dependent Ca^{2+} influx via TRPV4 and BK-dependent K^+ secretion is a physiologically relevant mechanism to maintain systemic K^+ balance (Fig. 2). It was also shown that stimulation of TRPV4 by dietary K^+ intake occurs largely in an aldosterone-dependent manner (Mamenko et al., 2017). Similarly, aldosterone increases TRPV4 activity and expression in cultured collecting duct cells (Tomilin et al., 2021). While the exact signaling mechanism is yet to be elucidated, pharmacological blockade of with-no-lysine(K)-kinase type 1 (WNK1) drastically diminished the stimulatory actions of aldosterone on TRPV4 (Tomilin et al., 2021). Furthermore, the WNK1 mutation K233M (devoid of kinase activity; Min, Lee, Cobb, & Goldsmith, 2004; Xu et al., 2005) not only eliminates stimulation of TRPV4 in over-expression systems, but also reduces TRPV4-mediated currents below the baseline, suggesting tonic regulation of the channel by endogenous WNK1 (Tomilin et al., 2021). It appears that WNK1 is also instrumental in stimulating BK with WNK1 deficient mice on high K^+ diet having elevated plasma K^+ levels, reduced urinary K^+ excretion and decreased BK channel activity in the collecting duct intercalated cells

(Liu et al., 2015; Ray et al., 2021; Webb et al., 2016). This indicates that the WNK1-dependent signaling mechanism operates to simultaneously stimulate TRPV4 and BK channels, augmenting FIKS and maintaining systemic K^+ balance.

TRPV4 also modulates the distal nephron K^+ transport in the state of dietary K^+ deficiency (Tomilin, Mamenko, Zaika, Wingo, & Pochynyuk, 2019). Interestingly, decreased dietary K^+ intake causes a notable elevation of tubular flow in the collecting duct, in part, due to resistance to the anti-diuretic actions of arginine vasopressin (Rieg et al., 2007; Wen et al., 2014). TRPV4 deletion or blockade with HC-067047 is protective against the development of hypokalemia in mice fed low K^+ diet (Tomilin, Mamenko, Zaika, Wingo, & Pochynyuk, 2019). This was manifested as a reduction of urinary K^+ levels and a greater activity of the apical H^+-K^+ ATPase, the major pathway for K^+ reabsorption in intercalated cells (Greenlee, Lynch, Gumz, Cain, & Wingo, 2010; Gumz, Lynch, Greenlee, Cain, & Wingo, 2010). The intracellular signaling mechanism coupling TRPV4-Ca^{2+} influx and the functional status of the pump has not been investigated. Altogether, the available experimental evidence suggests that TRPV4 in the distal nephron serves as a physiologically relevant kaliuretic factor, which favors K^+ redistribution out of cells directly by stimulating FIKS or indirectly by inhibiting K^+ reabsorption via H^+-K^+ ATPase (Fig. 2). It is plausible that TRPV4 activators/inhibitors could be an effective novel approach to correct hyperkalemia/hypokalemia, respectively. This possibility needs to be thoroughly tested in future translational studies.

6. Impaired TRPV4 activity in the setting of polycystic kidney disease

Polycystic kidney disease (PKD) encompasses a diverse group of hereditary disorders affecting over 12 million people worldwide. PKD is manifested by the development and progressive growth of fluid-filled cysts in the kidneys and often in other organs, most commonly in the liver (Chapman et al., 2015; Ecder, 2013). Enlargement of the cysts compromises the function of neighboring nephrons leading to progressive decline in renal function and development of end-stage renal disease (ESRD) (Harris & Torres, 2009; Torres, Harris, & Pirson, 2007). Abundant published evidence demonstrates abnormally low basal $[Ca^{2+}]_i$ levels and impaired mechano-sensitive Ca^{2+}-responses of cyst cells in both autosomal dominant and autosomal recessive PKD (ADPKD and ARPKD, respectively) (Hovater et al., 2008; Siroky et al., 2006; Xu et al., 2009; Yamaguchi, Hempson, Reif, Hedge, & Wallace, 2006; Zaika et al., 2013). Defective $[Ca^{2+}]_i$ homeostasis is one of the major causes for inappropriately elevated cAMP levels leading to Cl^--driven fluid secretion and enhanced proliferation in a B-Raf/ERK-dependent manner (Yamaguchi et al., 2006). Mutations in *PKD1* and *PKD2* genes encoding polycystin 1 (PC1) and polycystin 2 (PC2), respectively, underlie the overwhelming majority of ADPKD cases, the most common PKD form (occurrence 1:500–1000) (Harris & Torres, 2009; Torres et al., 2007). The TRP-like channel PC2 (also known as TRPP2) can form a tetrameric Ca^{2+}-permeable channel with PC1, an atypical G-protein coupled receptor (Parnell et al., 2018) with 3:1 stoichiometry (Su et al., 2018). However, direct activation of this complex by mechanical stimuli has not been demonstrated. It was proposed that flow-induced bending of the antenna-like non-motile primary cilium could lead to Ca^{2+} influx via PC2 (Nauli et al., 2003). However, high temporal resolution imaging demonstrated a complete lack of mechanically induced Ca^{2+} influxes in response to bending of the primary cilium by physiological and supra-physiological flow rates (Delling et al., 2016). Instead, it was shown that the flow-induced elevations of $[Ca^{2+}]_i$ originate from the stretching of the plasma membrane (Delling et al., 2016). Interestingly, PC2 can also form functional heteromeric channel with TRPV4 in a 2:2 stoichiometry (Kottgen et al., 2008; Stewart, Smith, Sandford, & Edwardson, 2010). Of note, TRPV4/PC2 produces a greater Ca^{2+} influx in response to mechanical stimulation than monomeric TRPV4 (Du et al., 2012; Kottgen et al., 2008). It is further demonstrated that TRPV4-mediated Ca^{2+} influx and TRPV4/PC2 single channel activity are reduced in freshly isolated monolayers of cystic cells in PCK 453 rats, a homologous model of human ARPKD (Zaika et al., 2013), and in primary cultures of human ADPKD kidneys after nephrectomy (Tomilin et al., 2018). This provides a mechanism for the disturbed $[Ca^{2+}]_i$ homeostasis in PKD cystic cells. Indeed, TRPV4 deletion similarly decreases $[Ca^{2+}]_i$ levels and abolishes flow-induced $[Ca^{2+}]_i$ elevations, as is observed in ADPKD and ARPKD cyst cells (Tomilin et al., 2018; Zaika et al., 2013). However, ablation of TRPV4 does not lead to the development of renal cysts in mice and zebrafish (Kottgen et al., 2008; Liedtke & Friedman, 2003), suggesting that the disrupted mechanosensitivity is a common consequence and an important driving force but not a cause of PKD development. While the exact mechanism of impaired TRPV4 activity in cyst cells is not established, dramatic reduction in TRPV4 glycosylation has been reported in both ADPKD and ARPKD (Tomilin et al., 2018; Zaika et al., 2013). Blockade of TRPV4 glycosylation with tunicamycin markedly decreased channel activity and compromised mechanosensitive $[Ca^{2+}]_i$ signaling in primary cultures of human distal tubule cells to the levels observed in primary cultures of human ADPKD cystic cells (Tomilin et al., 2018). Consistently, disrupted glycosylation and decreased channel activity have also been observed in TRPV4 mutations associated with the development of the familial digital arthropathy-brachydactyly (FDAP) in humans (Lamande et al., 2011). Thus, it is reasonable to assume that restoration of deficient TRPV4 glycosylation and/or stimulation of TRPV4 activity might be beneficial in counteracting PKD progression. In agreement with this note, long term administration of TRPV4 agonist, GSK1016790A in drinking water significantly reduced renal cyst burden in ARPKD rats (Zaika et al., 2013).

7. Final remarks and conclusions

The intrinsic mechanosensitive properties of the Ca^{2+}-conducting TRPV4 channel are ideally suited for renal epithelial cells, which are in a permanent need of adjusting their transport rates and regulating metabolism in response to dynamic changes in filtrate delivery. Regardless of apical or basolateral localization, TRPV4 plays a central role as flow/pressure sensor along the renal nephron. At the same time, mechanosensitive TRPV4 activation is utilized to regulate specific processes in different segments. This includes albumin reabsorption in the proximal tubule, generation of reactive oxygen species in the ascending limb, and control of K^+ balance in the connecting tubule/collecting duct. In addition, impaired TRPV4 activity and deficient mechanosensitive $[Ca^{2+}]_i$ signaling have been involved in pathophysiology of the polycystic kidney disease. The available experimental evidence in preconditioned and genetically manipulated animal models shows the possibility and importantly feasibility of targeting TRPV4 to affect renal function and, by extension, whole body homeostasis. However, these proof-of-principle concepts have not been implemented at the clinical level despite availability of highly potent and selective pharmacological tools to either inhibit or stimulate TRPV4 activity. It is anticipated that investigation of the translational significance of TRPV4 for human kidney is the matter of a very near future.

Funding

The research program of our group is supported by NIH-NIDDK DK119170, DK117865, AHA EIA35260097 (to O. Pochynyuk) and AHA-19CDA34660148 (to V.N. Tomilin).

References

Berrout, J., Jin, M., Mamenko, M., Zaika, O., Pochynyuk, O., & O'Neil, R. G. (2012). Function of TRPV4 as a mechanical transducer in flow-sensitive segments of the renal collecting duct system. *The Journal of Biological Chemistry*, *287*(12), 8782–8791.

Cabral, P. D., Capurro, C., & Garvin, J. L. (2015). TRPV4 mediates flow-induced increases in intracellular ca in medullary thick ascending limbs. *Acta Physiologica*, *214*(3), 319–328. https://doi.org/10.1111/apha.12528. PubMed PMID: 25980432; PubMed Central PMCID: PMC5406130.

Cai, Z., Xin, J., Pollock, D. M., & Pollock, J. S. (2000). Shear stress-mediated NO production in inner medullary collecting duct cells. *American Journal of Physiology. Renal Physiology*, *279*(2), F270–F274. https://doi.org/10.1152/ajprenal.2000.279.2.F270. PubMed PMID: 10919845.

Chapman, A. B., Devuyst, O., Eckardt, K. U., Gansevoort, R. T., Harris, T., Horie, S., et al. (2015). Autosomal-dominant polycystic kidney disease (ADPKD): Executive summary from a kidney disease: Improving global outcomes (KDIGO) controversies conference. *Kidney International*, *88*(1), 17–27. https://doi.org/10.1038/ki.2015.59. PubMed PMID: 25786098; PubMed Central PMCID: PMC4913350.

Curthoys, N. P., & Moe, O. W. (2014). Proximal tubule function and response to acidosis. *Clinical Journal of the American Society of Nephrology: CJASN*, *9*(9), 1627–1638. https://doi.org/10.2215/CJN.10391012. PubMed PMID: 23908456; PubMed Central PMCID: PMC4152816.

Delling, M., Indzhykulian, A. A., Liu, X., Li, Y., Xie, T., Corey, D. P., et al. (2016). Primary cilia are not calcium-responsive mechanosensors. *Nature*, *531*(7596), 656–660. https://doi.org/10.1038/nature17426. PubMed PMID: 27007841; PubMed Central PMCID: PMC4851444.

Deng, Z., Paknejad, N., Maksaev, G., Sala-Rabanal, M., Nichols, C. G., Hite, R. K., et al. (2018). Cryo-EM and X-ray structures of TRPV4 reveal insight into ion permeation and gating mechanisms. *Nature Structural & Molecular Biology*, *25*(3), 252–260. https://doi.org/10.1038/s41594-018-0037-5. PubMed PMID: 29483651.

Du, J., Wong, W. Y., Sun, L., Huang, Y., & Yao, X. (2012). Protein kinase G inhibits flow-induced Ca2 + entry into collecting duct cells. *Journal of the American Society of Nephrology*, *23*(7), 1172–1180. PMID: 22518003.

Ecder, T. (2013). Cardiovascular complications in autosomal dominant polycystic kidney disease. *Current Hypertension Reviews*, *9*(1), 2–11. PubMed PMID: 23971638.

Everaerts, W., Nilius, B., & Owsianik, G. (2010). The vanilloid transient receptor potential channel TRPV4: From structure to disease. *Progress in Biophysics and Molecular Biology*, *103*(1), 2–17. https://doi.org/10.1016/j.pbiomolbio.2009.10.002. PubMed PMID: 19835908.

Everaerts, W., Zhen, X., Ghosh, D., Vriens, J., Gevaert, T., Gilbert, J. P., et al. (2010). Inhibition of the cation channel TRPV4 improves bladder function in mice and rats with cyclophosphamide-induced cystitis. *Proceedings of the National Academy of Sciences of the*

United States of America, *107*(44), 19084–19089. https://doi.org/10.1073/pnas.1005333107. PubMed PMID: 20956320; PubMed Central PMCID: PMC2973867.

Giebisch, G. H., & Wang, W. H. (2010). Potassium transport--an update. *Journal of Nephrology*, *23*(Suppl. 16), S97–104. PubMed PMID: 21170894.

Gonzalez-Vicente, A., Saez, F., Monzon, C. M., Asirwatham, J., & Garvin, J. L. (2019). Thick ascending limb sodium transport in the pathogenesis of hypertension. *Physiological Reviews*, *99*(1), 235–309. https://doi.org/10.1152/physrev.00055.2017. PubMed PMID: 30354966; PubMed Central PMCID: PMC6335098.

Greenlee, M. M., Lynch, I. J., Gumz, M. L., Cain, B. D., & Wingo, C. S. (2010). The renal H,K-ATPases. *Current Opinion in Nephrology and Hypertension*, *19*(5), 478–482. https://doi.org/10.1097/MNH.0b013e32833ce65f. PubMed PMID: 20616716.

Grimm, P. R., Irsik, D. L., Settles, D. C., Holtzclaw, J. D., & Sansom, S. C. (2009). Hypertension of Kcnmb1 −/− is linked to deficient K secretion and aldosteronism. *Proceedings of the National Academy of Sciences of the United States of America*, *106*(28), 11800–11805. https://doi.org/10.1073/pnas.0904635106. PubMed PMID: 19556540; PubMed Central PMCID: PMC2701967.

Gualdani, R., Seghers, F., Yerna, X., Schakman, O., Tajeddine, N., Achouri, Y., et al. (2020). Mechanical activation of TRPV4 channels controls albumin reabsorption by proximal tubule cells. *Science Signaling*, *13*(653). https://doi.org/10.1126/scisignal.abc6967. PubMed PMID: 33051259.

Guilak, F., Leddy, H. A., & Liedtke, W. (2010). Transient receptor potential vanilloid 4: The sixth sense of the musculoskeletal system? *Annals of the New York Academy of Sciences*, *1192*, 404–409. https://doi.org/10.1111/j.1749-6632.2010.05389.x. PubMed PMID: 20392266; PubMed Central PMCID: PMC3580043.

Gumz, M. L., Lynch, I. J., Greenlee, M. M., Cain, B. D., & Wingo, C. S. (2010). The renal H+-K+-ATPases: Physiology, regulation, and structure. *American Journal of Physiology. Renal Physiology*, *298*(1), F12–F21. https://doi.org/10.1152/ajprenal.90723.2008. PubMed PMID: 19640897; PubMed Central PMCID: PMC2806118.

Haque, M. Z., Ares, G. R., Caceres, P. S., & Ortiz, P. A. (2011). High salt differentially regulates surface NKCC2 expression in thick ascending limbs of Dahl salt-sensitive and salt-resistant rats. *American Journal of Physiology. Renal Physiology*, *300*(5), F1096–F1104. https://doi.org/10.1152/ajprenal.00600.2010. PubMed PMID: 21307126; PubMed Central PMCID: PMC3094058.

Harris, P. C., & Torres, V. E. (2009). Polycystic kidney disease. *Annual Review of Medicine*, *60*, 321–337.

Hebert, S. C. (2003). Bartter syndrome. *Current Opinion in Nephrology and Hypertension*, *12*(5), 527–532. https://doi.org/10.1097/00041552-200309000-00008. PubMed PMID: 12920401.

Hebert, S. C., Desir, G., Giebisch, G., & Wang, W. (2005). Molecular diversity and regulation of renal potassium channels. *Physiological Reviews*, *85*(1), 319–371. https://doi.org/10.1152/physrev.00051.2003. PubMed PMID: 15618483; PubMed Central PMCID: PMC2838721.

Holtzclaw, J. D., Grimm, P. R., & Sansom, S. C. (2011). Role of BK channels in hypertension and potassium secretion. *Current Opinion in Nephrology and Hypertension*, *20*(5), 512–517. https://doi.org/10.1097/MNH.0b013e3283488889. PubMed PMID: 21670674; PubMed Central PMCID: PMC3839424.

Hong, N. J., Silva, G. B., & Garvin, J. L. (2010). PKC-alpha mediates flow-stimulated superoxide production in thick ascending limbs. *American Journal of Physiology. Renal Physiology*, *298*(4), F885–F891. https://doi.org/10.1152/ajprenal.00543.2009. PubMed PMID: 20053794; PubMed Central PMCID: PMC2853312.

Hovater, M. B., Olteanu, D., Hanson, E. L., Cheng, N. L., Siroky, B., Fintha, A., et al. (2008). Loss of apical monocilia on collecting duct principal cells impairs ATP secretion across the apical cell surface and ATP-dependent and flow-induced calcium signals. *Purinergic Signal, 4*(2), 155–170. https://doi.org/10.1007/s11302-007-9072-0. PubMed PMID: 18368523; PubMed Central PMCID: PMC2377318.

Janas, S., Seghers, F., Schakman, O., Alsady, M., Deen, P., Vriens, J., et al. (2016). TRPV4 is associated with central rather than nephrogenic osmoregulation. *Pflügers Archiv/European Journal of Physiology, 468*(9), 1595–1607. https://doi.org/10.1007/s00424-016-1850-5. PubMed PMID: 27364478.

Kottgen, M., Buchholz, B., Garcia-Gonzalez, M. A., Kotsis, F., Fu, X., Doerken, M., et al. (2008). TRPP2 and TRPV4 form a polymodal sensory channel complex. *The Journal of Cell Biology, 182*(3), 437–447. https://doi.org/10.1083/jcb.200805124. PubMed PMID: 18695040; PubMed Central PMCID: PMC2500130.

Lamande, S. R., Yuan, Y., Gresshoff, I. L., Rowley, L., Belluoccio, D., Kaluarachchi, K., et al. (2011). Mutations in TRPV4 cause an inherited arthropathy of hands and feet. *Nature Genetics, 43*(11), 1142–1146. https://doi.org/10.1038/ng.945. PubMed PMID: 21964574.

Liedtke, W., Choe, Y., Marti-Renom, M. A., Bell, A. M., Denis, C. S., Sali, A., et al. (2000). Vanilloid receptor-related osmotically activated channel (VR-OAC), a candidate vertebrate osmoreceptor. *Cell, 103*(3), 525–535. PubMed PMID: 11081638; PubMed Central PMCID: PMC2211528.

Liedtke, W., & Friedman, J. M. (2003). Abnormal osmotic regulation in trpv4 −/− mice. *Proceedings of the National Academy of Sciences of the United States of America, 100*(23), 13698–13703.

Liu, W., Morimoto, T., Woda, C., Kleyman, T. R., & Satlin, L. M. (2007). Ca2+ dependence of flow-stimulated K secretion in the mammalian cortical collecting duct. *American Journal of Physiology. Renal Physiology, 293*(1), F227–F235. https://doi.org/10.1152/ajprenal.00057.2007. PubMed PMID: 17389680.

Liu, W., Murcia, N. S., Duan, Y., Weinbaum, S., Yoder, B. K., Schwiebert, E., et al. (2005). Mechanoregulation of intracellular Ca2+ concentration is attenuated in collecting duct of monocilium-impaired orpk mice. *American Journal of Physiology. Renal Physiology, 289*(5), F978–F988.

Liu, Y., Song, X., Shi, Y., Shi, Z., Niu, W., Feng, X., et al. (2015). WNK1 activates large-conductance Ca2+−activated K+ channels through modulation of ERK1/2 signaling. *Journal of the American Society of Nephrology: JASN, 26*(4), 844–854. https://doi.org/10.1681/ASN.2014020186. PubMed PMID: 25145935; PubMed Central PMCID: PMC4378107.

Liu, W., Wei, Y., Sun, P., Wang, W. H., Kleyman, T. R., & Satlin, L. M. (2009). Mechanoregulation of BK channel activity in the mammalian cortical collecting duct: Role of protein kinases A and C. *American Journal of Physiology. Renal Physiology, 297*(4), F904–F915. https://doi.org/10.1152/ajprenal.90685.2008. PubMed PMID: 19656909; PubMed Central PMCID: PMC2775580.

Liu, W., Xu, S., Woda, C., Kim, P., Weinbaum, S., & Satlin, L. M. (2003). Effect of flow and stretch on the [Ca2+]i response of principal and intercalated cells in cortical collecting duct. *American Journal of Physiology. Renal Physiology, 285*(5), F998–F1012. https://doi.org/10.1152/ajprenal.00067.2003. PubMed PMID: 12837680.

Mamenko, M. V., Boukelmoune, N., Tomilin, V. N., Zaika, O. L., Jensen, V. B., O'Neil, R. G., et al. (2017). The renal TRPV4 channel is essential for adaptation to increased dietary potassium. *Kidney International, 91*(6), 1398–1409. https://doi.org/10.1016/j.kint.2016.12.010. PubMed PMID: 28187982; PubMed Central PMCID: PMC5429991.

Mamenko, M., Zaika, O. L., Boukelmoune, N., Berrout, J., O'Neil, R. G., & Pochynyuk, O. (2013). Discrete control of TRPV4 channel function in the distal nephron by protein kinases A and C. *The Journal of Biological Chemistry, 288*(28),

20306–20314. https://doi.org/10.1074/jbc.M113.466797. PubMed PMID: 23709216; PubMed Central PMCID: PMC3711297.

Mamenko, M., Zaika, O., Jin, M., O'Neil, R. G., & Pochynyuk, O. (2011). Purinergic activation of Ca-permeable TRPV4 channels is essential for mechano-sensitivity in the aldosterone-sensitive distal nephron. *PLoS One*, *6*(8), e22824.

Mannaa, M., Marko, L., Balogh, A., Vigolo, E., N'Diaye, G., Kassmann, M., et al. (2018). Transient receptor potential vanilloid 4 channel deficiency aggravates tubular damage after acute renal ischaemia reperfusion. *Scientific Reports*, *8*(1), 4878. https://doi.org/10.1038/s41598-018-23165-0. PubMed PMID: 29559678; PubMed Central PMCID: PMC5861116.

Masuyama, R., Vriens, J., Voets, T., Karashima, Y., Owsianik, G., Vennekens, R., et al. (2008). TRPV4-mediated calcium influx regulates terminal differentiation of osteoclasts. *Cell Metabolism*, *8*(3), 257–265. https://doi.org/10.1016/j.cmet.2008.08.002. PubMed PMID: 18762026.

Min, X., Lee, B. H., Cobb, M. H., & Goldsmith, E. J. (2004). Crystal structure of the kinase domain of WNK1, a kinase that causes a hereditary form of hypertension. *Structure*, *12*(7), 1303–1311. https://doi.org/10.1016/j.str.2004.04.014. PubMed PMID: 15242606.

Muramatsu, S., Wakabayashi, M., Ohno, T., Amano, K., Ooishi, R., Sugahara, T., et al. (2007). Functional gene screening system identified TRPV4 as a regulator of chondrogenic differentiation. *The Journal of Biological Chemistry*, *282*(44), 32158–32167. https://doi.org/10.1074/jbc.M706158200. PubMed PMID: 17804410.

Nauli, S. M., Alenghat, F. J., Luo, Y., Williams, E., Vassilev, P., Li, X., et al. (2003). Polycystins 1 and 2 mediate mechanosensation in the primary cilium of kidney cells. *Nature Genetics*, *33*(2), 129–137. https://doi.org/10.1038/ng1076. PubMed PMID: 12514735.

Nilius, B., & Honore, E. (2012). Sensing pressure with ion channels. *Trends in Neurosciences*, *35*(8), 477–486. https://doi.org/10.1016/j.tins.2012.04.002. PubMed PMID: 22622029.

Nilius, B., & Voets, T. (2013). The puzzle of TRPV4 channelopathies. *EMBO Reports*, *14*(2), 152–163. https://doi.org/10.1038/embor.2012.219. PubMed PMID: 23306656; PubMed Central PMCID: PMC3566843.

Nilius, B., Vriens, J., Prenen, J., Droogmans, G., & Voets, T. (2004). TRPV4 calcium entry channel: A paradigm for gating diversity. *American Journal of Physiology. Cell Physiology*, *286*(2), C195–C205. https://doi.org/10.1152/ajpcell.00365.2003. PubMed PMID: 14707014.

O'Neil, R. G., & Heller, S. (2005). The mechanosensitive nature of TRPV channels. *Pflügers Archiv/European Journal of Physiology*, *451*(1), 193–203. https://doi.org/10.1007/s00424-005-1424-4. PubMed PMID: 15909178.

Palmer, B. F. (2015). Regulation of potassium homeostasis. *Clinical Journal of the American Society of Nephrology: CJASN*, *10*(6), 1050–1060. https://doi.org/10.2215/CJN.08580813. PubMed PMID: 24721891; PubMed Central PMCID: PMC4455213.

Parnell, S. C., Magenheimer, B. S., Maser, R. L., Pavlov, T. S., Havens, M. A., Hastings, M. L., et al. (2018). A mutation affecting polycystin-1 mediated heterotrimeric G-protein signaling causes PKD. *Human Molecular Genetics*, *27*(19), 3313–3324. https://doi.org/10.1093/hmg/ddy223. PubMed PMID: 29931260; PubMed Central PMCID: PMC6140781.

Pearce, D., Soundararajan, R., Trimpert, C., Kashlan, O. B., Deen, P. M., & Kohan, D. E. (2015). Collecting duct principal cell transport processes and their regulation. *Clinical Journal of the American Society of Nephrology: CJASN*, *10*(1), 135–146. https://doi.org/10.2215/CJN.05760513. PubMed PMID: 24875192; PubMed Central PMCID: PMC4284417.

Praetorius, H. A., & Leipziger, J. (2010). Intrarenal purinergic signaling in the control of renal tubular transport. *Annual Review of Physiology*, *72*, 377–393. https://doi.org/10.1146/annurev-physiol-021909-135825. PubMed PMID: 20148681.

Ray, E. C., Carrisoza-Gaytan, R., Al-Bataineh, M., Marciszyn, A. L., Nkashama, L. J., Chen, J., et al. (2021). L-WNK1 is required for BK channel activation in intercalated cells. *American Journal of Physiology. Renal Physiology*, *321*(2), F245–F254. https://doi.org/10.1152/ajprenal.00472.2020. PubMed PMID: 34229479; PubMed Central PMCID: PMC8424664.

Rieg, T., Vallon, V., Sausbier, M., Sausbier, U., Kaissling, B., Ruth, P., et al. (2007). The role of the BK channel in potassium homeostasis and flow-induced renal potassium excretion. *Kidney International*, *72*(5), 566–573. https://doi.org/10.1038/sj.ki.5002369. PubMed PMID: 17579662.

Rosenbaum, T., Benitez-Angeles, M., Sanchez-Hernandez, R., Morales-Lazaro, S. L., Hiriart, M., Morales-Buenrostro, L. E., et al. (2020). TRPV4: A physio and pathophysiologically significant ion channel. *International Journal of Molecular Sciences*, *21*(11). https://doi.org/10.3390/ijms21113837. 32481620. PubMed Central PMCID: PMC7312103.

Rothberg, B. S. (2012). The BK channel: A vital link between cellular calcium and electrical signaling. *Protein & Cell*, *3*(12), 883–892. https://doi.org/10.1007/s13238-012-2076-8. PubMed PMID: 22996175.

Roy, A., Al-bataineh, M. M., & Pastor-Soler, N. M. (2015). Collecting duct intercalated cell function and regulation. *Clinical Journal of the American Society of Nephrology: CJASN*, *10*(2), 305–324. https://doi.org/10.2215/CJN.08880914. PubMed PMID: 25632105; PubMed Central PMCID: PMC4317747.

Saez, F., Hong, N. J., Cabral, P. D., & Garvin, J. L. (2020). Stretch-induced increases in intracellular Ca stimulate thick ascending limb O2(−) production and are enhanced in dahl salt-sensitive rats. *Hypertension*, *75*(2), 431–438. https://doi.org/10.1161/HYPERTENSIONAHA.119.13765. PubMed PMID: 31865796; PubMed Central PMCID: PMC7058965.

Satlin, L. M., Carattino, M. D., Liu, W., & Kleyman, T. R. (2006). Regulation of cation transport in the distal nephron by mechanical forces. *American Journal of Physiology. Renal Physiology*, *291*(5), F923–F931. https://doi.org/10.1152/ajprenal.00192.2006. PubMed PMID: 16849691.

Servin-Vences, M. R., Moroni, M., Lewin, G. R., & Poole, K. (2017). Direct measurement of TRPV4 and PIEZO1 activity reveals multiple mechanotransduction pathways in chondrocytes. *eLife*, *6*. https://doi.org/10.7554/eLife.21074. PubMed PMID: 28135189; PubMed Central PMCID: PMC5279942.

Silva, G. B., Ortiz, P. A., Hong, N. J., & Garvin, J. L. (2006). Superoxide stimulates NaCl absorption in the thick ascending limb via activation of protein kinase C. *Hypertension*, *48*(3), 467–472. https://doi.org/10.1161/01.HYP.0000236646.83354.51. PubMed PMID: 16894053.

Sipos, A., Vargas, S., & Peti-Peterdi, J. (2010). Direct demonstration of tubular fluid flow sensing by macula densa cells. *American Journal of Physiology. Renal Physiology*, *299*(5), F1087–F1093. https://doi.org/10.1152/ajprenal.00469.2009. PubMed PMID: 20719981; PubMed Central PMCID: PMC2980403.

Siroky, B. J., Ferguson, W. B., Fuson, A. L., Xie, Y., Fintha, A., Komlosi, P., et al. (2006). Loss of primary cilia results in deregulated and unabated apical calcium entry in ARPKD collecting duct cells. *American Journal of Physiology. Renal Physiology*, *290*(6), F1320–F1328. https://doi.org/10.1152/ajprenal.00463.2005. PubMed PMID: 16396941.

Stewart, A. P., Smith, G. D., Sandford, R. N., & Edwardson, J. M. (2010). Atomic force microscopy reveals the alternating subunit arrangement of the TRPP2-TRPV4 heterotetramer. *Biophysical Journal*, *99*(3), 790–797. https://doi.org/10.1016/j.bpj.2010.05.012. PubMed PMID: 20682256; PubMed Central PMCID: PMC2913176.

Su, Q., Hu, F., Ge, X., Lei, J., Yu, S., Wang, T., et al. (2018). Structure of the human PKD1-PKD2 complex. *Science*, *361*(6406). https://doi.org/10.1126/science.aat9819. PubMed PMID: 30093605.

Taniguchi, J., & Imai, M. (1998). Flow-dependent activation of maxi K+ channels in apical membrane of rabbit connecting tubule. *The Journal of Membrane Biology*, *164*(1), 35–45. PubMed PMID: 9636242.

Taniguchi, J., Tsuruoka, S., Mizuno, A., Sato, J., Fujimura, A., & Suzuki, M. (2007). TRPV4 as a flow sensor in flow-dependent K+ secretion from the cortical collecting duct. *American Journal of Physiology. Renal Physiology*, *292*(2), F667–F673.

Taylor, N. E., Glocka, P., Liang, M., & Cowley, A. W., Jr. (2006). NADPH oxidase in the renal medulla causes oxidative stress and contributes to salt-sensitive hypertension in dahl S rats. *Hypertension*, *47*(4), 692–698. https://doi.org/10.1161/01.HYP.0000203161.02046.8d. PubMed PMID: 16505210.

Tian, W., Salanova, M., Xu, H., Lindsley, J. N., Oyama, T. T., Anderson, S., et al. (2004). Renal expression of osmotically responsive cation channel TRPV4 is restricted to water-impairment nephron segments. *American Journal of Physiology. Renal Physiology*, *287*(1), F17–F24.

Toft-Bertelsen, T. L., & MacAulay, N. (2021). TRPing to the point of clarity: Understanding the function of the complex TRPV4 ion channel. *Cell*, *10*(1). https://doi.org/10.3390/cells10010165. PubMed PMID: 33467654; PubMed Central PMCID: PMC7830798.

Tomilin, V. N., Mamenko, M., Zaika, O., Ren, G., Marrelli, S. P., Birnbaumer, L., et al. (2019). TRPC3 determines osmosensitive [Ca2+]i signaling in the collecting duct and contributes to urinary concentration. *PLoS One*, *14*(12), e0226381. https://doi.org/10.1371/journal.pone.0226381. PubMed PMID: 31851715.

Tomilin, V., Mamenko, M., Zaika, O., Wingo, C. S., & Pochynyuk, O. (2019). TRPV4 deletion protects against hypokalemia during systemic K(+) deficiency. *American Journal of Physiology. Renal Physiology*, *316*(5), F948–F956. https://doi.org/10.1152/ajprenal.00043.2019. PubMed PMID: 30838874; PubMed Central PMCID: PMC6580258.

Tomilin, V. N., Pyrshev, K., Khayyat, N. H., Zaika, O., & Pochynyuk, O. (2021). With-no-lysine kinase 1 (WNK1) augments TRPV4 function in the aldosterone-sensitive distal nephron. *Cell*, *10*(6). https://doi.org/10.3390/cells10061482. PubMed PMID: 34204757; PubMed Central PMCID: PMC8231605.

Tomilin, V., Reif, G. A., Zaika, O., Wallace, D. P., & Pochynyuk, O. (2018). Deficient transient receptor potential vanilloid type 4 function contributes to compromised [Ca(2+)]i homeostasis in human autosomal-dominant polycystic kidney disease cells. *FASEB Journal: Official Publication of the Federation of American Societies for Experimental Biology*, *32*(8), 4612–4623. https://doi.org/10.1096/fj.201701535RR. PubMed PMID: 29553832; PubMed Central PMCID: PMC6044056.

Torres, V. E., Harris, P. C., & Pirson, Y. (2007). Autosomal dominant polycystic kidney disease. *Lancet*, *369*(9569), 1287–1301.

Venkatachalam, K., & Montell, C. (2007). TRP channels. *Annual Review of Biochemistry*, *76*, 387–417. https://doi.org/10.1146/annurev.biochem.75.103004.142819. PubMed PMID: 17579562.

Voets, T., Prenen, J., Vriens, J., Watanabe, H., Janssens, A., Wissenbach, U., et al. (2002). Molecular determinants of permeation through the cation channel TRPV4. *The Journal of Biological Chemistry*, *277*(37), 33704–33710. https://doi.org/10.1074/jbc.M204828200. PubMed PMID: 12093812.

Vriens, J., Watanabe, H., Janssens, A., Droogmans, G., Voets, T., & Nilius, B. (2004). Cell swelling, heat, and chemical agonists use distinct pathways for the activation of the cation channel TRPV4. *Proceedings of the National Academy of Sciences of the United States of America*, *101*(1), 396–401. https://doi.org/10.1073/pnas.0303329101. PubMed PMID: 14691263; PubMed Central PMCID: PMC314196.

Wang, W. H., & Giebisch, G. (2009). Regulation of potassium (K) handling in the renal collecting duct. *Pflügers Archiv/European Journal of Physiology*, *458*(1), 157–168. https://doi.org/10.1007/s00424-008-0593-3. PubMed PMID: 18839206; PubMed Central PMCID: PMC2730119.

Wang, W. H., Schwab, A., & Giebisch, G. (1990). Regulation of small-conductance K+ channel in apical membrane of rat cortical collecting tubule. *The American Journal of Physiology*, *259*(3 Pt 2), F494–F502. PubMed PMID: 2396675.

Watanabe, H., Davis, J. B., Smart, D., Jerman, J. C., Smith, G. D., Hayes, P., et al. (2002). Activation of TRPV4 channels (hVRL-2/mTRP12) by phorbol derivatives. *The Journal of Biological Chemistry*, *277*(16), 13569–13577. https://doi.org/10.1074/jbc.M200062200. PubMed PMID: 11827975.

Watanabe, H., Vriens, J., Janssens, A., Wondergem, R., Droogmans, G., & Nilius, B. (2003). Modulation of TRPV4 gating by intra- and extracellular Ca2+. *Cell Calcium*, *33*(5–6), 489–495. PubMed PMID: 12765694.

Watanabe, H., Vriens, J., Prenen, J., Droogmans, G., Voets, T., & Nilius, B. (2003). Anandamide and arachidonic acid use epoxyeicosatrienoic acids to activate TRPV4 channels. *Nature*, *424*(6947), 434–438. https://doi.org/10.1038/nature01807. PubMed PMID: 12879072.

Webb, T. N., Carrisoza-Gaytan, R., Montalbetti, N., Rued, A., Roy, A., Socovich, A. M., et al. (2016). Cell-specific regulation of L-WNK1 by dietary K. *American Journal of Physiology. Renal Physiology*, *310*(1), F15–F26. https://doi.org/10.1152/ajprenal.00226.2015. PubMed PMID: 26662201; PubMed Central PMCID: PMC4675801.

Weinbaum, S., Duan, Y., Satlin, L. M., Wang, T., & Weinstein, A. M. (2010). Mechanotransduction in the renal tubule. *American Journal of Physiology. Renal Physiology*, *299*(6), F1220–F1236. https://doi.org/10.1152/ajprenal.00453.2010. PubMed PMID: 20810611; PubMed Central PMCID: PMC3006307.

Wen, D., Cornelius, R. J., & Sansom, S. C. (2014). Interacting influence of diuretics and diet on BK channel-regulated K homeostasis. *Current Opinion in Pharmacology*, *15*, 28–32. https://doi.org/10.1016/j.coph.2013.11.001. PubMed PMID: 24721651; PubMed Central PMCID: PMC3984455.

Woda, C. B., Bragin, A., Kleyman, T. R., & Satlin, L. M. (2001). Flow-dependent K+ secretion in the cortical collecting duct is mediated by a maxi-K channel. *American Journal of Physiology. Renal Physiology*, *280*(5), F786–F793.

Woda, C. B., Leite, M., Jr., Rohatgi, R., & Satlin, L. M. (2002). Effects of luminal flow and nucleotides on [Ca(2+)](i) in rabbit cortical collecting duct. *American Journal of Physiology. Renal Physiology*, *283*(3), F437–F446. https://doi.org/10.1152/ajprenal.00316.2001. PubMed PMID: 12167594.

Wu, L., Gao, X., Brown, R. C., Heller, S., & O'Neil, R. G. (2007). Dual role of the TRPV4 channel as a sensor of flow and osmolality in renal epithelial cells. *American Journal of Physiology. Renal Physiology*, *293*(5), F1699–F1713.

Xu, B. E., Lee, B. H., Min, X., Lenertz, L., Heise, C. J., Stippec, S., et al. (2005). WNK1: Analysis of protein kinase structure, downstream targets, and potential roles in hypertension. *Cell Research*, *15*(1), 6–10. https://doi.org/10.1038/sj.cr.7290256. PubMed PMID: 15686619.

Xu, C., Shmukler, B. E., Nishimura, K., Kaczmarek, E., Rossetti, S., Harris, P. C., et al. (2009). Attenuated, flow-induced ATP release contributes to absence of flow-sensitive, purinergic Cai2+ signaling in human ADPKD cyst epithelial cells. *American Journal of Physiology. Renal Physiology*, *296*(6), F1464–F1476. https://doi.org/10.1152/ajprenal.90542.2008. PubMed PMID: 19244404; PubMed Central PMCID: PMC2692447.

Yamaguchi, T., Hempson, S. J., Reif, G. A., Hedge, A. M., & Wallace, D. P. (2006). Calcium restores a normal proliferation phenotype in human polycystic kidney disease epithelial cells. *Journal of the American Society of Nephrology: JASN*, *17*(1), 178–187. https://doi.org/10.1681/ASN.2005060645. PubMed PMID: 16319189.

Youn, J. H., & McDonough, A. A. (2009). Recent advances in understanding integrative control of potassium homeostasis. *Annual Review of Physiology*, *71*, 381–401. https://doi.org/10.1146/annurev.physiol.010908.163241. PubMed PMID: 18759636; PubMed Central PMCID: PMC4946439.

Zaika, O., Mamenko, M., Berrout, J., Boukelmoune, N., O'Neil, R. G., & Pochynyuk, O. (2013). TRPV4 dysfunction promotes renal cystogenesis in autosomal recessive polycystic kidney disease. *Journal of the American Society of Nephrology: JASN*, *24*(4), 604–616. https://doi.org/10.1681/ASN.2012050442. PubMed PMID: 23411787; PubMed Central PMCID: PMC3609133.

CHAPTER FIVE

Transient receptor potential vanilloid type 4 (TRPV4) in urinary bladder structure and function

Megan Elizabeth Perkins and Margaret A. Vizzard*
Department of Neurological Sciences, The Larner College of Medicine, The University of Vermont, Burlington, VT, United States
*Corresponding author: e-mail address: margaret.vizzard@uvm.edu

Contents

Current Topics in Membranes, Volume 89
ISSN 1063-5823
https://doi.org/10.1016/bs.ctm.2022.06.002

Abstract

Bladder pain syndrome (BPS)/interstitial cystitis (IC) is a urologic, chronic pelvic pain syndrome characterized by pelvic pain, pressure, or discomfort with urinary symptoms. Symptom exacerbation (flare) is common with multiple, perceived triggers including stress. Multiple transient receptor potential (TRP) channels (TRPA1, TRPV1, TRPV4) expressed in the bladder have specific tissue distributions in the lower urinary tract (LUT) and are implicated in bladder disorders including overactive bladder (OAB) and BPS/IC. TRPV4 channels are strong candidates for mechanosensors in the urinary bladder and TRPV4 antagonists are promising therapeutic agents for OAB. In this perspective piece, we address the current knowledge of TRPV4 distribution and function in the LUT and its plasticity with injury or disease with an emphasis on BPS/IC. We review our studies that extend the knowledge of TRPV4 in urinary bladder function by focusing on (i) TRPV4 involvement in voiding dysfunction, pelvic pain, and non-voiding bladder contractions in NGF-OE mice; (ii) distention-induced luminal ATP release mechanisms and (iii) involvement of TRPV4 and vesicular release mechanisms. Finally, we review our lamina propria studies in postnatal rat studies that demonstrate: (i) the predominance of the TRPV4+ and PDGFRα+ lamina propria cellular network in early postnatal rats; (ii) the ability of exogenous mediators (i.e., ATP, TRPV4 agonist) to activate and increase the number of lamina propria cells exhibiting active Ca^{2+} events; and (iii) the ability of ATP and TRPV4 agonist to increase the rate of integrated Ca^{2+} activity corresponding to coupled lamina propria network events and the formation of propagating wavefronts.

1. Introduction

Micturition relies on the components of the lower urinary tract (LUT) including the bladder and urethra, working together in a reciprocal manner to store and eliminate urine. In the storage phase, the bladder remains relaxed to act as a reservoir to collect and store urine produced by the kidneys, while the urethra remains contracted to prevent leakage. Upon elimination, the bladder contracts to expel urine through the relaxed urethra. The LUT consists of 3 main tissue layers: the mucosa (closest to the lumen), the muscularis propria, and the surrounding adventitia and serosa (Birder & Andersson, 2013).

2. LUT tissues

2.1 Mucosa

The mucosa is composed of the urothelium, basement membrane, lamina propria and smooth muscle cells (i.e., muscularis mucosae) (Birder & Andersson, 2013).

2.2 Urothelium

The urothelium (lining the bladder lumen) consists of transitional epithelial cells organized in three layers including a basal, intermediate, and superficial cell layer. The basal cell layer adheres to the basement membrane. The intermediate and basal urothelial cell layers play important roles in proliferation and differentiation during injury. The umbrella cell, a specialized epithelial cell, forms the superficial urothelial layer. The bladder must be inherently pliable to properly store accumulating urine, expanding, and contracting, when necessary, while maintaining bladder function. Large (25–250 μm and hexagonally shaped, umbrella cells greatly contribute to this ability as they change shape in response to bladder filling, due to their unique protein expression profile and vesicular trafficking mechanisms. Umbrella cells express uroplakin, specialized lipid, and tight junction (e.g., occludins, claudins) proteins which aid in maintaining the urothelial barrier (Birder & Andersson, 2013). A polysaccharide glycosaminoglycan (GAG) layer lines the luminal surface of the umbrella cells and provides additional barrier support by preventing passage of microorganisms and toxic substances. The urothelial layers and properties remain relatively consistent throughout the bladder (i.e., detrusor, trigone regions); however, the urothelial layer of the bladder neck/proximal urethra and renal pelvis are notably different. The urethral urothelium exhibits stratified, columnar epithelium and lacks urothelial specific protein expression (i.e., uroplakin, occludins, claudins) (Birder & Andersson, 2013).

The urothelium was once thought to be a relatively passive component of the LUT, acting solely as a barrier, to prevent toxins, urinary metabolites, and microorganisms from penetrating the bladder wall. However, it is now well known that the urothelium is a biologically active layer. Numerous studies demonstrate that the urothelium plays critical modulatory roles in LUT sensory signaling (Birder & Andersson, 2013). Mechanical (i.e., stretch, pressure) or chemical (i.e., urine composition, heat, mediators released from nerves and surrounding cells) stimuli may induce the release of neuroactive chemicals (e.g., adenosine triphosphate (ATP), acetylcholine (ACh), nitric oxide (NO), growth factors) from urothelial cells. Additionally, urothelial cells express diverse channels and receptors (e.g., noradrenaline, ACh, ATP, uridine-5′-triphosphate (UTP), adenosine diphosphate (ADP), adenosine, CRH, neurotrophins, endothelins, estrogen, bradykinin, tachykinin, amiloride, ENaC, and TRP channels). Many of these channels and receptors are also expressed on afferent nerves. Compounds released from the urothelium may act by autocrine or paracrine mechanisms. Furthermore, they may modulate sensory information via interactions with urothelial cells, lamina propria cells, detrusor smooth muscle cells (SMCs), blood vessels, inflammatory cells, and afferent nerves (Birder & Andersson, 2013).

2.3 Lamina propria

The lamina propria lies between the most external layer of the bladder, muscularis detrusor, and the urothelial basement membrane. This tissue may also be referred to as the submucosa or suburothelium. The lamina propria is an extracellular matrix consisting of many fiber types (e.g., elastic, collagen), cell types (e.g., adipocytes, fibroblasts, interstitial cells), blood vessels, lymphatic vessels, smooth muscle fascicles (i.e., muscularis mucosae) and nerve terminal endings. The lamina propria is one continuous tissue; however, many investigators distinguish between a deep, more fibrous region near the detrusor and a superficial or upper region near the urothelium that is densely populated with LP cells (Gabella & Davis, 1998). A dense collection of afferent nerve fibers terminates within the lamina propria, with some endings extending into the urothelium (Birder & Andersson, 2013; Fowler, Griffiths, & de Groat, 2008). Afferent nerves primarily terminate in the lower region of the bladder body and neck (Birder & Andersson, 2013). Efferent nerve terminals are present but sparse in the LP and primarily terminate within smooth muscle (Andersson & Arner, 2004). Small, parasympathetic intramural ganglia are present in the LP and within detrusor smooth muscle bundles in the human bladder, but are absent in rats, cats, guinea pigs and rabbits (Birder & Andersson, 2013; Gosling & Dixon, 1974). Like the urothelium, investigators have recently begun to re-examine the involvement of the LP in afferent signaling as numerous studies suggest that urinary, LP interstitial cells may act as intermediaries or amplifiers of bladder sensations (Birder & Andersson, 2013; Koh, Lee, Ward, & Sanders, 2018; Neuhaus, Gonsior, Cheng, Stolzenburg, & Berger, 2020).

2.4 Muscularis propria

The detrusor muscle is the muscularis propria component of the bladder, which primary function is to contract during voiding events to expel urine in the bladder, into urethra. The detrusor is made of smooth muscle cells (SMC) organized in three sheets. Two longitudinal smooth muscle sheets surround a middle, circular smooth muscle sheet. A urinary interstitial cell (IC) sub-population exists in the detrusor and the urethral muscularis that is separate and distinct from LP ICs (Koh et al., 2012; Sancho, Triguero, Lafuente-Sanchis, & Garcia-Pascual, 2017). These cells lie alongside and in between smooth muscle bundles and may contribute to regulating SMCs excitability (Koh et al., 2018). The detrusor muscle extends over much of the bladder body and base but is notably different than the smooth musculature of the ureter orifices/distal ureters, trigone, and urethra.

2.5 Urethra

Trigonal and detrusor smooth muscle both extend into the proximal urethra. Additional layers of urethral smooth muscle include a thick, inner longitudinal layer and a thin, outer circular layer (Mistry, Klarskov, DeLancey, & Lose, 2020). An internal urethral sphincter does not exist in the true, anatomical sense in males or females; however, the bladder base, neck and proximal urethral musculature all contribute to create sphincteric-like action (Elbadawi, 1996).

3. Neural control of micturition

Neural circuits in the brain and spinal cord coordinate the actions of the bladder and urethra to switch between the storage phase and the elimination phase (Fig. 1). This coordination is mediated by three sets of peripheral nerves: (i) Hypogastric sympathetic nerves that inhibit the detrusor muscle and contract the bladder base and urethra for proper storage; (ii) Pelvic parasympathetic nerves that contract the detrusor muscle and relax the urethra for voiding; and (iii) Pudendal somatic nerves that excite the striated external urethral sphincter to close the bladder outlet (Yoshimura & Chancellor, 2003). All three peripheral nerves also contain afferent nerves that convey sensory information related to bladder filling (Andersson & Michel, 2011; Walsh, 1986). Finally, higher order brain areas provide input to the micturition circuit to support voluntary control of voiding.

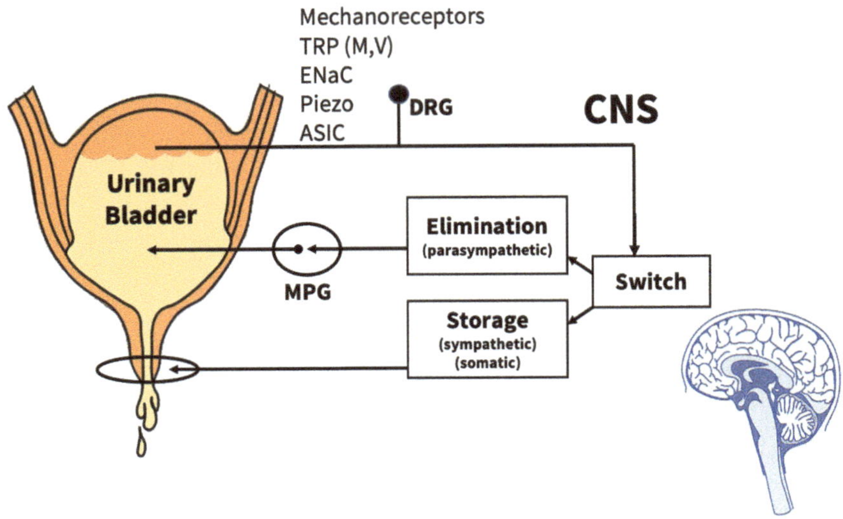

Fig. 1 Storage and elimination (voiding) of urine. LUT function is controlled by neural pathways that maintain a reciprocal relationship between the urinary bladder and the urethral outlet. During bladder filling, storage reflexes, organized primarily in the spinal cord, are activated whereas voiding is mediated by reflex mechanisms that are organized in the brain. During bladder filling and storage, the parasympathetic innervation of the detrusor is inhibited, and the urethral sphincter is activated, preventing involuntary urine release. During bladder filling, the parasympathetic efferent pathway to the bladder, including a population of CNS (e.g., pontine micturition center) neurons, is turned off. As bladder filling continues and a threshold level of bladder distension (i.e., stretch) is achieved, the afferent activity from urinary bladder mechanoreceptors switches the pathway to the elimination mode. During elimination (voiding), parasympathetic activity is activated resulting in urinary bladder contraction, whereas sympathetic activity and somatomotor activity is withdrawn. DRG, dorsal root ganglion; MPG, major pelvic ganglion. TRP, transient receptor potential channel; M, melastatin; V, vanilloid; ENaC, epithelial sodium channel; ASIC, acid-sensing ion channels. *Modified from Vizzard, M. A. (2006). Neurochemical plasticity and the role of neurotrophic factors in bladder reflex pathways after spinal cord injury.* Progress in Brain Research, 152, *97–115. https://doi.org/10.1016/S0079-6123(05)52007-7.*

Bladder storage (i.e., continence) is, in part, mediated by the sympathetic nervous system (Fig. 1). Sympathetic preganglionic neurons that innervate the bladder originate in the intermediolateral nucleus of spinal cord levels T11–L2 in humans and L1–L2 in rodents (Nadelhaft, Vera, Card, & Miselis, 1992). After leaving the spinal cord, these sympathetic fibers form lumbar splanchnic nerves and some synapse in the inferior mesenteric ganglion, while others travel farther to make their first synapse in the ganglia of the pelvic plexus. The postganglionic fibers then travel along the hypogastric nerve or pelvic nerve, respectively, to the bladder where they release norepinephrine to provide inhibitory input to the detrusor to promote bladder accommodation via β-adrenergic receptors, and excitatory input to the bladder base and smooth muscle of the urethra to facilitate storage via α-adrenergic receptors (Andersson & Arner, 2004; de Groat, Griffiths, & Yoshimura, 2015; Fowler et al., 2008).

Parasympathetic preganglionic neurons that innervate the bladder and urethra are located in the sacral parasympathetic nucleus in the lateral horn of the sacral spinal cord (S2–S4 in humans, L6–S1 in rodents) (Walsh, 1986). Axons from these neurons leave the spinal cord and immediately separate to become pelvic splanchnic nerves (i.e., pelvic nerve). Axons within this nerve either synapse with postganglionic neurons in the pelvic plexus or postganglionic neurons located in terminal ganglia within the bladder wall (e.g., intramural ganglia or major pelvic ganglia) (de Groat, 1990; de Groat et al., 2015; DeGroat & Saum, 1976; Elbadawi, 1996; Fowler et al., 2008; Girard, Tooke, & Vizzard, 2017). During bladder filling, the parasympathetic nervous system is inhibited; however, when the micturition threshold is exceeded and it a socially appropriate time to urinate, the parasympathetic nervous system is activated alone and leads to the contraction of the detrusor smooth muscle through the release of acetylcholine and ATP onto muscarinic (M3) and purinergic (P2X) receptors, respectively (Fig. 1). Additionally parasympathetic fibers from the pelvic nerve relax the internal urethral sphincter through the release of nitric oxide (Bennett et al., 1995) (Fig. 1).

The external urethral sphincter is striated muscle controlled by the somatic nervous system. Motor neurons responsible for innervating the external urethral sphincter reside in Onuf's nucleus (humans) or the dorsolateral motor nucleus (rodents), both of which are a collection of motor neurons in the ventral horn of the sacral spinal cord. These neurons send projections to contract the external urethral sphincter via the pudendal nerve to maintain continence in synchronized fashion with the sympathetic nervous system (de Groat, 1990; Elbadawi, 1996; Fowler et al., 2008; Girard et al., 2017).

Afferent nerves containing sensory information related to bladder fullness travel to the spinal cord via the pelvic and hypogastric nerves, whereas sensory input from the bladder base and urethra travel to the spinal cord via the pudendal and hypogastric nerves (Fowler et al., 2008). During normal bladder filling, sensory information travels via myelinated Aδ-fibers that respond to mechanosensory channels that become activated during passive distension of the bladder. As the bladder nears capacity, tension increases and mechanosensitive channels that respond to pressure open and lead to increased firing of Aδ-fibers (Fowler et al., 2008). Under pathological conditions, small diameter, unmyelinated, C-fibers are recruited to carry sensory signals related to noxious stimuli from the LUT to the spinal cord. The cell bodies of both Aδ- and C-fibers are in dorsal root ganglia in spinal levels T11–L2 and S2–S4 in humans (de Groat & Yoshimura, 2010; Fowler et al., 2008). Bladder afferents enter the spinal cord through Lissauer's tract and terminate on neurons in the superficial dorsal horn, dorsal commissure, and sacral parasympathetic nucleus. Some neurons then project rostrally to supraspinal micturition integration centers in the brain.

Supraspinal neural substrates play a fundamental role in the initiation of voiding as well as maintaining continence. In humans, the pontine micturition center, also called Barrington's nucleus, is in the pontine medial tegmentum within the brainstem and is the most prominent efferent nucleus within the micturition reflex circuit (Andersson & Michel, 2011; de Groat, 1990; de Groat et al., 2015; Elbadawi, 1996; Fowler et al., 2008). Chemical or electrical activation of the pontine micturition center (PMC) causes a contraction of the detrusor smooth muscle and a relaxation of the urethra that results in micturition (Hou et al., 2016; Sugaya, Matsuyama, Takakusaki, & Mori, 1987). To do this, the PMC sends descending fibers to the sacral parasympathetic nucleus in the spinal cord to synapse on parasympathetic neurons that inhibit the detrusor (Andersson & Michel, 2011) as well as GABAergic interneurons that then inhibit alpha motor neurons in Onuf's nucleus to induce relaxation of the external urethral sphincter (Andersson & Michel, 2011). Though the PMC is considered the micturition control center, in that it switches the LUT from a state of storage to a state of elimination, it does not directly receive afferent input from the LUT in humans (Andersson & Michel, 2011) (Fig. 1). After the Aδ- and C-fibers synapse in the spinal cord, their projections join ascending spinal projections to synapse in the periaqueductal gray (PAG), a region in the mesencephalon of the brainstem (Andersson & Michel, 2011). The PAG projects to the PMC and serves as an integration center of LUT sensory signals along with inputs from other higher order brain regions such as the hypothalamus, preoptic region, central nucleus of the amygdala, bed nucleus of the stria terminalis (BNST), and prefrontal cortex (Beckel & Holstege, 2011). During bladder filling, input from these brain areas prevent the PAG from exciting neurons in the PMC to maintain conscious and voluntary control of micturition (Beckel & Holstege, 2011; Fowler et al., 2008).

4. Neurochemistry of LUT afferent neurons

Afferent fibers traveling in the pelvic, hypogastric (lumbar splanchnic), and pudendal nerves convey information from the LUT to the sacral spinal cord. The sensations transmitted vary from degree of bladder distension to bladder pain and the activation stimuli (e.g., stretch/volume/chemicals) of bladder afferents are varied (Andersson & Michel, 2011). To communicate this wide range of information, LUT afferent nerves, namely Aδ- and C-fibers have diverse properties. LUT afferent neurons in the lumbosacral dorsal root ganglia (DRG) are not a focus of this perspectives piece but a thorough review of this topic can be found here (Andersson & Michel, 2011).

5. Urothelial signaling

Although historically viewed as a passive barrier, research now suggests an integral role of the urothelial layer of the LUT as a sensory and signaling organ that receives, amplifies, and transmits both mechanical and chemical signals to cells deeper in the bladder wall. A variety of receptors and channels are expressed on afferent neurons in the bladder are also found on urothelial cells, including: bradykinin receptors (Chopra et al., 2005), neurotrophin receptors (TrkA, TrkB, and p75) (Murray, Malley, Qiao, Hu, & Vizzard, 2004), purinergic receptors (P_2X and P_2Y) (Birder et al., 2004; Chopra et al., 2008; Lee, Bardini, & Burnstock, 2000; Tempest et al., 2004), VEGF receptors (Kopparapu et al., 2013; Saban et al., 2008), adrenergic receptors (both α and β) (Birder et al., 2002; Birder, Apodaca, De Groat, & Kanai, 1998; Kullmann et al., 2011; Limberg et al., 2010), muscarinic and nicotinic acetylcholine receptors (Beckel & Birder, 2012; Beckel, Kanai, Lee, de Groat, & Birder, 2006; Birder et al., 1998; Kullmann et al., 2008; Kullmann, Artim, Birder, & de Groat, 2008), sodium channels (Araki et al., 2004; Smith et al., 1998; Wang et al., 2003), a variety of TRP channels (TRPV1, TRPV2, TRPV4, TRPM8, and TRPA1) (Birder et al., 2001, 2002; Du et al., 2008; Gevaert et al., 2007; Girard et al., 2019; Kullmann, Shah, Birder, & de Groat, 2009; Mochizuki et al., 2009; Stein et al., 2004; Yamada et al., 2009), and Piezo1/2 channels (Dalghi et al., 2019; Liu et al., 2018; Marshall et al., 2020).

Activation of these channels and receptors occurs due to various stimuli such as increases in bladder wall tension from bladder filling, soluble chemical mediators in urine, and chemical mediators released from afferent neurons, inflammatory cells, and blood vessels (Birder & Andersson, 2013; Birder & de Groat, 2007; Hanna-Mitchell et al., 2007; LaBerge, Malley, Zvarova, & Vizzard, 2006). In response to activation of these channels, urothelial cells release an array of signaling molecules including NO, ATP, nerve growth factor (NGF), and substance P onto both afferent and efferent nerves located in or near the urothelial cell layer (Birder et al., 1998, 2003; Birder & Andersson, 2013; Burnstock, 2001; Ferguson, Kennedy, & Burton, 1997; Girard et al., 2019). ATP is one of the main urothelial signaling mediators and its release from urothelial cells and subsequent binding to P_2X receptors on afferent nerve terminals is essential for a proper functioning micturition reflex. As the bladder fills, the apical surface of the urothelial cells begins to stretch, causing an activation of stretch activated channels (SACs). This activation leads to an influx of calcium and ultimately results in a robust release of ATP from the urothelial cells (Ferguson et al., 1997). The precise mechanism of this release is still debated and is not limited to a singular mechanism (Birder & Andersson, 2013; Winder, Tobin, Zupančič, & Romih, 2014). Research supports a combination of both ATP release via vesicular exocytosis and non-vesicular release through channels like pannexin and connexin, among other transporters (Lazarowski, 2012; Sui et al., 2014; Winder et al., 2014). For vesicular release, ATP is typically co-packaged into vesicles and released with other neurotransmitters like acetylcholine (Dahl, 2015; Unsworth & Johnson, 1990). For non-vesicular release in urothelial cells, ATP can leave the cell through non-junctional connexin or pannexin hemichannels and signal through both autocrine and paracrine mechanisms (Beckel et al., 2015).

6. Interstitial cystitis/bladder pain syndrome (IC/BPS)

IC/BPS is a chronic pelvic pain condition in which patients experience pain, pressure or discomfort perceived to be related to the urinary bladder, in addition to LUT symptoms, particularly increased urinary frequency and/or urgency (Fig. 2). IC/BPS presents with very diverse clinical phenotypes (Akiyama et al., 2021; Hanno, 2015). Despite this diversity, IC/BPS's hallmark feature, distinguishing the syndrome from related LUT conditions, is pain related to bladder filling. IC/BPS is diagnosable after symptoms have persisted for 6 weeks and other identifiable causes, such as infection, or related conditions have been excluded (Hanno, 2015) (Fig. 2). There are currently few treatment options for IC/BPS patients. Most treatments that are available aim to manage symptoms, rather than targeting the underlying

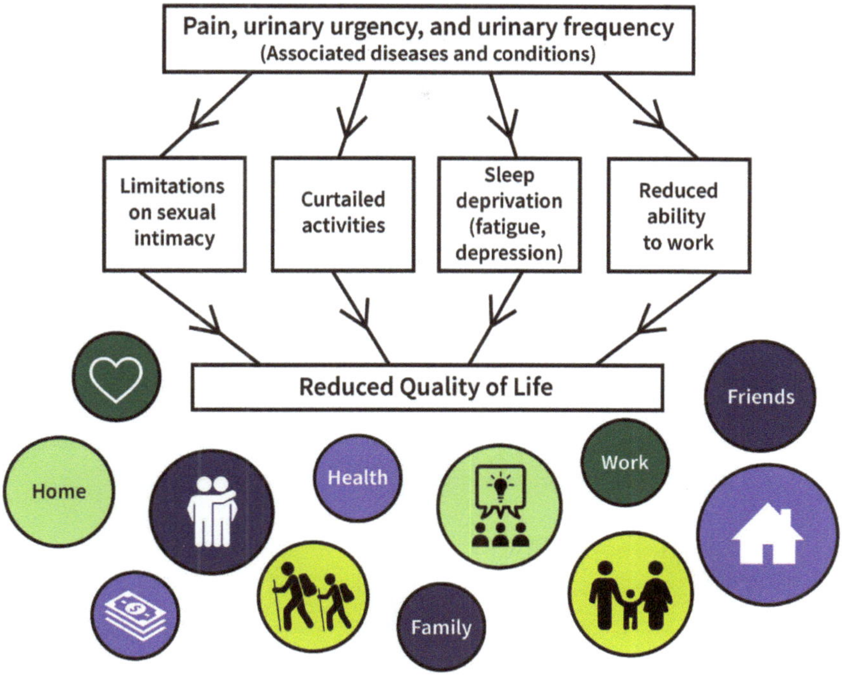

Fig. 2 IC/BPS reduces quality of life (QoL). Symptoms of IC/BPS can negatively affect sexual relationships and relationships in general and limit one's ability to work and to participate in other activities, resulting in sleep deprivation, fatigue and depression which collectively reduce QoL.

mechanisms of IC/BPS and many are used off-label. Additionally, treatment outcomes vary greatly among patients and even with short-term treatment success, long-term success may not be achieved. Pentosan polysulfate sodium, or Elmiron, is currently the only FDA-approved medication specifically for IC/BPS, yet with modest success (Hanno, 2015; Marcu, Campian, & Tu, 2018).

6.1 Epidemiology

Prevalence estimates have increased over the past 50 years. This may be due to increased awareness of IC/BPS symptoms and recognition of IC/BPS as a condition amongst physicians, leading to fewer misdiagnoses. In a nationwide study conducted via telephone surveys with statistically modeled predictors for screening, Berry et al. (2011) estimated 2.7–6.5% of US women (3.3–7.9 million) met IC/BPS diagnosis criteria, depending on a more or a less stringent definition of IC/BPS. Historically, it has been thought that

IC/BPS has a higher prevalence in women compared to men, with a ratio of 5:1 (Akiyama et al., 2021; Berry et al., 2011). However, this notion has been challenged in more recent years with some studies suggesting the accepted prevalence amongst men is higher than originally estimated due to misdiagnoses and underreporting (Forrest & Schmidt, 2004; Suskind et al., 2013). Using a similar methodological approach to Berry et al. (2011), Suskind et al. (2013) estimated the prevalence of IC/BPS in US men at 1.9–4.2% (2.1–4.6 million).

7. Pathophysiology

The etiology of IC/BPS is still not well understood. However, there are consistent pathological features associated with IC/BPS that may contribute to its development and/or persistence.

7.1 Urothelial alterations

The GAG layer, specialized lipids, uroplakin proteins, and tight junction proteins create a high transepithelial resistance, allowing the urothelium to act as a fantastic barrier against irritative substances (Birder & Andersson, 2013). However, the urothelium exhibits drastic changes in patients with and without Hunner's ulcers, such as reduced and aberrant expression of urothelial barrier proteins and deficient urothelial repair mechanisms, leading to increased urothelial permeability, urothelial thinning, and denudation (Birder & Andersson, 2013; Hauser et al., 2008; Keay et al., 2004; Parsons, Lilly, & Stein, 1991) (Fig. 3). Cationic substances (e.g., potassium), metabolites from diet, medication, or supplements, hormonal changes, and bacterial infections can exert cytotoxic effects on urothelial cells and impair barrier function (Akiyama et al., 2021). Furthermore, increased urothelial permeability allows irritative substances that would normally be contained and expelled, to permeate the urothelium, depolarize bladder wall constituents (e.g., interstitial cells, nerve terminal endings, SMCs) and trigger secondary inflammatory processes, in response to released endogenous pathogens from damaged tissue (Akiyama et al., 2021; Birder & Andersson, 2013). General inflammatory changes are also observed in IC/BPS patients including immune cell infiltration, MC degranulation, angiogenesis, edema, and stromal fibrosis (Akiyama et al., 2021; Arms, Girard, Malley, & Vizzard, 2013; Arms, Girard, & Vizzard, 2010; Gonzalez, Girard, & Vizzard, 2013; Guo et al., 2018; Malley & Vizzard, 2002). B cell clonal expansion is specifically noted in IC/BPS patients with Hunner's ulcers (Moldwin et al., 2022).

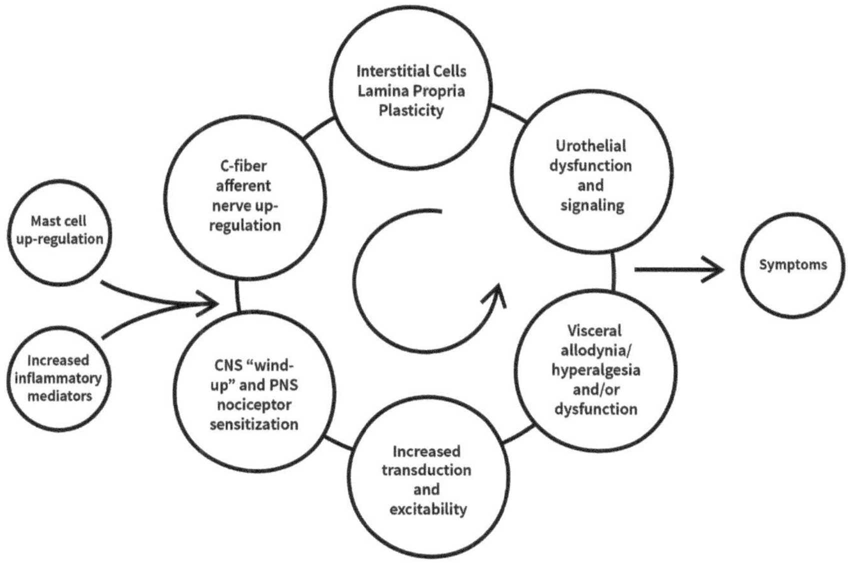

Fig. 3 Potential etiologic cascade and pathogenesis underlying painful bladder syndrome (BPS)/interstitial cystitis (IC). BPS/IC likely has a multifactorial etiology. There is no consensus regarding the etiology or pathogenesis of BPS/IC, but a number of proposals include: a "leaky epithelium," mast cell activation and release of neuroactive compounds at the level of the urinary bladder, "awakening" of C-fiber bladder afferents, and upregulation of inflammatory mediators (e.g., cytokines and chemokines). Inflammatory mediators can produce CNS "wind-up" and peripheral nociceptor sensitization resulting in chronic bladder pain and voiding dysfunction. *Adapted from Gonzalez, E. J., Merrill, L., & Vizzard, M. A. (2014). Bladder sensory physiology: Neuroactive compounds and receptors, sensory transducers, and target-derived growth factors as targets to improve function.* American Journal of Physiology. Regulatory, Integrative and Comparative Physiology, 306*(12), R869–R878. https://doi.org/10.1152/ajpregu.00030.2014; Sant, G. R., Kempuraj, D., Marchand, J. E., & Theoharides, T. C. (2007). The mast cell in interstitial cystitis: role in pathophysiology and pathogenesis.* Urology, 69*(4 Suppl.), 34–40. https://doi.org/10.1016/j.urology.2006.08.1109.*

When inflammation becomes chronic, the exposure and activation of nerve terminals to inflammatory mediators induces afferent nerve sensitization that, if left untreated, can lead to hyperalgesia and allodynia because of nerve hyperexcitability induced by neurogenic inflammation (Birder & Kullmann, 2018; Grundy, Caldwell, & Brierley, 2018) (Fig. 3). Neurogenic inflammation is a component of the positive feedback loop triggered when increased stimulation of afferent nerves causes them to release inflammatory peptides including Substance P, CGRP, and neurokinin A (Birder & Kullmann, 2018). This release of neuropeptides causes a cascade of events involving the immune system, the vasculature, and the nervous system in the following ways: (i) the vasculature becomes angiogenic which can increase edema thereby perpetuating the inflammatory response (Birder & Kullmann, 2018); (ii) immune system activation induces recruitment of additional pro-inflammatory mediators like mast cells and leukocytes to the site of inflammation (Birder & Kullmann, 2018); and (iii) underlying nerves become hyperexcitable due to increased nerve density (Pang, Marchand, Sant, Kream, & Theoharides, 1995), increased expression of receptors such as TRPs, and purinergic receptors (Birder & Kullmann, 2018; Brady, Apostolidis, Harper, et al., 2004; Brady, Apostolidis, Yiangou, et al., 2004), and increased activation and expression of free ions like K^+, Ca^{2+} that lower the action potential threshold in sensory neurons (de Groat & Yoshimura, 2006, 2010). This persistent activation of peripheral nociceptive signals creates an increased drive of C-fiber afferent input to the spinal cord which, over time, leads to aberrant functioning of CNS circuits. Specifically, nociceptive neurons in the brain and spinal cord will increase their firing given normal or sub-threshold stimuli (Baron, Hans, & Dickenson, 2013). This phenomenon is called central sensitization and is the driver of the hyperalgesia (increased sensitivity to pain) and allodynia (pain response to non-noxious stimuli) symptoms in patients with IC/BPS (Reynolds, Dmochowski, Wein, & Bruehl, 2016) (Fig. 3).

7.2 Stress-induced bladder dysfunction

Stress may also contribute to neural changes and upregulation in IC/BPS. Stress can exacerbate, and even induce, symptoms in IC/BPS patients (Nazif, Teichman, & Gebhart, 2007; Rothrock, Lutgendorf, Kreder, Ratliff, & Zimmerman, 2001a, 2001b) and animal models (Birder & Andersson, 2018; Charrua, Pinto, Birder, & Cruz, 2015; Girard et al., 2017; Matos et al., 2017; Merrill, Malley, & Vizzard, 2013; Merrill & Vizzard, 2014; Mingin et al., 2015; Mingin, Peterson, Erickson, Nelson, & Vizzard, 2014; Nazif et al., 2007; Rothrock et al., 2001a, 2001b) (Fig. 3). The hypothalamic-pituitary-adrenal (HPA) axis is a neuroendocrine feedback loop that acts to maintain physiological homeostasis. However, abnormal, or chronic HPA activation is associated with dysregulated responses to stress and inflammation and may result in aberrant sympathetic activation (Godoy, Rossignoli, Delfino-Pereira, Garcia-Cairasco, & de Lima Umeoka, 2018; Nazif et al., 2007; Sheng et al., 2020). Symptom exacerbation due to stress is prevalent in many disease states, including functional disorders of the urinary bladder (e.g., overactive bladder (OAB), BPS/IC) (Nickel et al., 2016; Sutcliffe et al., 2014, 2018); however, the mechanisms underlying the effects of stress on the micturition reflex function are unclear. The repeated variate stress (RVS) model has been used previously to examine stress-induced changes in bladder function and pelvic sensation in rats (Girard et al., 2017; Gonzalez, Girard, Braas, May, & Vizzard, 2016; Merrill et al., 2013; Merrill & Vizzard, 2014). The RVS paradigm (Hammack et al., 2009; King, Toufexis, & Hammack, 2017) has multiple advantages including: (i) lack of habituation with novel stressor exposure; (ii) reproducible and robust changes in urinary bladder function (Merrill et al., 2013; Merrill & Vizzard, 2014); and (iii) reproducible decrease (~10–25%) in weight gain during RVS as demonstrated in rats and mice (Merrill et al., 2013; Merrill & Vizzard, 2014). We have previously used the RVS protocol to characterize effects on bladder function and somatic sensitivity (Merrill et al., 2013; Merrill & Vizzard, 2014) in rats. RVS in rats (Merrill et al., 2013; Merrill & Vizzard, 2014) produced similar changes in bladder function and somatic sensation observed in the present study, including increased urinary frequency, and somatic sensitivity of the hindpaw and pelvic region as well as increased NGF content in the urinary bladder. In addition, RVS in rats also caused additional changes in the inflammatory milieu of the urinary bladder including, changes in histamine, myeloperoxidase, and the chemokine, CXCL12 (Merrill et al., 2013; Merrill & Vizzard, 2014).

The contribution of other conditions and organ systems should be considered, as IC/BPS is commonly comorbid with psychological (e.g., depression, generalized anxiety), GI-related (e.g., irritable bowel syndrome, inflammatory bowel disease), auto-immune (e.g., Sjogren's syndrome, Lupus) or somatoform syndromes (e.g., fibromyalgia, chronic fatigue syndrome, vulvodynia) (Grundy & Brierley, 2018; Grundy et al., 2018; Yoshikawa et al., 2015). Non-ulcerative IC/BPS patients have higher rates of comorbidities than those with Hunner's ulcers (Akiyama et al., 2021). Patients report that certain drugs (e.g., alcohol, nicotine) and foods (e.g., caffeine, high potassium, high acid) aggravate IC/BPS symptoms; however, a direct link between diet and symptomology has not been established (Patnaik et al., 2017).

7.3 Multiple contributing factors

IC/BPS pathophysiology and etiology is likely a product of multiple factors mentioned above (Birder & Andersson, 2018). Immune cell infiltration, urothelial and GAG alterations, angiogenesis, afferent, and central sensitization work in concert to promote damaging positive feedback cycles (Fig. 3).

Comorbid conditions and/or stress may further exacerbate or induce IC/BPS features. The combination of these factors and the feedback cycles they create influence IC/BPS progression and further complicates understanding, diagnosing, and managing/treating underlying IC/BPS pathophysiology (Fig. 3).

8. IC/BPS animal models

There are multiple animal models of IC/BPS use to mimic the clinical signs and symptoms of IC/BPS. Birder and Andersson (Birder & Andersson, 2018) broadly group IC/BPS animal models into three categories: (i) bladder-centric models; (ii) models with complex mechanisms (i.e., CNS, immune system); and (iii) psychological and physical stressors, and natural models. For this perspectives piece, we focus on two bladder centric models and stress models. For additional information, please see (Birder & Andersson, 2018; Bjorling, Wang, & Bushman, 2011; Bjorling et al., 2011).

8.1 Cyclophosphamide (CYP)-induced cystitis

The most widely used bladder-centric model in the literature is IC/BPS induced by CYP. CYP is an anti-neoplastic alkylating agent used in chemotherapy to treat several types of cancer (Olivar & Laird, 1999). Though CYP itself is inert, it is metabolized in the liver to phosphamide mustard and acrolein, which are excreted into the urine and accumulate in the bladder (Olivar & Laird, 1999). This accumulation of toxic metabolites, especially acrolein, causes intense changes that to the urinary bladder that mimic changes seen in patients with IC/BPS (Olivar & Laird, 1999). CYP-induced cystitis can be used as an acute bladder inflammation model, induced with one intraperitoneal injection of CYP, or a chronic model, induced with a series of lower dose intraperitoneal CYP injections over the course of days (Birder & Andersson, 2018). Both models yield the hallmark symptoms of IC/BPS (pain and increased voiding frequency) while also inducing localized bladder specific pathological changes like mucosal edema, urothelium breakdown, inflammatory cell infiltration, petechial hemorrhaging, and afferent nerve hyperexcitability (Birder & Andersson, 2018; Olivar & Laird, 1999). Although CYP induces a greater inflammatory response than seen in some IC/BPS patients, it is an appealing model because it is a non-invasive and reproducible model that produces pathological and functional bladder-specific changes mimicking many of the signs/symptoms of the clinical syndrome (Bjorling et al., 2011; Olivar & Laird, 1999).

8.2 NGF-overexpression (OE)

We have characterized a transgenic mouse model of chronic, urothelium-specific NGF-overexpression that represents a novel approach to exploring the role of NGF in urinary bladder inflammation and sensory function (Schnegelsberg et al., 2010). Functionally, NGF-OE mice exhibit frequent voiding, the presence of NVCs during the bladder filling phase and referred somatic pelvic hypersensitivity (Schnegelsberg et al., 2010). In addition, NGF-OE mice exhibit changes in the morphology and neurochemistry of the urinary bladder and increases in inflammatory cell infiltrates to the urinary bladder (Schnegelsberg et al., 2010). Neurochemical changes include increased expression of Sub P and CGRP in the suburothelial nerve plexus of the urinary bladder (Schnegelsberg et al., 2010). Our findings support and extend many previous studies in rodents demonstrating involvement of NGF in altered bladder sensory function and the development of referred hyperalgesia in response to bladder inflammation (Dmitrieva & McMahon, 1996; Guerios, Wang, & Bjorling, 2006; Guerios, Wang, Boldon, Bushman, & Bjorling, 2008; Hu et al., 2005; Jaggar, Scott, & Rice, 1999; Lamb, Gebhart, & Bielefeldt, 2004; Zvara & Vizzard, 2007).

8.3 Stress-induced bladder dysfunction

Stress models have been shown to induce pain behaviors in rodents along with increased voiding frequency (Merrill & Vizzard, 2014; Smith et al., 2011; Wang et al., 2017). Experiments with the WAS model of stress have shown increased inflammatory cell infiltration, urothelial breakdown, evidence of peripheral and central sensitization, and increased blood flow to areas in the brain associated with voiding urgency (Birder & Andersson, 2018; Wang et al., 2017). Although underlying mechanisms are unknown, research suggests that it is a consequence of an aberrantly functioning sympathetic nervous system, more specifically the HPA axis. As evidence for this, chronic stress has been shown to increase sympathetic nerve activity which results in increased adrenergic stimulation due to dysfunctional HPA feedback mechanisms to various stress hormones (Birder & Andersson, 2018).

9. Mechanotransduction

Mechanotransduction, the conversion of mechanical forces into biological signals, is an essential physiological process used throughout the body; however, understanding the mechanism of mechano-sensing is still largely unknown in multiple organ systems (Nilius, 2010). In the urinary bladder, mechanotransduction is thought to be the mechanism used to transmit signals related to bladder distention to sensory signals that relay the level of bladder fullness to the spinal cord and higher order brain centers that integrate that information to determine an appropriate time to void. Increasing evidence suggests mechano-sensing channels on urothelial cells are also activated by numerous diverse signals including bladder stretch, temperature changes, pH changes, and chemical irritation. Once activated, the urothelial cells transduce these signals to bladder afferent nerves by releasing chemical mediators, such as ATP (Birder & Andersson, 2013; Olsen, Stover, & Nagatomi, 2011).

Research is ongoing to determine which mechano-sensing channels are responsible for transducing this signal and the mechanism(s) by which they carry out this function. Calcium-permeable stretch-activated cation channels (SACs) were discovered over 20 years ago and research suggests they are the main drivers of mechanotransduction. SACs can be classified into channels that are directly gated or indirectly gated. An ion channel that has direct gating is the primary transducer of mechanical force, meaning that it alone senses the force and passes on a receptor current. A channel with indirect gating plays a necessary role in mechanical transduction but does so by second messenger signaling via G-proteins, phospholipases, and the activation of downstream kinases (Christensen & Corey, 2007; Nilius, 2010). For a channel to be considered directly mechano-sensitive, the candidate ion channel must meet multiple functional criteria (Christensen & Corey, 2007).

Research suggests the presence of several mechanosensors on urothelial cells that respond to mechanical stretch of the bladder, among other signals.

10. Transient receptor potential (TRP) channels

The TRP channels are a novel superfamily of channels that contribute to changes in the intracellular concentration of calcium ([Ca_i^{2+}]) by either acting as a Ca^{2+} entry pathway in the plasma membrane or by changing membrane polarization and thus modulating the driving force for Ca^{2+} entry into the cell (Nilius, Owsianik, Voets, & Peters, 2007). There are more than 50 TRP channels described in species from yeast to human and 28 known mammalian TRP channels (Everaerts, Gevaert, Nilius, & De Ridder, 2008). The TRP channel superfamily consists of 7 subfamilies: TRPC (canonical), TRPM (melastatin), TRPV (vanilloid), TRPA (ankyrin), TRPP (polycystin), TRPML (mucolipin), and TRPN (no mechanopotential).

Each TRP channel consists of 4 subunits, and each subunit consists of 6 transmembrane spanning segments (S1–S6), a cation-permeable pore-forming loop between S5 and S6, and intracellularly located NH_2 and COOH termini (Everaerts et al., 2008; Vriens, Appendino, & Nilius, 2009).

The TRPV subfamily consists of six members (TRPV1-6). TRPV1-4 are all heat-activated and nonselective for cations. These channels are modestly permeable to Ca^{2+}, with a PCa: PNa of ~6:1 and are also chemosensors for many endogenous and synthetic ligands (Angelico & Testa, 2010; Nilius et al., 2007) TRPV5 and TRPV6 differ greatly; they are the only highly selective Ca^{2+} channels in the TRP superfamily (Angelico & Testa, 2010). TRPV channels can act as signal integrators, in that their stimuli have an additive effect on the gating of these channels, which becomes important for pathological states (Nilius et al., 2007).

The TRPV4 channel is a Ca^{2+}-permeable, stretch-activated, nonselective cation channel that is widely expressed throughout the body. In the brain, TRPV4 is found in neurons in the lamina terminalis, anterior hypothalamus, and choroid plexus where it acts as an osmosensor (Liedtke et al., 2000). The TRPV4 channel is also found in motor, spinal ventral root and DRG neurons, where it is thought to play a role in nociception (Nilius & Voets, 2013). TRPV4 is also expressed and has mainly a sensory role in auditory hair cells, skin keratinocytes, endothelial cells, cardiac fibroblasts, myocytes, skeletal muscle fibers, and renal and urinary bladder epithelium (Andersson, Gratzke, & Hedlund, 2010; Angelico & Testa, 2010; Nilius & Voets, 2013; Vriens et al., 2009). It was originally identified as a channel activated by hypotonic cell swelling (Liedtke et al., 2000; Vriens et al., 2009), but has since been found to be activated by other physical and chemical stimuli, including mechanical or shear stress, moderate heat (>27 °C), arachidonic acid, the endogenous ligand anandamide, and synthetic ligands such as 4alpha-phorbol 12,13-didecanoate (4α-PDD) and GSK1016790A (Andersson et al., 2010; Angelico & Testa, 2010; Nilius & Voets, 2013; Vriens et al., 2009).

11. TRPV4 channels and the urinary bladder

Studies indicate that several TRP channels, including TRPV1/2/4, TRPM7/8, and TRPA1, are expressed in the urinary bladder and may act as sensors of stretch and/or chemical irritation in the lower urinary tract (Andersson et al., 2010; Yu, Hill, Apodaca, & Zeidel, 2011). Most of these channels are also implicated in bladder diseases such as OAB and BPS/IC

(Everaerts et al., 2008; Nilius et al., 2007). Therefore, it is important to understand the expression as well as the mechanisms by which these channels act to create potential therapeutic targets for these diseases.

TRPV4 expression was first shown in the urinary bladder urothelium of mice and rats in 2007, where it was mainly found in bladder basal and intermediate urothelial cells (Gevaert et al., 2007). Since then, other groups have confirmed the expression of TRPV4 in urothelium (Birder, 2007; Kullmann et al., 2009; Mochizuki et al., 2009; Yamada et al., 2009; Yu et al., 2011). In addition, the TRPV4 agonist 4α-PDD increased $[Ca^{2+}]_i$ when administered to cultured urothelial cells, and this increase was blocked by the relatively nonselective TRP antagonist ruthenium red (RR), providing evidence that functional TRPV4 channels are in fact expressed in urothelium (Birder, 2007). Other groups found TRPV4 expression in the detrusor muscle in addition to the urothelium; however, expression levels were found to be approximately 20-fold higher (Thorneloe et al., 2008) to 36-fold higher (Xu et al., 2009) in the urothelium compared to detrusor smooth muscle.

12. TRPV4 and bladder function

The role of TRPV4 in bladder function was first reported in 2007 when an abnormal urine voiding pattern in the knockout (KO) mouse was observed (Angelico & Testa, 2010; Gevaert et al., 2007). Using cystometrograms (CMG), a decreased frequency of voiding contractions and an increased frequency of nonvoiding contractions, intermicturition interval (IMI), and total urine volume per micturition were observed in TRPV4 KO mice (Gevaert et al., 2007). Later studies replicated these results, finding a decrease in voiding frequency and an increase in bladder capacity and void volumes in the TRPV4 KO (Everaerts, Nilius, & Owsianik, 2010; Everaerts, Zhen, et al., 2010; Thorneloe et al., 2008). With the creation of a TRPV4 KO mouse model, the role of TRPV4 in bladder function has been further explored. TRPV4 KO mice are viable and fertile and have no obvious behavioral abnormalities (Everaerts, Nilius, & Owsianik, 2010; Everaerts, Zhen, et al., 2010). Several studies have reported various mild phenotypes including an abnormal response to osmotic and somatosensory mechanical stimuli, deficits in osmotic homeostasis, increased blood osmolarity, increased bone mass, aging-related hearing impairment, and they drink significantly less compared to their WT counterpart (Everaerts, Nilius, & Owsianik, 2010; Everaerts, Zhen, et al., 2010; Nilius et al., 2007).

In addition to the TRPV4 KO mouse model used to study bladder function, agonists and antagonists of this channel have been used to study its role.

Agonists of the TRPV4 channel appear to increase bladder hyperactivity, while antagonists show a decrease, making it a promising target for OAB and possibly other bladder dysfunction disorders (Janssen et al., 2011). Administration of 4α-PDD to conscious rats caused an increase in amplitude of reflex bladder contractions during CMG (Birder, 2007). GSK1016790A, a highly selective and potent TRPV4 agonist that is approximately 300-fold more potent than 4α-PDD, induced hyperactivity *in vivo* by decreasing infused and voided volume in normal mice (Thorneloe et al., 2008) and rats (Aizawa, Wyndaele, Homma, & Igawa, 2012). HC-067047, a potent and selective TRPV4 antagonist, provoked a decrease in micturition frequency and an increase in bladder capacity when administered systemically to mice and rats with CYP-induced cystitis (Everaerts, Nilius, & Owsianik, 2010; Everaerts, Zhen, et al., 2010). This TRPV4 blocker has advantages over other antagonists like RR, which is relatively nonselective, as well as Gd^{3+} (RN-1734) and La^{3+} (RN-9893), of which little is known (Vriens et al., 2009).

Current research demonstrates that TRPV4 has an important role in normal bladder function due to the regulation of ATP release from the urothelium in response to increases in intravesical pressure (Everaerts, Nilius, & Owsianik, 2010; Everaerts, Zhen, et al., 2010; Gevaert et al., 2007). It is widely accepted that the urothelium participates in the initiation of the micturition reflex (Birder, 2007; Yamada et al., 2009). Evidence shows that mechanical stretch evokes the release of ATP from urothelial cells to then activate P2X3 receptors on sensory nerve terminals, leading to membrane depolarization, action potential generation, and signals to the spinal cord to initiate the micturition reflex (Birder, 2007). This process may occur, in part, to activation of TRPV4 channels (and possibly TRPV1 channels and other SACs) present on the urothelial cell surface. Administration of 4α-PDD to cultured urothelial cells resulted in an increase in ATP release, which was blocked by RR (Birder, 2007). Using calcium imaging in urothelial cells from WT and TRPV4 KO mice, 4α-PDD evoked an increase in $[Ca^{2+}]_i$ in WT cells that was absent in cells from KO mice. This increase in $[Ca^{2+}]_i$ was minimal in the absence of extracellular Ca^{2+}, indicating that the increase is due to Ca^{2+} influx through TRPV4 channels (Mochizuki et al., 2009). In an *in vitro* cell-stretch system, WT urothelial cells show a transient increase in $[Ca^{2+}]_i$ after stretching followed by a sustained increase, which is blocked by RR. Also, after stretch, there is a prominent increase in ATP release in these cells, which is significantly decreased in cells from TRPV4 KO mice and suppressed by RR and depletion of extracellular Ca^{2+} (Gevaert et al., 2007; Mochizuki et al., 2009).

Furthermore, P2X3 receptor KO mice lack stretch-evoked reflex bladder contractions (Birder, 2007). Taken together, these results provide evidence that TRPV4 channels, located on urothelial cells in the urinary bladder, release ATP upon activation that binds to receptors on bladder sensory neurons, sending signals to the spinal cord to initiate micturition.

13. TRPV channels and pain

Evidence indicates that the TRPV channel subfamily is involved in the detection of acute noxious thermal, mechanical, and chemical stimuli. Hypersensitivity and pain in various pathological conditions are often due to upregulated expression and/or increases in sensitivity of TRPV channels (Nilius et al., 2007). TRPV1 is expressed in >50% of DRG neurons and is activated by heat, acid, and vanilloids, indicating an important role in the detection of noxious stimuli. In fact, TRPV1 KO mice show a decreased response to noxious thermal stimuli and are defective in developing inflammatory thermal hyperalgesia (Nilius et al., 2007). In addition, TRPV4 KO mice have a decreased sensitivity to tail pressure, show deficits in mechanically evoked paw withdrawal responses (Birder, 2007), and thermal hyperalgesia is also absent in these mice. Therefore, TRPV1 and TRPV4 are implicated in the etiology of thermal allodynia and hyperalgesia. Further, taxol-induced painful neuropathy can be treated by gene silencing of TRPV4 (Alessandri-Haber et al., 2004).

Our major contributions to understanding the role(s) of TRPV4 in micturition have focused on (i) TRPV4 function in the NGF-OE mouse that exhibits increased voiding frequency and somatic sensitivity; (ii) TRPV4 expression and function in platelet-derived growth factor receptor-α (PDGFRα)$^+$ ICs in the lamina propria of the postnatal urinary bladder; and (iii) the role of TRPV4 in ATP release via vesicular exocytosis. The following sections will summarize these contributions and provide suggestions for additional areas of study.

14. TRPV4 function in NGF-OE mice that exhibit increased voiding frequency and somatic sensitivity

We previously demonstrated (Girard, Merrill, Malley, & Vizzard, 2013) increased TRPV4 transcript and protein expression in the LUT tissues (e.g., urothelium, suburothelium, suburothelial nerve plexus, lumbosacral DRG) in NGF-OE consistent with TRP channel function and modulation being dependent upon target tissue expression of growth factors (e.g., NGF, glial-derived neurotrophic factor (GDNF) (Ernsberger, 2009; Homma et al., 2013; Malin et al., 2011; Shimizu et al., 2018)). Interestingly, NGF-OE mice display opposite phenotypic changes in bladder function compared to TRPV4 KO (Gevaert et al., 2007) mice including reduced bladder capacity, increased frequency of voiding and somatic pelvic hypersensitivity (Schnegelsberg et al., 2010). The role of NGF in bladder sensory function and inflammation-induced referred hyperalgesia and in IC/BPS is well documented. We used three complementary approaches (i.e., conscious cystometry, natural voiding assays and void spot assays) to characterize the effects of intravesical TRPV4 blockade on urinary bladder function in NGF-OE and wildtype (WT) mice.

We demonstrated that intravesical TRPV4 blockade with HC-067047 significantly reduced voiding frequency and NVCs using conscious cystometry with continuous saline instillation in NGF-OE mice with an open outlet. Intravesical instillation of HC-067047 in littermate WT mice also significantly reduced voiding frequency but the reduction was less than that observed in NGF-OE mice. Using void spot assays, we showed that intravesical HC-067047 in NGF-OE mice reduced the number void spots (small and large) but no effects were observed in WT mice. Natural voiding assessments using the Urovoid approach demonstrated that intravesical instillation of HC-067047 significantly increased the void mass and IMI in NGF-OE mice studied over multiple 12h light: dark cycles. In addition, pelvic sensitivity determination, using von Frey filament testing, was significantly reduced in NGF-OE mice following intravesical infusion of HC-067047. These studies continue to suggest that TRPV4 blockade at the level of the urinary bladder may be a useful target to reduce urinary frequency (Everaerts, Nilius, & Owsianik, 2010; Everaerts, Zhen, et al., 2010; Gevaert et al., 2007; Girard et al., 2013; Merrill & Vizzard, 2014) and somatic sensitivity.

15. TRPV4 has a functional role in ATP release via vesicular exocytosis

ATP release from urothelial cells and subsequent binding to purinergic receptors regulate a variety of actions in the urinary bladder and micturition reflexes including sensory nerve activity, detrusor smooth muscle contraction, synaptic transmission, and nociception (Chen et al., 1995; Sui et al., 2014; Wang et al., 2005). Intravesical instillation of ATP enhanced spinal bladder neuron excitability (Munoz, Somogyi, Boone, & Smith, 2011) and increased bladder activity (Pandita & Andersson, 2002). In contrast, attenuation of luminal ATP release (Beckel et al., 2015; Smith, Vemulakonda, Kiss, Boone, & Somogyi, 2005; Timoteo et al., 2014) reduced NVCs and voiding frequency. Altered ATP signaling may play a significant role in several functional LUT pathologies (e.g., OAB, IC/BPS) (Ruggieri, 2006; Silva-Ramos et al., 2013; Sun & Chai, 2006). Pathological conditions of the bladder that elevate luminal ATP presumably activate mucosal purinoceptors to increase nerve excitability (Beckel et al., 2015; Gonzalez, Girard, et al., 2016; Gonzalez, Heppner, Nelson, & Vizzard, 2016; Vlaskovska et al., 2001; Yu & de Groat, 2008). One mechanism that may account for this effect is autocrine purinoceptor activation on the urothelium stimulating continual ATP release (Sun & Chai, 2006). This aberrant signaling of ATP could be limited by controlling extracellular release from the urothelium and/or its binding to purinergic receptors within deeper layers of the bladder wall. Previous studies suggested a functional role for TRPV4 in ATP release such that TRPV4 activation (e.g., stretch or agonist application) leads to an increase in Ca^{2+} and ATP that is attenuated in TRPV4 KO mice or by TRPV4 antagonist (Gevaert et al., 2007; Mochizuki et al., 2009). Several mechanisms of urothelial ATP release including vesicular exocytosis (Gonzalez, Girard, et al., 2016; Gonzalez, Heppner, et al., 2016; McLatchie & Fry, 2015; Sui et al., 2014; Wang, Olivecrona, et al., 2005), pannexin and/or connexin ion channels (Beckel et al., 2015; McLatchie & Fry, 2015; Sui et al., 2014), and nucleoside transporters (Lazarowski, Boucher, & Harden, 2003; Wang et al., 2005; Wang, Olivecrona, et al., 2005) have been demonstrated. Our studies focused on the pharmacological manipulation of vesicular exocytosis and pannexin-1 channels (Fig. 4). We intravesically instilled the general secretory inhibitor, BFA, to inhibit ATP vesicular release (Fig. 4). BFA has previously been shown to decrease mechanical- and stretch-evoked ATP release from the urothelium (Gonzalez, Girard, et al., 2016; Gonzalez, Heppner, et al., 2016; McLatchie & Fry, 2015; Sui et al., 2014; Wang, Lee, et al., 2005). We also intravesically instilled the inhibitory peptide, 10Panx, to block pannexin-1 channels, implicated in distention-evoked ATP release from the urothelium (Gonzalez, Girard, et al., 2016; Gonzalez, Heppner, et al., 2016; Negoro et al., 2014; Timoteo et al., 2014; Wang, Olivecrona, et al., 2005) (Fig. 4).

Our studies with BFA, 10Panx or co-administration of both in NGF-OE and WT mice with urinary bladder distention to 25 cm H_2O demonstrated that vesicular exocytosis contributed to most of the distention-evoked

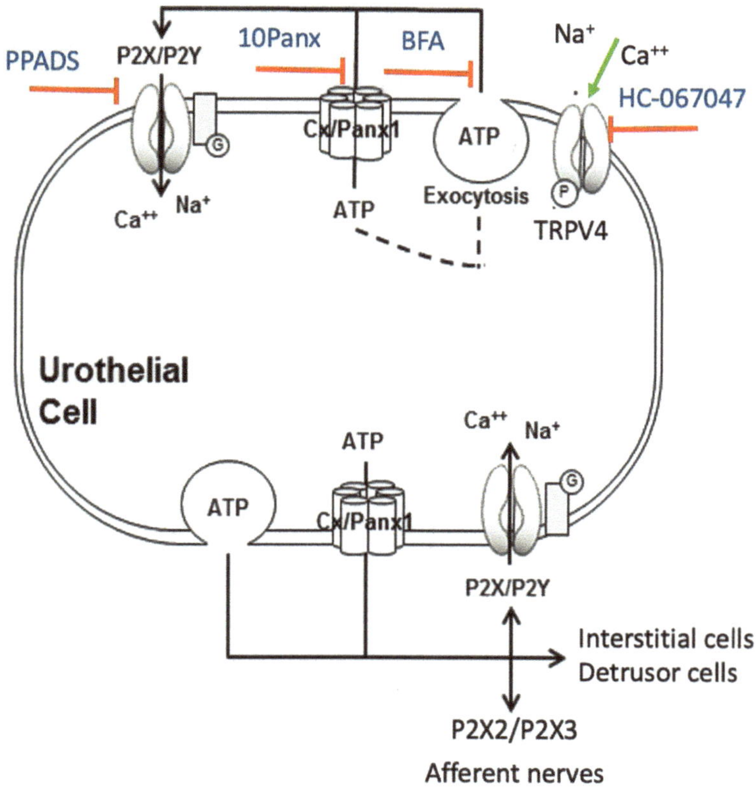

Fig. 4 ATP is released by vesicular exocytosis in the urinary bladder and TRPV4 plays a functional role in the urothelium. Diagram illustrating the possible mechanisms and underlying key components contributing to ATP release in urothelial cells with TRPV4 playing a functional role. Our studies focused on the pharmacological manipulation of vesicular exocytosis and pannexin-1 channels using the general secretory inhibitor, brefeldin A (10 μM) (BFA), to inhibit ATP vesicular release and/or the inhibitory peptide, 10Panx (50 μM), to block pannexin-1 channels. We quantified luminal ATP after functionally inhibiting known mechanisms of ATP release via vesicular exocytosis and pannexin channels as well as through TRPV4 (HC-067047) (1 μM) and P2X (PPADS) (300 μM) blockade. With intravesical inhibition of TRPV4 or inhibition of vesicular release in the urinary bladder, distention-induced luminal ATP release was decreased in both WT and NGF-OE mice although the magnitude of reduction was greater in NGF-OE mice. Inhibition of pannexin channels locally in the urinary bladder did not decrease the luminal release of ATP. Sites of inhibition are indicated with red lines. PPADS, pyridoxalphosphate-6-azophenyl-2′,4′-disulfonic acid.

luminal ATP release consistent with previous studies (Gonzalez, Girard, et al., 2016; Gonzalez, Heppner, et al., 2016) (Fig. 4). We quantified luminal ATP after functionally inhibiting known mechanisms of ATP release via vesicular exocytosis and pannexin channels as well as through TRPV4

(Gevaert et al., 2007; Mochizuki et al., 2009) and P2X blockade (Sui et al., 2014; Sun, Keay, Lehrfeld, & Chai, 2009; Vlaskovska et al., 2001) (Fig. 4). Upon intravesical inhibition of TRPV4 or inhibition of vesicular release in the urinary bladder, distention-induced luminal ATP release was decreased in both WT and NGF-OE mice although the magnitude of reduction was greater in NGF-OE mice. Inhibition of pannexin channels locally in the urinary bladder did not decrease the luminal release of ATP in WT or NGF-OE mice. These data suggest ATP is released by vesicular exocytosis in the urinary bladder and TRPV4 plays a functional role (Fig. 4). Although these results suggest a vesicular secretory mechanism, the effects of BFA may also generalize to inhibit the transport of cell surface proteins, like hemichannels, to attenuate release (Wang, Olivecrona, et al., 2005). Future studies are needed to differentiate the effects of vesicle release and hemichannel inhibition.

16. ATP and TRPV4 activate cells in the laminar propria network, leading to the appearance of organized, propagating wavefronts

As discussed above, the lamina propria is located between the urothelium and the detrusor and is composed of loose connective tissue, ICs, vasculature, lymphatic vessels, nerve fibers and nerve terminals and may serve to integrate epithelial and smooth muscle input to maintain normal bladder function (Andersson, 2002, 2004; Birder & Andersson, 2013). The physical proximity of the structural components of the urinary bladder suggests that reciprocal communication among urothelial cells, the lamina propria and detrusor smooth muscle may be possible (Birder & Andersson, 2013).

Lamina propria ICs may interact with an extensive network of nerve fibers (subepithelial plexus), including afferents, that course through the lamina propria (Gabella & Davis, 1998). Immunohistochemistry as well as electron microscopy imaging reveal close physical proximity between lamina propria cells and nerve fibers (Andersson & McCloskey, 2014; Davidson & McCloskey, 2005; Wiseman, Fowler, & Landon, 2003). A dense network of kit+ lamina propria cells that expressed platelet-derived growth factor receptor-α (PDGFRα) and exhibited stellate or spindle-shaped morphology were also found in murine bladder (Koh et al., 2012). Although the lamina propria is heterogeneous, consisting of diverse cell types, certain cells may be important modulators of neural activity and form a communication link between the urothelium and detrusor; the importance of which may depend upon age or presence of pathology.

We demonstrated that the lamina propria cell network in early postnatal rat pups ($P \leq 21$) exhibited a predominance of PDGFRα- and TRPV4-IR. Application of exogenous mediators (i.e., ATP, TRPV4 agonist) activated and increased the number of lamina propria cells that exhibited active Ca^{2+} events. Further, lamina propria cell activity was not random as demonstrated with spatio-temporal maps and calcium transient particle analyses that demonstrated varying degrees of coupling (e.g., tight, partial, or loose organization). This coupling of cellular Ca^{2+} activity of cells could be modified to generate organized, propagating bands of activity (wavefronts) with ATP in the lamina propria network. These findings suggest that lamina propria Ca^{2+} signaling may facilitate communication through this syncytial network to other cell types or tissue layers of the urinary bladder. Spatio-temporal patterning in the lamina propria network may be a means to affect change in sensory processing and/or detrusor contractility.

Using lamina propria wholemounts from postnatal rats, we identified Ca^{2+} events that had long durations with slow upstroke and downstroke phases. These Ca^{2+} events in the lamina propria network were largely dependent on Ca^{2+} release from internal stores because cyclopiazonic acid (CPA), a specific and reversible blocker of the sarcoplasmic or endoplasmic reticulum Ca^{2+}-ATPase (SERCA) pump, significantly reduced the (i) overall duration of cell activation and (ii) the number of cells exhibiting Ca^{2+} events. In lamina propria wholemounts from postnatal rats, we demonstrated Ca^{2+} events spreading along processes between cells. The lamina propria network exhibited a range of Ca^{2+} behaviors from ongoing, loosely coupled individual cellular activity to near synchronous cell activation in a particular direction to form a Ca^{2+} wavefront that propagated substantial distances throughout the lamina propria. The range of Ca^{2+} behaviors is indicative of flexible coupling between lamina propria cells (Neuhaus et al., 2020). Although the mechanism underlying the formation of a Ca^{2+} wavefront is unclear, the lamina propria exhibits substantial Cx43 expression, consistent with the presence of gap junctions (Ikeda et al., 2007).

Purinergic signaling in the LUT involves multiple tissues and cell types, including: suburothelial nerve plexus, smooth muscle cells, lamina propria cells including ICs as well as inflammatory cells (Birder, 2005; Birder & Andersson, 2013; Li et al., 2013). ICs express P2X and P2Y (i.e., P2X3 P2Y2, P2Y4, P2Y6) receptors and are proposed to form a functional syncytium with smooth muscle cells (Drumm, Koh, Andersson, & Ward, 2014; Sui, Wu, & Fry, 2006). Although the mechanism coupling ICs to sensory activity is unknown, the location of ICs and the responsiveness to ATP suggest they may have a regulatory role in the afferent limb of the micturition reflex (Wu, Sui, & Fry, 2004). We demonstrated that exogenous ATP increased Ca^{2+} transient activity in numerous lamina propria cells that resulted in multiple Ca^{2+} waves propagating through the lamina propria (Heppner, Hennig, Nelson, & Vizzard, 2017) (Figs. 5 and 6). In addition, we showed that exogenous ATP converted lamina propria cells from a loosely coupled pattern to a highly organized pattern with the formation of a Ca^{2+} wavefront (Heppner et al., 2017) (Figs. 5 and 6). This finding is consistent with the hypothesis that ATP released from urothelium or other tissues (i.e., suburothelial nerves) increases lamina propria Ca^{2+} signaling and facilitates communication through this syncytial network to other cell types or tissue layers (Andersson & McCloskey, 2014; Heppner et al., 2017) (Figs. 5 and 6). A large area of the lamina propria network undergoing near-synchronous activity may provide a much more potent signal/stimulus to other cell types compared to loosely coupled patterns of activation of lamina propria cells.

As described above, multiple TRP channels from different subfamilies are expressed in the urinary bladder and have specific LUT tissue distributions. We demonstrated functional TRPV4 expression in lamina propria cells in postnatal rat pups (Heppner et al., 2017). In wholemount tissue preparations from postnatal rats ($\leq$P21), TRPV4-IR was observed in lamina propria cells that exhibited similar morphology to those expressing PDGFRα-IR. Application of the TRPV4 agonist, GSK1016790, increased the number of cells that exhibited active Ca^{2+} events and time active as shown by the rate of integrated Ca^{2+} activity (Heppner et al., 2017). The TRPV4 antagonist, GSK2193874, did not affect the number of lamina propria cells exhibiting Ca^{2+} transients or the duration of Ca^{2+} transients suggesting that TRPV4 channels are not consistently active in lamina propria cells. Activation of TRPV4 channels via proteolytic activation of protease-activated receptor 2 under certain condition including inflammation (Sostegni et al., 2015) has been demonstrated, but the background level of TRPV4 activation in normal bladder function is apparently quite minimal.

In wholemount preparations from rat pups ($\leq$P21), we consistently identified a dense network of PDGFRα- or TRPV4-IR cells in the lamina propria; however, few such cells were present in adult preparations (Heppner et al., 2017). Differences between preparations from adult and postnatal rats may reflect the continuing maturation of the micturition reflexes during the early postnatal period. The functional syncytium between ICs and the detrusor smooth muscle may have greater importance

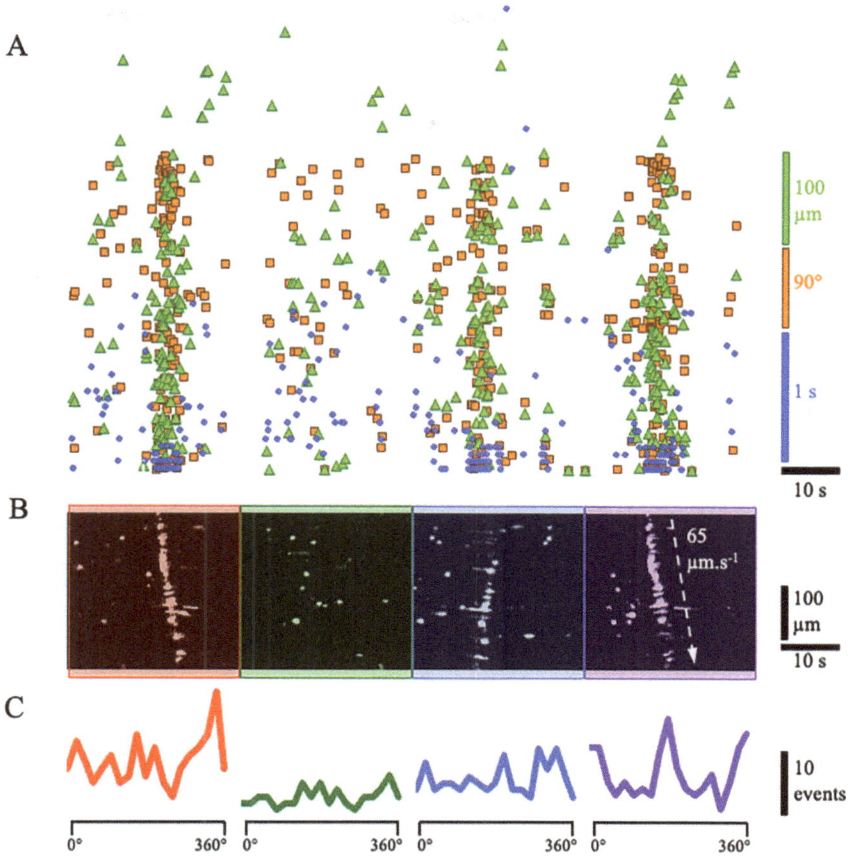

Fig. 5 Tightly and loosely coupled Ca^{2+} network activity in the lamina propria syncytium. (A) Plot of the firing characteristics of lamina propria cells showing (i) the distance (green triangles), (ii) the angle (orange squares) and (iii) the time delay (blue dots) between lamina propria cells in the forward firing sequence. (B) Spatio-temporal map of lamina propria cells firing showing 4 network firing events denoted by red, green, blue and purple overlays. (C) Frequency of angles histograms between next-to--fire cells (forward sequence) during the 4 network firing events. The first and last network firing events show a high degree organization (B: red and purple overlays) with a cluster of small delays (A: blue dots) between firing of cells and a strong bias for next-to-fire cells to occur at specific angles (C: red & purple lines; 160 & 340°) corresponding to the direction of the wavefront (90° to the propagation direction). The second network event (B: green overlay) does not show a tightly organized network firing sequence, with variable delays (A: blue dots) and little bias in the angle between next-to-fire cells (C: green line). The third event (B: blue overlay) in which the wavefront propagates in the opposite direction, shows partially coupled network activity with some bias in the angle between next-to-fire cells but does not involve all the activatable cells in the field of view. *Modified from Heppner, T. J., Hennig, G. W., Nelson, M. T., & Vizzard, M. A. (2017). Rhythmic calcium events in the lamina propria network of the urinary bladder of rat pups.* Frontiers in Systems Neuroscience, 11, *87. https:/doi.org/10.3389/fnsys.2017.00087.*

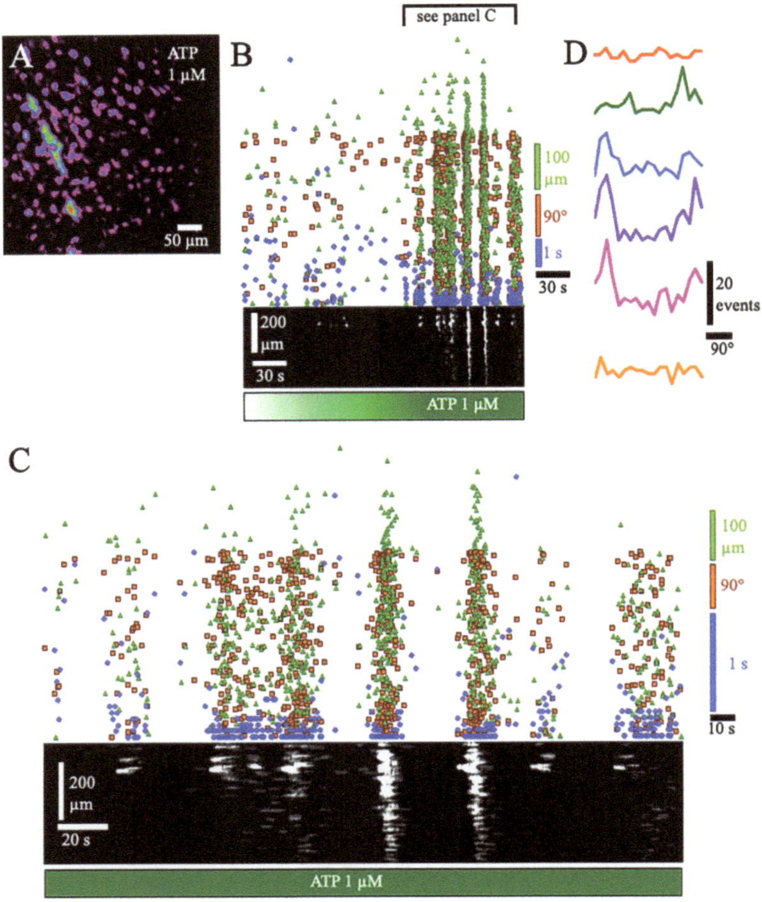

Fig. 6 Lamina propria cell network response to ATP. (A) Prevalence map of lamina propria syncytium. (B) Gradual perfusion of ATP to 1 µM (green bar) resulted in increasing number of cells firing, eventually leading to the development of propagating wavefronts. (C) Zoomed-in area from (B) showing greater detail of altered activity resulting in wavefront development. (D) Angle histograms showing the progression from unorganized activity to well-defined wavefronts based on angle bias. In preparations with the presence of propagating network Ca^{2+} events, the velocity (60–70 µm/s) of propagation was consistent, even though the direction of propagation was often variable. The bias in angle (10.65 ± 1.63 (n = 8)) was determined to evaluate the overall degree of coupling between active lamina propria cells and indicated that lamina propria cell activity was not random and had a defined direction (loosely/tightly coupled wavefront). *Modified from Heppner, T. J., Hennig, G. W., Nelson, M. T., & Vizzard, M. A. (2017). Rhythmic calcium events in the lamina propria network of the urinary bladder of rat pups.* Frontiers in Systems Neuroscience, 11, *87. https://doi.org/10.3389/fnsys.2017.00087.*

for coordinating bladder emptying in the early postnatal period due to the absence of mature and functional neural input needed to coordinate the activities of the CNS and PNS including the urinary bladder (Kanai et al., 2007). The current studies in rat pups ($\leq$P21) are consistent with this suggestion as demonstrated by: (i) the predominance of the lamina propria cellular network in early postnatal rat pups; (ii) the ability of exogenous mediators (i.e., ATP, TRPV4 agonist) to activate and increase the number of lamina propria cells that exhibited active Ca^{2+} events; and (iii) the ability of ATP and TRPV4 agonist to increase the rate of integrated Ca^{2+} activity corresponding to coupled lamina propria network events and the formation of propagating wavefronts (Heppner et al., 2017) (Figs. 5 and 6).

In future studies, it would be of interest to examine the lamina propria network in wholemount preparations from preclinical animal models of bladder dysfunction or SCI to determine if the lamina propria network at the urothelial-lamina propria junction is changed and/or if spontaneous and evoked Ca^{2+} transients and network activity is altered. A change in the number of cells exhibiting Ca^{2+} events, their pattern of activation, the duration of Ca^{2+} events and/or the ability of exogenous mediators (e.g., ATP, TRPV4 agonists) to increase the rate of integrated Ca^{2+} activity corresponding to coupled lamina propria network events could contribute to altered sensory processing (e.g., mechanosensation, pain) and altered urinary bladder function (i.e., hyperactivity). Plasticity in lamina propria cellular network has been previously demonstrated. For example, ICs in the detrusor are decreased 5 weeks following SCI, target organ innervation is reduced, and the smooth muscle is hypertrophied (Johnston et al., 2012). However, there are variable and inconsistent changes in ICs from individuals with bladder dysfunction (e.g., neurogenic and idiopathic detrusor overactivity) (McCloskey, 2010). Thus, future studies should examine the lamina propria network at the urothelial-lamina propria junction following neural injury, disease, or urinary bladder dysfunction to fully understand its functional significance.

Disclaimers

The National Institutes of Health had no role in the experiments described, including the design, data collection, and analysis of studies performed in the Vizzard laboratory, decision to publish, or preparation of the manuscript. The contents are solely the responsibility of the authors and do not necessarily represent the official views of the National Institutes of Health.

Conflict of interest

The authors declare that the research was conducted in the absence of any commercial or financial relationships that could be construed as a potential conflict of interest.

Author contributions

The authors thank all current and former Vizzard laboratory members who have contributed to the studies described in this perspective.

Funding

This work was funded by National Institute of Diabetes and Digestive and Kidney Diseases Grants 5R01DK120108 and 1R01DK124580 to M. A. Vizzard.

References

Aizawa, N., Wyndaele, J. J., Homma, Y., & Igawa, Y. (2012). Effects of TRPV4 cation channel activation on the primary bladder afferent activities of the rat. *Neurourology and Urodynamics*, *31*(1), 148–155. https://doi.org/10.1002/nau.21212.

Akiyama, Y., Yao, J. R., Kreder, K. J., O'Donnell, M. A., Lutgendorf, S. K., Lyu, D., et al. (2021). Autoimmunity to urothelial antigen causes bladder inflammation, pelvic pain, and voiding dysfunction: A novel animal model for Hunner-type interstitial cystitis. *American Journal of Physiology. Renal Physiology*, *320*(2), F174–F182. https://doi.org/10.1152/ajprenal.00290.2020.

Alessandri-Haber, N., Dina, O. A., Yeh, J. J., Parada, C. A., Reichling, D. B., & Levine, J. D. (2004). Transient receptor potential vanilloid 4 is essential in chemotherapy-induced neuropathic pain in the rat. *The Journal of Neuroscience*, *24*(18), 4444–4452. https://doi.org/10.1523/jneurosci.0242-04.2004.

Andersson, K. E. (2002). Bladder activation: afferent mechanisms [Review]. *Urology*, *59*(5 Suppl. 1), 43–50. http://www.ncbi.nlm.nih.gov/pubmed/12007522.

Andersson, K. E. (2004). Mechanisms of disease: Central nervous system involvement in overactive bladder syndrome [Research Support, Non-U.S. Gov't Review]. *Nature Clinical Practice. Urology*, *1*(2), 103–108. https://doi.org/10.1038/ncpuro0021.

Andersson, K. E., & Arner, A. (2004). Urinary bladder contraction and relaxation: Physiology and pathophysiology. *Physiological Reviews*, *84*(3), 935–986. https://doi.org/10.1152/physrev.00038.2003.

Andersson, K. E., Gratzke, C., & Hedlund, P. (2010). The role of the transient receptor potential (TRP) superfamily of cation-selective channels in the management of the overactive bladder. *BJU International*, *106*(8), 1114–1127. https://doi.org/10.1111/j.1464-410X.2010.09650.x.

Andersson, K. E., & McCloskey, K. D. (2014). Lamina propria: The functional center of the bladder? *Neurourology and Urodynamics*, *33*(1), 9–16. https://doi.org/10.1002/nau.22465.

Andersson, K. E., & Michel, M. C. (2011). *Urinary tract*. Berlin Heidelberg: Springer. https://books.google.com/books?id=jnfQ5VbxJ_AC.

Angelico, P., & Testa, R. (2010). TRPV4 as a target for bladder overactivity. *F1000 Biology Reports*, *2*. https://doi.org/10.3410/B2-12.

Araki, I., Du, S., Kamiyama, M., Mikami, Y., Matsushita, K., Komuro, M., et al. (2004). Overexpression of epithelial sodium channels in epithelium of human urinary bladder with outlet obstruction. *Urology*, *64*(6), 1255–1260. https://doi.org/10.1016/j.urology.2004.06.064.

Arms, L., Girard, B. M., Malley, S. E., & Vizzard, M. A. (2013). Expression and function of CCL2/CCR2 in rat micturition reflexes and somatic sensitivity with urinary bladder inflammation. *American Journal of Physiology. Renal Physiology, 305*(1), F111–F122. https://doi.org/10.1152/ajprenal.00139.2013.

Arms, L., Girard, B. M., & Vizzard, M. A. (2010). Expression and function of CXCL12/CXCR4 in rat urinary bladder with cyclophosphamide-induced cystitis. *American Journal of Physiology. Renal Physiology, 298*(3), F589–F600. https://doi.org/10.1152/ajprenal.00628.2009.

Baron, R., Hans, G., & Dickenson, A. H. (2013). Peripheral input and its importance for central sensitization. *Annals of Neurology, 74*(5), 630–636. https://doi.org/10.1002/ana.24017.

Beckel, J. M., & Birder, L. A. (2012). Differential expression and function of nicotinic acetylcholine receptors in the urinary bladder epithelium of the rat. *The Journal of Physiology, 590*(6), 1465–1480. https://doi.org/10.1113/jphysiol.2011.226860.

Beckel, J. M., Daugherty, S. L., Tyagi, P., Wolf-Johnston, A. S., Birder, L. A., Mitchell, C. H., et al. (2015). Pannexin 1 channels mediate the release of ATP into the lumen of the rat urinary bladder. *The Journal of Physiology, 593*(8), 1857–1871. https://doi.org/10.1113/jphysiol.2014.283119.

Beckel, J. M., & Holstege, G. (2011). Neuroanatomy of the lower urinary tract. *Handbook of Experimental Pharmacology, 202*, 99–116. https://doi.org/10.1007/978-3-642-16499-6_6.

Beckel, J. M., Kanai, A., Lee, S. J., de Groat, W. C., & Birder, L. A. (2006). Expression of functional nicotinic acetylcholine receptors in rat urinary bladder epithelial cells. *American Journal of Physiology. Renal Physiology, 290*(1), F103–F110. https://doi.org/10.1152/ajprenal.00098.2005.

Bennett, B. C., Kruse, M. N., Roppolo, J. R., Flood, H. D., Fraser, M., & de Groat, W. C. (1995). Neural control of urethral outlet activity in vivo: Role of nitric oxide. *The Journal of Urology, 153*(6), 2004–2009.

Berry, S. H., Elliott, M. N., Suttorp, M., Bogart, L. M., Stoto, M. A., Eggers, P., et al. (2011). Prevalence of symptoms of bladder pain syndrome/interstitial cystitis among adult females in the United States. *The Journal of Urology, 186*(2), 540–544. https://doi.org/10.1016/j.juro.2011.03.132.

Birder, L. A. (2005). More than just a barrier: Urothelium as a drug target for urinary bladder pain [Research Support, N.I.H., Extramural Research Support, U.S. Gov't, P.H.S. Review]. *American Journal of Physiology. Renal Physiology, 289*(3), F489–F495. https://doi.org/10.1152/ajprenal.00467.2004.

Birder, L. A. (2007). TRPs in bladder diseases. *Biochimica et Biophysica Acta, 1772*(8), 879–884. https://doi.org/10.1016/j.bbadis.2007.04.003.

Birder, L., & Andersson, K.-E. (2013). Urothelial signaling. *Physiological Reviews, 93*(2), 653–680. https://doi.org/10.1152/physrev.00030.2012.

Birder, L., & Andersson, K.-E. (2018). Animal modelling of interstitial cystitis/bladder pain syndrome. *International Neurourology Journal, 22*(Suppl. 1), S3–S9. https://doi.org/10.5213/inj.1835062.531.

Birder, L. A., Apodaca, G., De Groat, W. C., & Kanai, A. J. (1998). Adrenergic- and capsaicin-evoked nitric oxide release from urothelium and afferent nerves in urinary bladder. *The American Journal of Physiology, 275*(2), F226–F229. https://doi.org/10.1152/ajprenal.1998.275.2.F226.

Birder, L. A., Barrick, S. R., Roppolo, J. R., Kanai, A. J., de Groat, W. C., Kiss, S., et al. (2003). Feline interstitial cystitis results in mechanical hypersensitivity and altered ATP release from bladder urothelium. *American Journal of Physiology. Renal Physiology, 285*(3), F423–F429. https://doi.org/10.1152/ajprenal.00056.2003.

Birder, L. A., & de Groat, W. C. (2007). Mechanisms of disease: involvement of the urothelium in bladder dysfunction [Research Support, N.I.H., Extramural Review]. *Nature Clinical Practice. Urology, 4*(1), 46–54. https://doi.org/10.1038/ncpuro0672.

Birder, L. A., Kanai, A. J., de Groat, W. C., Kiss, S., Nealen, M. L., Burke, N. E., et al. (2001). Vanilloid receptor expression suggests a sensory role for urinary bladder epithelial cells. *Proceedings of the National Academy of Sciences of the United States of America, 98*(23), 13396–13401. https://doi.org/10.1073/pnas.231243698.

Birder, L. A., & Kullmann, F. A. (2018). Role of neurogenic inflammation in local communication in the visceral mucosa. *Seminars in Immunopathology, 40*(3), 261–279. https://doi.org/10.1007/s00281-018-0674-0.

Birder, L. A., Nakamura, Y., Kiss, S., Nealen, M. L., Barrick, S., Kanai, A. J., et al. (2002). Altered urinary bladder function in mice lacking the vanilloid receptor TRPV1 [Research Support, Non-U.S. Gov't Research Support, U.S. Gov't, P.H.S.]. *Nature Neuroscience, 5*(9), 856–860. https://doi.org/10.1038/nn902.

Birder, L. A., Nealen, M. L., Kiss, S., de Groat, W. C., Caterina, M. J., Wang, E., et al. (2002). Beta-adrenoceptor agonists stimulate endothelial nitric oxide synthase in rat urinary bladder urothelial cells. *The Journal of Neuroscience, 22*(18), 8063–8070.

Birder, L. A., Ruan, H. Z., Chopra, B., Xiang, Z., Barrick, S., Buffington, C. A., et al. (2004). Alterations in P2X and P2Y purinergic receptor expression in urinary bladder from normal cats and cats with interstitial cystitis. *American Journal of Physiology. Renal Physiology, 287*(5), F1084–F1091. https://doi.org/10.1152/ajprenal.00118.2004.

Bjorling, D. E., Wang, Z.-Y., & Bushman, W. (2011). Models of inflammation of the lower urinary tract. *Neurourology and Urodynamics, 30*(5), 673–682. https://doi.org/10.1002/nau.21078.

Brady, C. M., Apostolidis, A. N., Harper, M., Yiangou, Y., Beckett, A., Jacques, T. S., et al. (2004). Parallel changes in bladder suburothelial vanilloid receptor TRPV1 and pan-neuronal marker PGP9.5 immunoreactivity in patients with neurogenic detrusor overactivity after intravesical resiniferatoxin treatment. *BJU International, 93*(6), 770–776. https://doi.org/10.1111/j.1464-410X.2003.04722.x.

Brady, C. M., Apostolidis, A., Yiangou, Y., Baecker, P. A., Ford, A. P., Freeman, A., et al. (2004). P2X3-immunoreactive nerve fibres in neurogenic detrusor overactivity and the effect of intravesical resiniferatoxin [Research Support, Non-U.S. Gov't]. *European Urology, 46*(2), 247–253. https://doi.org/10.1016/j.eururo.2003.12.017.

Burnstock, G. (2001). Purine-mediated signalling in pain and visceral perception. *Trends in Pharmacological Sciences, 22*(4), 182–188. https://doi.org/10.1016/s0165-6147(00)01643-6.

Charrua, A., Pinto, R., Birder, L. A., & Cruz, F. (2015). Sympathetic nervous system and chronic bladder pain: a new tune for an old song. *Translational Andrology and Urology, 4*(5), 534–542. https://doi.org/10.3978/j.issn.2223-4683.2015.09.06.

Chen, C. C., Akopian, A. N., Sivilotti, L., Colquhoun, D., Burnstock, G., & Wood, J. N. (1995). A P2X purinoceptor expressed by a subset of sensory neurons. *Nature, 377*(6548), 428–431. https://doi.org/10.1038/377428a0.

Chopra, B., Barrick, S. R., Meyers, S., Beckel, J. M., Zeidel, M. L., Ford, A. P., et al. (2005). Expression and function of bradykinin B1 and B2 receptors in normal and inflamed rat urinary bladder urothelium. *The Journal of Physiology, 562*(Pt. 3), 859–871. https://doi.org/10.1113/jphysiol.2004.071159.

Chopra, B., Gever, J., Barrick, S. R., Hanna-Mitchell, A. T., Beckel, J. M., Ford, A. P., et al. (2008). Expression and function of rat urothelial P2Y receptors. *American Journal of Physiology. Renal Physiology, 294*(4), F821–F829. https://doi.org/10.1152/ajprenal.00321.2006.

Christensen, A. P., & Corey, D. P. (2007). TRP channels in mechanosensation: Direct or indirect activation? *Nature Reviews. Neuroscience, 8*(7), 510–521. https://doi.org/10.1038/nrn2149.

Dahl, G. (2015). ATP release through pannexon channels. *Philosophical Transactions of the Royal Society of London. Series B, Biological Sciences, 370*(1672). https://doi.org/10.1098/rstb.2014.0191.

Dalghi, M. G., Clayton, D. R., Ruiz, W. G., Al-Bataineh, M. M., Satlin, L. M., Kleyman, T. R., et al. (2019). Expression and distribution of PIEZO1 in the mouse urinary tract. *American Journal of Physiology. Renal Physiology, 317*(2), F303–F321. https://doi.org/10.1152/ajprenal.00214.2019.

Davidson, R. A., & McCloskey, K. D. (2005). Morphology and localization of interstitial cells in the guinea pig bladder: structural relationships with smooth muscle and neurons [Research Support, Non-U.S. Gov't]. *The Journal of Urology, 173*(4), 1385–1390. https://doi.org/10.1097/01.ju.0000146272.80848.37.

de Groat, W. C. (1990). Central neural control of the lower urinary tract [Research Support, U.S. Gov't, Non-P.H.S. Research Support, U.S. Gov't, P.H.S. Review]. *CIBA Foundation Symposium, 151*, 27–44. discussion 44–56 http://www.ncbi.nlm.nih.gov/pubmed/2226064.

de Groat, W. C., Griffiths, D., & Yoshimura, N. (2015). Neural control of the lower urinary tract. *Comprehensive Physiology, 5*(1), 327–396. https://doi.org/10.1002/cphy.c130056.

de Groat, W. C., & Yoshimura, N. (2006). Mechanisms underlying the recovery of lower urinary tract function following spinal cord injury. *Progress in Brain Research, 152*, 59–84. https://doi.org/10.1016/S0079-6123(05)52005-3.

de Groat, W. C., & Yoshimura, N. (2010). Changes in afferent activity after spinal cord injury. *Neurourology and Urodynamics, 29*(1), 63–76. https://doi.org/10.1002/nau.20761.

DeGroat, W. C., & Saum, W. R. (1976). Synaptic transmission in parasympathetic ganglia in the urinary bladder of the cat. *The Journal of Physiology, 256*(1), 137–158. https://doi.org/10.1113/jphysiol.1976.sp011316.

Dmitrieva, N., & McMahon, S. B. (1996). Sensitisation of visceral afferents by nerve growth factor in the adult rat [Research Support, Non-U.S. Gov't]. *Pain, 66*(1), 87–97. http://www.ncbi.nlm.nih.gov/pubmed/8857635.

Drumm, B. T., Koh, S. D., Andersson, K. E., & Ward, S. M. (2014). Calcium signalling in cajal-like interstitial cells of the lower urinary tract. *Nature Reviews. Urology, 11*(10), 555–564. https://doi.org/10.1038/nrurol.2014.241.

Du, S., Araki, I., Kobayashi, H., Zakoji, H., Sawada, N., & Takeda, M. (2008). Differential expression profile of cold (TRPA1) and cool (TRPM8) receptors in human urogenital organs. *Urology, 72*(2), 450–455. https://doi.org/10.1016/j.urology.2007.11.127.

Elbadawi, A. (1996). Functional anatomy of the organs of micturition. *The Urologic Clinics of North America, 23*(2), 177–210. https://www.ncbi.nlm.nih.gov/pubmed/8659020.

Ernsberger, U. (2009). Role of neurotrophin signalling in the differentiation of neurons from dorsal root ganglia and sympathetic ganglia. *Cell and Tissue Research, 336*(3), 349–384. https://doi.org/10.1007/s00441-009-0784-z.

Everaerts, W., Gevaert, T., Nilius, B., & De Ridder, D. (2008). On the origin of bladder sensing: Tr(i)ps in urology. *Neurourology and Urodynamics, 27*(4), 264–273. https://doi.org/10.1002/nau.20511.

Everaerts, W., Nilius, B., & Owsianik, G. (2010). The vanilloid transient receptor potential channel TRPV4: From structure to disease [Research Support, Non-U.S. Gov't Review]. *Progress in Biophysics and Molecular Biology, 103*(1), 2–17. https://doi.org/10.1016/j.pbiomolbio.2009.10.002.

Everaerts, W., Zhen, X., Ghosh, D., Vriens, J., Gevaert, T., Gilbert, J. P., et al. (2010). Inhibition of the cation channel TRPV4 improves bladder function in mice and rats with cyclophosphamide-induced cystitis [Research Support, Non-U.S. Gov't]. *Proceedings of the National Academy of Sciences of the United States of America, 107*(44), 19084–19089. https://doi.org/10.1073/pnas.1005333107.

Ferguson, D. R., Kennedy, I., & Burton, T. J. (1997). ATP is released from rabbit urinary bladder epithelial cells by hydrostatic pressure changes- -a possible sensory mechanism? [Research Support, Non-U.S. Gov't]. *The Journal of Physiology, 505*(Pt. 2), 503–511. http://www.ncbi.nlm.nih.gov/pubmed/9423189.

Forrest, J. B., & Schmidt, S. (2004). Interstitial cystitis, chronic nonbacterial prostatitis and chronic pelvic pain syndrome in men: a common and frequently identical clinical entity. *The Journal of Urology*, *172*(6 Pt. 2), 2561–2562. https://doi.org/10.1097/01.ju.0000145102.40773.43.

Fowler, C. J., Griffiths, D., & de Groat, W. C. (2008). The neural control of micturition. *Nature Reviews. Neuroscience*, *9*(6), 453–466. https://doi.org/10.1038/nrn2401.

Gabella, G., & Davis, C. (1998). Distribution of afferent axons in the bladder of rats [Research Support, Non-U.S. Gov't]. *Journal of Neurocytology*, *27*(3), 141–155. http://www.ncbi.nlm.nih.gov/pubmed/10640174.

Gevaert, T., Vriens, J., Segal, A., Everaerts, W., Roskams, T., Talavera, K., et al. (2007). Deletion of the transient receptor potential cation channel TRPV4 impairs murine bladder voiding. *The Journal of Clinical Investigation*, *117*(11), 3453–3462. https://doi.org/10.1172/JCI31766.

Girard, B. M., Campbell, S. E., Perkins, M., Hsiang, H., Tooke, K., Drescher, C., et al. (2019). TRPV4 blockade reduces voiding frequency, ATP release, and pelvic sensitivity in mice with chronic urothelial overexpression of NGF. *American Journal of Physiology. Renal Physiology*, *317*(6), F1695–F1706. https://doi.org/10.1152/ajprenal.00147.2019.

Girard, B. M., Merrill, L., Malley, S., & Vizzard, M. A. (2013). Increased TRPV4 expression in urinary bladder and lumbosacral dorsal root ganglia in mice with chronic overexpression of NGF in urothelium [Research Support, N.I.H., Extramural Research Support, Non-U.S. Gov't]. *Journal of Molecular Neuroscience*, *51*(2), 602–614. https://doi.org/10.1007/s12031-013-0033-5.

Girard, B. M., Tooke, K., & Vizzard, M. A. (2017). PACAP/receptor system in urinary bladder dysfunction and pelvic pain following urinary bladder inflammation or stress. *Frontiers in Systems Neuroscience*, *11*, 90. https://doi.org/10.3389/fnsys.2017.00090.

Godoy, L. D., Rossignoli, M. T., Delfino-Pereira, P., Garcia-Cairasco, N., & de Lima Umeoka, E. H. (2018). A comprehensive overview on stress neurobiology: Basic concepts and clinical implications. *Frontiers in Behavioral Neuroscience*, *12*, 127. https://doi.org/10.3389/fnbeh.2018.00127.

Gonzalez, E. J., Girard, B., Braas, K. M., May, V., & Vizzard, M. A. (2016). Neuroplasticity of PACAP expression and function in micturition reflex pathways. In D. Reglodi, & A. Tamas (Eds.), *Pituitary adenylate cyclase activating polypeptide—PACAP* (pp. 313–334). Springer International Publishing. https://doi.org/10.1007/978-3-319-35135-3_19.

Gonzalez, E. J., Girard, B. M., & Vizzard, M. A. (2013). Expression and function of transforming growth factor-beta isoforms and cognate receptors in the rat urinary bladder following cyclophosphamide-induced cystitis. *American Journal of Physiology. Renal Physiology*, *305*(9), F1265–F1276. https://doi.org/10.1152/ajprenal.00042.2013.

Gonzalez, E. J., Heppner, T. J., Nelson, M. T., & Vizzard, M. A. (2016). Purinergic signalling underlies transforming growth factor-beta-mediated bladder afferent nerve hyperexcitability. *The Journal of Physiology*, *594*(13), 3575–3588. https://doi.org/10.1113/JP272148.

Gosling, J. A., & Dixon, J. S. (1974). Species variation in the location of upper urinary tract pacemaker cells. *Investigative Urology*, *11*(5), 418–423. https://www.ncbi.nlm.nih.gov/pubmed/4856100.

Grundy, L., & Brierley, S. M. (2018). Cross-organ sensitization between the colon and bladder: To pee or not to pee? *American Journal of Physiology. Gastrointestinal and Liver Physiology*, *314*(3), G301–G308. https://doi.org/10.1152/ajpgi.00272.2017.

Grundy, L., Caldwell, A., & Brierley, S. M. (2018). Mechanisms underlying overactive bladder and interstitial cystitis/painful bladder syndrome. *Frontiers in Neuroscience*, *12*, 931. https://doi.org/10.3389/fnins.2018.00931.

Guerios, S. D., Wang, Z. Y., & Bjorling, D. E. (2006). Nerve growth factor mediates peripheral mechanical hypersensitivity that accompanies experimental cystitis in mice [Comparative Study]. *Neuroscience Letters, 392*(3), 193–197.

Guerios, S. D., Wang, Z. Y., Boldon, K., Bushman, W., & Bjorling, D. E. (2008). Blockade of NGF and trk receptors inhibits increased peripheral mechanical sensitivity accompanying cystitis in rats [Research Support, N.I.H., Extramural Research Support, Non-U. S. Gov't]. *American Journal of Physiology. Regulatory, Integrative and Comparative Physiology, 295*(1), R111–R122. https://doi.org/10.1152/ajpregu.00728.2007.

Guo, M., Chang, P., Hauke, E., Girard, B. M., Tooke, K., Ojala, J., et al. (2018). Expression and function of chemokines CXCL9-11 in micturition pathways in cyclophosphamide (CYP)-induced cystitis and somatic sensitivity in mice. *Frontiers in Systems Neuroscience, 12*, 9. https://doi.org/10.3389/fnsys.2018.00009.

Hammack, S. E., Cheung, J., Rhodes, K. M., Schutz, K. C., Falls, W. A., Braas, K. M., et al. (2009). Chronic stress increases pituitary adenylate cyclase-activating peptide (PACAP) and brain-derived neurotrophic factor (BDNF) mRNA expression in the bed nucleus of the stria terminalis (BNST): Roles for PACAP in anxiety-like behavior. *Psychoneuroendocrinology, 34*(6), 833–843. https://doi.org/10.1016/j.psyneuen.2008.12.013.

Hanna-Mitchell, A. T., Beckel, J. M., Barbadora, S., Kanai, A. J., de Groat, W. C., & Birder, L. A. (2007). Non-neuronal acetylcholine and urinary bladder urothelium. *Life Sciences, 80*(24-25), 2298–2302. https://doi.org/10.1016/j.lfs.2007.02.010.

Hanno, P. M. (2015). *Afferent clinical data for lead candidate, AF-219, Demonstrate Improvements in Pain and Urinary Urgency in Interstitial Cystitis/Bladder Pain Syndrome (IC/BPS).*

Hauser, P. J., Dozmorov, M. G., Bane, B. L., Slobodov, G., Culkin, D. J., & Hurst, R. E. (2008). Abnormal expression of differentiation related proteins and proteoglycan core proteins in the urothelium of patients with interstitial cystitis. *The Journal of Urology, 179*(2), 764–769. https://doi.org/10.1016/j.juro.2007.09.022.

Heppner, T. J., Hennig, G. W., Nelson, M. T., & Vizzard, M. A. (2017). Rhythmic calcium events in the lamina propria network of the urinary bladder of rat pups. *Frontiers in Systems Neuroscience, 11*, 87. https://doi.org/10.3389/fnsys.2017.00087.

Homma, Y., Nomiya, A., Tagaya, M., Oyama, T., Takagaki, K., Nishimatsu, H., et al. (2013). Increased mRNA expression of genes involved in pronociceptive inflammatory reactions in bladder tissue of interstitial cystitis. *The Journal of Urology, 190*(5), 1925–1931. https://doi.org/10.1016/j.juro.2013.05.049.

Hou, X. H., Hyun, M., Taranda, J., Huang, K. W., Todd, E., Feng, D., et al. (2016). Central control circuit for context-dependent micturition. *Cell, 167*(1), 73–86.e12. https://doi.org/10.1016/j.cell.2016.08.073.

Hu, V. Y., Zvara, P., Dattilio, A., Redman, T. L., Allen, S. J., Dawbarn, D., et al. (2005). Decrease in bladder overactivity with REN1820 in rats with cyclophosphamide induced cystitis [Research Support, U.S. Gov't, P.H.S.]. *The Journal of Urology, 173*(3), 1016–1021. https://doi.org/10.1097/01.ju.0000155170.15023.e5.

Ikeda, Y., Fry, C., Hayashi, F., Stolz, D., Griffiths, D., & Kanai, A. (2007). Role of gap junctions in spontaneous activity of the rat bladder. *American Journal of Physiology. Renal Physiology, 293*(4), F1018–F1025. https://doi.org/10.1152/ajprenal.00183.2007.

Jaggar, S. I., Scott, H. C., & Rice, A. S. (1999). Inflammation of the rat urinary bladder is associated with a referred thermal hyperalgesia which is nerve growth factor dependent [Research Support, Non-U.S. Gov't]. *British Journal of Anaesthesia, 83*(3), 442–448. http://www.ncbi.nlm.nih.gov/pubmed/10655916.

Janssen, D. A., Hoenderop, J. G., Jansen, K. C., Kemp, A. W., Heesakkers, J. P., & Schalken, J. A. (2011). The mechanoreceptor TRPV4 is localized in adherence junctions of the human bladder urothelium: A morphological study. *The Journal of Urology, 186*(3), 1121–1127. https://doi.org/10.1016/j.juro.2011.04.107.

Johnston, L., Cunningham, R. M., Young, J. S., Fry, C. H., McMurray, G., Eccles, R., et al. (2012). Altered distribution of interstitial cells and innervation in the rat urinary bladder following spinal cord injury. *Journal of Cellular and Molecular Medicine*, *16*(7), 1533–1543. https://doi.org/10.1111/j.1582-4934.2011.01410.x.

Kanai, A., Roppolo, J., Ikeda, Y., Zabbarova, I., Tai, C., Birder, L., et al. (2007). Origin of spontaneous activity in neonatal and adult rat bladders and its enhancement by stretch and muscarinic agonists. *American Journal of Physiology. Renal Physiology*, *292*(3), F1065–F1072. https://doi.org/10.1152/ajprenal.00229.2006.

Keay, S. K., Szekely, Z., Conrads, T. P., Veenstra, T. D., Barchi, J. J., Jr., Zhang, C. O., et al. (2004). An antiproliferative factor from interstitial cystitis patients is a frizzled 8 protein-related sialoglycopeptide. *Proceedings of the National Academy of Sciences of the United States of America*, *101*(32), 11803–11808. https://doi.org/10.1073/pnas.0404509101.

King, S. B., Toufexis, D. J., & Hammack, S. E. (2017). Pituitary adenylate cyclase activating polypeptide (PACAP), stress, and sex hormones. *Stress*, *1–11*. https://doi.org/10.1080/10253890.2017.1336535.

Koh, S. D., Lee, H., Ward, S. M., & Sanders, K. M. (2018). The mystery of the interstitial cells in the urinary bladder. *Annual Review of Pharmacology and Toxicology*, *58*, 603–623. https://doi.org/10.1146/annurev-pharmtox-010617-052615.

Koh, B. H., Roy, R., Hollywood, M. A., Thornbury, K. D., McHale, N. G., Sergeant, G. P., et al. (2012). Platelet-derived growth factor receptor-alpha cells in mouse urinary bladder: A new class of interstitial cells. *Journal of Cellular and Molecular Medicine*, *16*(4), 691–700. https://doi.org/10.1111/j.1582-4934.2011.01506.x.

Kopparapu, P. K., Boorjian, S. A., Robinson, B. D., Downes, M., Gudas, L. J., Mongan, N. P., et al. (2013). Expression of VEGF and its receptors VEGFR1/VEGFR2 is associated with invasiveness of bladder cancer. *Anticancer Research*, *33*(6), 2381–2390.

Kullmann, F. A., Artim, D., Beckel, J., Barrick, S., de Groat, W. C., & Birder, L. A. (2008). Heterogeneity of muscarinic receptor-mediated Ca^{2+} responses in cultured urothelial cells from rat. *American Journal of Physiology. Renal Physiology*, *294*(4), F971–F981. https://doi.org/10.1152/ajprenal.00313.2007.

Kullmann, F. A., Artim, D. E., Birder, L. A., & de Groat, W. C. (2008). Activation of muscarinic receptors in rat bladder sensory pathways alters reflex bladder activity. *The Journal of Neuroscience*, *28*(8), 1977–1987. https://doi.org/10.1523/jneurosci.4694-07.2008.

Kullmann, F. A., Downs, T. R., Artim, D. E., Limberg, B. J., Shah, M., Contract, D., et al. (2011). Urothelial beta-3 adrenergic receptors in the rat bladder. *Neurourology and Urodynamics*, *30*(1), 144–150. https://doi.org/10.1002/nau.20965.

Kullmann, F. A., Shah, M. A., Birder, L. A., & de Groat, W. C. (2009). Functional TRP and ASIC-like channels in cultured urothelial cells from the rat. *American Journal of Physiology. Renal Physiology*, *296*(4), F892–F901. https://doi.org/10.1152/ajprenal.90718.2008.

LaBerge, J., Malley, S. E., Zvarova, K., & Vizzard, M. A. (2006). Expression of corticotropin-releasing factor and CRF receptors in micturition pathways after cyclophosphamide-induced cystitis [Research Support, N.I.H., Extramural]. *American Journal of Physiology. Regulatory, Integrative and Comparative Physiology*, *291*(3), R692–R703. https://doi.org/10.1152/ajpregu.00086.2006.

Lamb, K., Gebhart, G. F., & Bielefeldt, K. (2004). Increased nerve growth factor expression triggers bladder overactivity [Research Support, U.S. Gov't, P.H.S.]. *The Journal of Pain*, *5*(3), 150–156. http://www.ncbi.nlm.nih.gov/pubmed/15106127.

Lazarowski, E. R. (2012). Vesicular and conductive mechanisms of nucleotide release. *Purinergic Signal*, *8*(3), 359–373. https://doi.org/10.1007/s11302-012-9304-9.

Lazarowski, E. R., Boucher, R. C., & Harden, T. K. (2003). Mechanisms of release of nucleotides and integration of their action as P2X- and P2Y-receptor activating molecules. *Molecular Pharmacology*, *64*(4), 785–795. https://doi.org/10.1124/mol.64.4.785.

Lee, H. Y., Bardini, M., & Burnstock, G. (2000). Distribution of P2X receptors in the urinary bladder and the ureter of the rat. *The Journal of Urology*, *163*(6), 2002–2007. https://www.ncbi.nlm.nih.gov/pubmed/10799247.

Li, Y., Xue, L., Miao, Q., Mao, F., Yao, L., Yuan, J., et al. (2013). Expression and electrophysiological characteristics of P2X3 receptors in interstitial cells of Cajal in rats with partial bladder outlet obstruction [Comparative Study Research Support, Non-U.S. Gov't]. *BJU International*, *111*(5), 843–851. https://doi.org/10.1111/j.1464-410X.2012.11408.x.

Liedtke, W., Choe, Y., Marti-Renom, M. A., Bell, A. M., Denis, C. S., Sali, A., et al. (2000). Vanilloid receptor-related osmotically activated channel (VR-OAC), a candidate vertebrate osmoreceptor. *Cell*, *103*(3), 525–535. https://doi.org/10.1016/s0092-8674(00)00143-4.

Limberg, B. J., Andersson, K. E., Aura Kullmann, F., Burmer, G., de Groat, W. C., & Rosenbaum, J. S. (2010). beta-Adrenergic receptor subtype expression in myocyte and non-myocyte cells in human female bladder. *Cell and Tissue Research*, *342*(2), 295–306. https://doi.org/10.1007/s00441-010-1053-x.

Liu, Q., Sun, B., Zhao, J., Wang, Q., An, F., Hu, X., et al. (2018). Increased Piezo1 channel activity in interstitial Cajal-like cells induces bladder hyperactivity by functionally interacting with NCX1 in rats with cyclophosphamide-induced cystitis. *Experimental & Molecular Medicine*, *50*(5), 60. https://doi.org/10.1038/s12276-018-0088-z.

Malin, S., Molliver, D., Christianson, J. A., Schwartz, E. S., Cornuet, P., Albers, K. M., et al. (2011). TRPV1 and TRPA1 function and modulation are target tissue dependent. *The Journal of Neuroscience*, *31*(29), 10516–10528. https://doi.org/10.1523/jneurosci.2992-10.2011.

Malley, S. E., & Vizzard, M. A. (2002). Changes in urinary bladder cytokine mRNA and protein after cyclophosphamide-induced cystitis. *Physiological Genomics*, *9*(1), 5–13. https://doi.org/10.1152/physiolgenomics.00117.2001.

Marcu, I., Campian, E. C., & Tu, F. F. (2018). Interstitial cystitis/bladder pain syndrome. *Seminars in Reproductive Medicine*, *36*(2), 123–135. https://doi.org/10.1055/s-0038-1676089.

Marshall, K. L., Saade, D., Ghitani, N., Coombs, A. M., Szczot, M., Keller, J., et al. (2020). PIEZO2 in sensory neurons and urothelial cells coordinates urination. *Nature*, *588*(7837), 290–295. https://doi.org/10.1038/s41586-020-2830-7.

Matos, R., Serrao, P., Rodriguez, L., Birder, L. A., Cruz, F., & Charrua, A. (2017). The water avoidance stress induces bladder pain due to a prolonged alpha1A adrenoceptor stimulation. *Naunyn-Schmiedeberg's Archives of Pharmacology*, *390*(8), 839–844. https://doi.org/10.1007/s00210-017-1384-1.

McCloskey, K. D. (2010). Interstitial cells in the urinary bladder—Localization and function. *Neurourology and Urodynamics*, *29*(1), 82–87. https://doi.org/10.1002/nau.20739.

McLatchie, L. M., & Fry, C. H. (2015). ATP release from freshly isolated guinea-pig bladder urothelial cells: A quantification and study of the mechanisms involved. *BJU International*, *115*(6), 987–993. https://doi.org/10.1111/bju.12954.

Merrill, L., Malley, S., & Vizzard, M. A. (2013). Repeated variate stress in male rats induces increased voiding frequency, somatic sensitivity, and urinary bladder nerve growth factor expression. *American Journal of Physiology. Regulatory, Integrative and Comparative Physiology*, *305*(2), R147–R156. https://doi.org/10.1152/ajpregu.00089.2013.

Merrill, L., & Vizzard, M. A. (2014). Intravesical TRPV4 blockade reduces repeated variate stress-induced bladder dysfunction by increasing bladder capacity and decreasing voiding frequency in male rats [Research Support, N.I.H., Extramural]. *American Journal of Physiology. Regulatory, Integrative and Comparative Physiology*, *307*(4), R471–R480. https://doi.org/10.1152/ajpregu.00008.2014.

Mingin, G. C., Heppner, T. J., Tykocki, N. R., Erickson, C. S., Vizzard, M. A., & Nelson, M. T. (2015). Social stress in mice induces urinary bladder overactivity and increases TRPV1 channel-dependent afferent nerve activity. *American Journal of Physiology. Regulatory, Integrative and Comparative Physiology*, *309*(6), R629–R638. https://doi.org/10.1152/ajpregu.00013.2015.

Mingin, G. C., Peterson, A., Erickson, C. S., Nelson, M. T., & Vizzard, M. A. (2014). Social stress induces changes in urinary bladder function, bladder NGF content, and generalized bladder inflammation in mice. *American Journal of Physiology. Regulatory, Integrative and Comparative Physiology*, *307*(7), R893–R900. https://doi.org/10.1152/ajpregu.00500.2013.

Mistry, M. A., Klarskov, N., DeLancey, J. O., & Lose, G. (2020). A structured review on the female urethral anatomy and innervation with an emphasis on the role of the urethral longitudinal smooth muscle. *International Urogynecology Journal*, *31*(1), 63–71. https://doi.org/10.1007/s00192-019-04104-7.

Mochizuki, T., Sokabe, T., Araki, I., Fujishita, K., Shibasaki, K., Uchida, K., et al. (2009). The TRPV4 cation channel mediates stretch-evoked Ca2+ influx and ATP release in primary urothelial cell cultures. *The Journal of Biological Chemistry*, *284*(32), 21257–21264. https://doi.org/10.1074/jbc.M109.020206.

Moldwin, R. M., Nursey, V., Yaskiv, O., Dalvi, S., Macdonald, E. J., Funaro, M., et al. (2022). Immune cell profiles of patients with interstitial cystitis/bladder pain syndrome. *Journal of Translational Medicine*, *20*(1), 97. https://doi.org/10.1186/s12967-022-03236-7.

Munoz, A., Somogyi, G. T., Boone, T. B., & Smith, C. P. (2011). Lumbosacral sensory neuronal activity is enhanced by activation of urothelial purinergic receptors. *Brain Research Bulletin*, *86*(5-6), 380–384. https://doi.org/10.1016/j.brainresbull.2011.09.001.

Murray, E., Malley, S. E., Qiao, L. Y., Hu, V. Y., & Vizzard, M. A. (2004). Cyclophosphamide induced cystitis alters neurotrophin and receptor tyrosine kinase expression in pelvic ganglia and bladder [Research Support, N.I.H., Extramural]. *The Journal of Urology*, *172*(6 Pt 1), 2434–2439. http://www.ncbi.nlm.nih.gov/pubmed/15538286.

Nadelhaft, I., Vera, P. L., Card, J. P., & Miselis, R. R. (1992). Central nervous system neurons labelled following the injection of pseudorabies virus into the rat urinary bladder. *Neuroscience Letters*, *143*(1–2), 271–274. https://doi.org/10.1016/0304-3940(92)90281-b.

Nazif, O., Teichman, J. M., & Gebhart, G. F. (2007). Neural upregulation in interstitial cystitis. *Urology*, *69*(4 Suppl), 24–33. https://doi.org/10.1016/j.urology.2006.08.1108.

Negoro, H., Urban-Maldonado, M., Liou, L. S., Spray, D. C., Thi, M. M., & Suadicani, S. O. (2014). Pannexin 1 channels play essential roles in urothelial mechanotransduction and intercellular signaling. *PLoS One*, *9*(8), e106269. https://doi.org/10.1371/journal.pone.0106269.

Neuhaus, J., Gonsior, A., Cheng, S., Stolzenburg, J. U., & Berger, F. P. (2020). Mechanosensitivity is a characteristic feature of cultured suburothelial interstitial cells of the human bladder. *International Journal of Molecular Sciences*, *21*(15). https://doi.org/10.3390/ijms21155474.

Nickel, J. C., Stephens, A., Landis, J. R., Mullins, C., van Bokhoven, A., Lucia, M. S., et al. (2016). Assessment of the lower urinary tract microbiota during symptom flare in women with urologic chronic pelvic pain syndrome: A MAPP network study. *The Journal of Urology*, *195*(2), 356–362. https://doi.org/10.1016/j.juro.2015.09.075.

Nilius, B. (2010). Pressing and squeezing with Piezos. *EMBO Reports*, *11*(12), 902–903. https://doi.org/10.1038/embor.2010.181.

Nilius, B., Owsianik, G., Voets, T., & Peters, J. A. (2007). Transient receptor potential cation channels in disease. *Physiological Reviews*, *87*(1), 165–217. https://doi.org/10.1152/physrev.00021.2006.

Nilius, B., & Voets, T. (2013). The puzzle of TRPV4 channelopathies. *EMBO Reports*, *14*(2), 152–163. https://doi.org/10.1038/embor.2012.219.

Olivar, T., & Laird, J. M. (1999). Cyclophosphamide cystitis in mice: behavioural characterisation and correlation with bladder inflammation. *European Journal of Pain (London, England)*, *3*(2), 141–149. https://doi.org/10.1053/eujp.1998.0105.

Olsen, S. M., Stover, J. D., & Nagatomi, J. (2011). Examining the role of mechanosensitive ion channels in pressure mechanotransduction in rat bladder urothelial cells. *Annals of Biomedical Engineering*, *39*(2), 688–697. https://doi.org/10.1007/s10439-010-0203-3.

Pandita, R. K., & Andersson, K. E. (2002). Intravesical adenosine triphosphate stimulates the micturition reflex in awake, freely moving rats. *The Journal of Urology*, *168*(3), 1230–1234. https://doi.org/10.1097/01.ju.0000023398.35234.ea.

Pang, X., Marchand, J., Sant, G. R., Kream, R. M., & Theoharides, T. C. (1995). Increased number of substance P positive nerve fibres in interstitial cystitis. *British Journal of Urology*, *75*(6), 744–750. https://doi.org/10.1111/j.1464-410x.1995.tb07384.x.

Parsons, C. L., Lilly, J. D., & Stein, P. (1991). Epithelial dysfunction in nonbacterial cystitis (interstitial cystitis). *The Journal of Urology*, *145*(4), 732–735. https://doi.org/10.1016/s0022-5347(17)38437-9.

Patnaik, S. S., Lagana, A. S., Vitale, S. G., Buttice, S., Noventa, M., Gizzo, S., et al. (2017). Etiology, pathophysiology and biomarkers of interstitial cystitis/painful bladder syndrome. *Archives of Gynecology and Obstetrics*, *295*(6), 1341–1359. https://doi.org/10.1007/s00404-017-4364-2.

Reynolds, W. S., Dmochowski, R., Wein, A., & Bruehl, S. (2016). Does central sensitization help explain idiopathic overactive bladder? *Nature Reviews. Urology*, *13*(8), 481–491. https://doi.org/10.1038/nrurol.2016.95.

Rothrock, N. E., Lutgendorf, S. K., Kreder, K. J., Ratliff, T., & Zimmerman, B. (2001a). Stress and symptoms in patients with interstitial cystitis: A life stress model. *Urology*, *57*(3), 422–427. https://www.ncbi.nlm.nih.gov/pubmed/11248609.

Rothrock, N. E., Lutgendorf, S. K., Kreder, K. J., Ratliff, T. L., & Zimmerman, B. (2001b). Daily stress and symptom exacerbation in interstitial cystitis patients. *Urology*, *57*(6 Suppl. 1), 122. https://www.ncbi.nlm.nih.gov/pubmed/11378108.

Ruggieri, M. R., Sr. (2006). Mechanisms of disease: Role of purinergic signaling in the pathophysiology of bladder dysfunction. *Nature Clinical Practice. Urology*, *3*(4), 206–215. https://doi.org/10.1038/ncpuro0456.

Saban, M. R., Backer, J. M., Backer, M. V., Maier, J., Fowler, B., Davis, C. A., et al. (2008). VEGF receptors and neuropilins are expressed in the urothelial and neuronal cells in normal mouse urinary bladder and are upregulated in inflammation. *American Journal of Physiology. Renal Physiology*, *295*(1), F60–F72. https://doi.org/10.1152/ajprenal.00618.2007.

Sancho, M., Triguero, D., Lafuente-Sanchis, A., & Garcia-Pascual, A. (2017). Proliferation of interstitial cells in the cyclophosphamide-induced cystitis and the preventive effect of imatinib. *BioMed Research International*, *2017*, 3457093. https://doi.org/10.1155/2017/3457093.

Schnegelsberg, B., Sun, T. T., Cain, G., Bhattacharya, A., Nunn, P. A., Ford, A. P., et al. (2010). Overexpression of NGF in mouse urothelium leads to neuronal hyperinnervation, pelvic sensitivity, and changes in urinary bladder function. *American Journal of Physiology. Regulatory, Integrative and Comparative Physiology*, *298*(3), R534–R547. https://doi.org/10.1152/ajpregu.00367.2009.

Sheng, J. A., Bales, N. J., Myers, S. A., Bautista, A. I., Roueinfar, M., Hale, T. M., et al. (2020). The hypothalamic-pituitary-adrenal axis: Development, programming actions of hormones, and maternal-fetal interactions. *Frontiers in Behavioral Neuroscience*, *14*, 601939. https://doi.org/10.3389/fnbeh.2020.601939.

Shimizu, N., Wada, N., Shimizu, T., Suzuki, T., Takaoka, E. I., Kanai, A. J., et al. (2018). Effects of nerve growth factor neutralization on TRP channel expression in laser-captured bladder afferent neurons in mice with spinal cord injury. *Neuroscience Letters*, *683*, 100–103. https://doi.org/10.1016/j.neulet.2018.06.049.

Silva-Ramos, M., Silva, I., Oliveira, O., Ferreira, S., Reis, M. J., Oliveira, J. C., et al. (2013). Urinary ATP may be a dynamic biomarker of detrusor overactivity in women with overactive bladder syndrome. *PLoS One*, *8*(5), e64696. https://doi.org/10.1371/journal.pone.0064696.

Smith, A. L., Leung, J., Kun, S., Zhang, R., Karagiannides, I., Raz, S., et al. (2011). The effects of acute and chronic psychological stress on bladder function in a rodent model. *Urology*, *78*(4), 967.e961-967. https://doi.org/10.1016/j.urology.2011.06.041.

Smith, P. R., Mackler, S. A., Weiser, P. C., Brooker, D. R., Ahn, Y. J., Harte, B. J., et al. (1998). Expression and localization of epithelial sodium channel in mammalian urinary bladder. *The American Journal of Physiology*, *274*(1), F91–F96. https://doi.org/10.1152/ajprenal.1998.274.1.F91.

Smith, C. P., Vemulakonda, V. M., Kiss, S., Boone, T. B., & Somogyi, G. T. (2005). Enhanced ATP release from rat bladder urothelium during chronic bladder inflammation: effect of botulinum toxin A. *Neurochemistry International*, *47*(4), 291–297. https://doi.org/10.1016/j.neuint.2005.04.021.

Sostegni, S., Diakov, A., McIntyre, P., Bunnett, N., Korbmacher, C., & Haerteis, S. (2015). Sensitisation of TRPV4 by PAR2 is independent of intracellular calcium signalling and can be mediated by the biased agonist neutrophil elastase. *Pflügers Archiv*, *467*(4), 687–701. https://doi.org/10.1007/s00424-014-1539-6.

Stein, R. J., Santos, S., Nagatomi, J., Hayashi, Y., Minnery, B. S., Xavier, M., et al. (2004). Cool (TRPM8) and hot (TRPV1) receptors in the bladder and male genital tract. *The Journal of Urology*, *172*(3), 1175–1178. https://doi.org/10.1097/01.ju.0000134880.55119.cf.

Sugaya, K., Matsuyama, K., Takakusaki, K., & Mori, S. (1987). Electrical and chemical stimulations of the pontine micturition center. *Neuroscience Letters*, *80*(2), 197–201. https://doi.org/10.1016/0304-3940(87)90653-7.

Sui, G., Fry, C. H., Montgomery, B., Roberts, M., Wu, R., & Wu, C. (2014). Purinergic and muscarinic modulation of ATP release from the urothelium and its paracrine actions. *American Journal of Physiology. Renal Physiology*, *306*(3), F286–F298. https://doi.org/10.1152/ajprenal.00291.2013.

Sui, G. P., Wu, C., & Fry, C. H. (2006). Characterization of the purinergic receptor subtype on guinea-pig suburothelial myofibroblasts. *BJU International*, *97*(6), 1327–1331. https://doi.org/10.1111/j.1464-410X.2006.06200.x.

Sun, Y., & Chai, T. C. (2006). Augmented extracellular ATP signaling in bladder urothelial cells from patients with interstitial cystitis. *American Journal of Physiology. Cell Physiology*, *290*(1), C27–C34. https://doi.org/10.1152/ajpcell.00552.2004.

Sun, Y., Keay, S., Lehrfeld, T. J., & Chai, T. C. (2009). Changes in adenosine triphosphate-stimulated ATP release suggest association between cytokine and purinergic signaling in bladder urothelial cells. *Urology*, *74*(5), 1163–1168. https://doi.org/10.1016/j.urology.2009.02.066.

Suskind, A. M., Berry, S. H., Ewing, B. A., Elliott, M. N., Suttorp, M. J., & Clemens, J. Q. (2013). The prevalence and overlap of interstitial cystitis/bladder pain syndrome and chronic prostatitis/chronic pelvic pain syndrome in men: Results of the RAND Interstitial Cystitis Epidemiology male study. *The Journal of Urology*, *189*(1), 141–145. https://doi.org/10.1016/j.juro.2012.08.088.

Sutcliffe, S., Colditz, G. A., Goodman, M. S., Pakpahan, R., Vetter, J., Ness, T. J., et al. (2014). Urological chronic pelvic pain syndrome symptom flares: Characterisation of the full range of flares at two sites in the Multidisciplinary Approach to the Study of Chronic Pelvic Pain (MAPP) Research Network. *BJU International*, *114*(6), 916–925. https://doi.org/10.1111/bju.12778.

Sutcliffe, S., Jemielita, T., Lai, H. H., Andriole, G. L., Bradley, C. S., Clemens, J. Q., et al. (2018). A case-crossover study of urological chronic pelvic pain syndrome flare triggers in the MAPP research network. *The Journal of Urology*, *199*(5), 1245–1251. https://doi.org/10.1016/j.juro.2017.12.050.

Tempest, H. V., Dixon, A. K., Turner, W. H., Elneil, S., Sellers, L. A., & Ferguson, D. R. (2004). P2X and P2X receptor expression in human bladder urothelium and changes in interstitial cystitis. *BJU International, 93*(9), 1344–1348. https://doi.org/10.1111/j.1464-410X.2004.04858.x.

Thorneloe, K. S., Sulpizio, A. C., Lin, Z., Figueroa, D. J., Clouse, A. K., McCafferty, G. P., et al. (2008). N-((1S)-1-{[4-((2S)-2-{[(2,4-dichlorophenyl)-sulfonyl]amino}-3-hydroxypropanoyl)-1 -piperazinyl]carbonyl}-3-methylbutyl)-1-benzothiophene-2-carboxamide (GSK1016790A), a novel and potent transient receptor potential vanilloid 4 channel agonist induces urinary bladder contraction and hyperactivity: Part I. *The Journal of Pharmacology and Experimental Therapeutics, 326*(2), 432–442. https://doi.org/10.1124/jpet.108.139295.

Timoteo, M. A., Carneiro, I., Silva, I., Noronha-Matos, J. B., Ferreirinha, F., Silva-Ramos, M., et al. (2014). ATP released via pannexin-1 hemichannels mediates bladder overactivity triggered by urothelial P2Y6 receptors. *Biochemical Pharmacology, 87*(2), 371–379. https://doi.org/10.1016/j.bcp.2013.11.007.

Unsworth, C. D., & Johnson, R. G. (1990). Acetylcholine and ATP are coreleased from the electromotor nerve terminals of Narcine brasiliensis by an exocytotic mechanism. *Proceedings of the National Academy of Sciences of the United States of America, 87*(2), 553–557. https://doi.org/10.1073/pnas.87.2.553.

Vlaskovska, M., Kasakov, L., Rong, W., Bodin, P., Bardini, M., Cockayne, D. A., et al. (2001). P2X3 knock-out mice reveal a major sensory role for urothelially released ATP. *The Journal of Neuroscience, 21*(15), 5670–5677. https://www.ncbi.nlm.nih.gov/pubmed/11466438.

Vriens, J., Appendino, G., & Nilius, B. (2009). Pharmacology of vanilloid transient receptor potential cation channels. *Molecular Pharmacology, 75*(6), 1262–1279. https://doi.org/10.1124/mol.109.055624.

Walsh, P. C. (1986). *Campbell's urology. Vol. 2.* WB Saunders Co.

Wang, Z., Chang, H. H., Gao, Y., Zhang, R., Guo, Y., Holschneider, D. P., et al. (2017). Effects of water avoidance stress on peripheral and central responses during bladder filling in the rat: A multidisciplinary approach to the study of urologic chronic pelvic pain syndrome (MAPP) research network study. *PLoS One, 12*(9), e0182976. https://doi.org/10.1371/journal.pone.0182976.

Wang, E. C., Lee, J. M., Johnson, J. P., Kleyman, T. R., Bridges, R., & Apodaca, G. (2003). Hydrostatic pressure-regulated ion transport in bladder uroepithelium. *American Journal of Physiology. Renal Physiology, 285*(4), F651–F663. https://doi.org/10.1152/ajprenal.00403.2002.

Wang, E. C., Lee, J. M., Ruiz, W. G., Balestreire, E. M., von Bodungen, M., Barrick, S., et al. (2005). ATP and purinergic receptor-dependent membrane traffic in bladder umbrella cells. *The Journal of Clinical Investigation, 115*(9), 2412–2422. https://doi.org/10.1172/JCI24086.

Wang, L., Olivecrona, G., Gotberg, M., Olsson, M. L., Winzell, M. S., & Erlinge, D. (2005). ADP acting on P2Y13 receptors is a negative feedback pathway for ATP release from human red blood cells. *Circulation Research, 96*(2), 189–196. https://doi.org/10.1161/01.Res.0000153670.07559.E4.

Winder, M., Tobin, G., Zupančič, D., & Romih, R. (2014). Signalling molecules in the urothelium. *BioMed Research International, 2014*, 297295. https://doi.org/10.1155/2014/297295.

Wiseman, O. J., Fowler, C. J., & Landon, D. N. (2003). The role of the human bladder lamina propria myofibroblast. *BJU International, 91*(1), 89–93. https://www.ncbi.nlm.nih.gov/pubmed/12614258.

Wu, C., Sui, G. P., & Fry, C. H. (2004). Purinergic regulation of guinea pig suburothelial myofibroblasts. *The Journal of Physiology, 559*(Pt 1), 231–243. https://doi.org/10.1113/jphysiol.2004.067934.

Xu, X., Gordon, E., Lin, Z., Lozinskaya, I. M., Chen, Y., & Thorneloe, K. S. (2009). Functional TRPV4 channels and an absence of capsaicin-evoked currents in freshly-isolated, guinea-pig urothelial cells. *Channels (Austin, TX)*, *3*(3), 156–160. https://www.ncbi.nlm.nih.gov/pubmed/19411839.

Yamada, T., Ugawa, S., Ueda, T., Ishida, Y., Kajita, K., & Shimada, S. (2009). Differential localizations of the transient receptor potential channels TRPV4 and TRPV1 in the mouse urinary bladder. *The Journal of Histochemistry and Cytochemistry*, *57*(3), 277–287. https://doi.org/10.1369/jhc.2008.951962.

Yoshikawa, S., Kawamorita, N., Oguchi, T., Funahashi, Y., Tyagi, P., Chancellor, M. B., et al. (2015). Pelvic organ cross-sensitization to enhance bladder and urethral pain behaviors in rats with experimental colitis. *Neuroscience*, *284*, 422–429. https://doi.org/10.1016/j.neuroscience.2014.08.064.

Yoshimura, N., & Chancellor, M. B. (2003). Neurophysiology of lower urinary tract function and dysfunction. *Reviews in Urology*, *5*(Suppl. 8), S3–S10. https://pubmed.ncbi.nlm.nih.gov/16985987 https://www.ncbi.nlm.nih.gov/pmc/articles/PMC1502389/.

Yu, Y., & de Groat, W. C. (2008). Sensitization of pelvic afferent nerves in the in vitro rat urinary bladder-pelvic nerve preparation by purinergic agonists and cyclophosphamide pretreatment. *American Journal of Physiology. Renal Physiology*, *294*(5), F1146–F1156. https://doi.org/10.1152/ajprenal.00592.2007.

Yu, W., Hill, W. G., Apodaca, G., & Zeidel, M. L. (2011). Expression and distribution of transient receptor potential (TRP) channels in bladder epithelium. *American Journal of Physiology. Renal Physiology*, *300*(1), F49–F59. https://doi.org/10.1152/ajprenal.00349.2010.

Zvara, P., & Vizzard, M. A. (2007). Exogenous overexpression of nerve growth factor in the urinary bladder produces bladder overactivity and altered micturition circuitry in the lumbosacral spinal cord [Research Support, N.I.H., Extramural]. *BMC Physiology*, *7*, 9. https://doi.org/10.1186/1472-6793-7-9.

CHAPTER SIX

The role of TRPV4 channels in cutaneous epithelia

Carlene Moore[a,*]
[a]Division of Headache and Division of Translational Brain Sciences, Department of Neurology, Duke University School of Medicine, Durham, NC, United States
*Corresponding author: e-mail address: carlene.moore@duke.edu

Contents

Abstract

Transient receptor potential vanilloid 4 (TRPV4) channels are multi-modally activated cation permeable channels that are expressed most organ tissues including the skin. TRPV4 is highly expressed in the skin and functions in skin resident cells such as epidermal keratinocytes, melanocytes, immune mast cells and macrophages, and cutaneous neurons. TRPV4 plays many crucial roles in skin homeostasis to affect an extensive range of processes such as temperature sensation, osmo-sensation, hair growth, cell apoptosis, skin barrier integrity, differentiation, nociception and itch. Since TRPV4 functions in a plenitude of pathological states, TRPV4 can become a versatile therapeutic target for diseases such as chronic pain, itch and skin cancer.

1. Introduction

The surface epithelium, the epidermis of skin provides barrier protection against dehydration and the potentially harmful external environment

Current Topics in Membranes, Volume 89
ISSN 1063-5823
https://doi.org/10.1016/bs.ctm.2022.06.003

(Fuchs, 2009; Patapoutian, 2005; Roosterman et al., 2006; Tominaga & Caterina, 2004). The epidermis also assumes an underappreciated sensory and sentinel function as a "frontline sensor" of a potentially harmful environment (Fuchs, 2009). TRPV4 belongs to a family of transmembrane non-selective calcium permeable ion channels, TRP channels. TRPV4 was initially identified as an osmolarity sensor, and is now known to be multi-modally activated by warm temperature, mechanical force (osmotic and shear stress), phorbol ester derivatives, UVB (type B ultraviolet rays), and a wide number of chemical agonists (Boudaka, Al-Yazeedi, & Al-Lawati, 2020; Liedtke et al., 2003; Moore et al., 2013). TRPV4 is abundant in the skin and mediates calcium influx into many resident skin cells including epidermal keratinocytes, melanocytes, immune cells such as mast cells and macrophages, endothelial cells of skin blood vessels and cutaneous neurons to affect wide range of processes such as hair growth, cell apoptosis, skin barrier integrity and differentiation, the release of paracrine and autocrine factors, nociception and itch as illustrated in Fig. 1 (Caterina & Pang, 2016; Chen et al., 2016; Liedtke et al., 2000; Liedtke & Friedman, 2003a; Moore & Liedtke, 2017; Olivan-Viguera et al., 2018; Sokabe et al., 2010; Szabó et al., 2019). TRPV4 might also be involved in skin aging and sexually dimorphic differences in skin. Skin inflammation is an

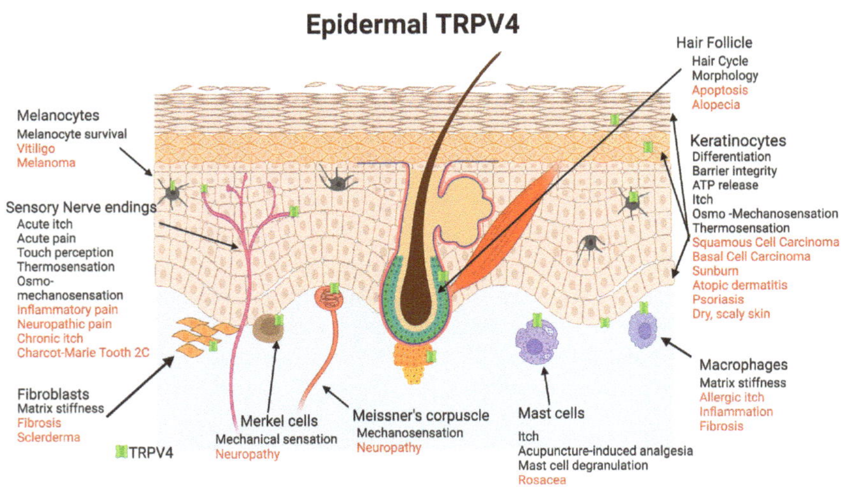

Fig. 1 The role of TRPV4 in skin homeostasis (black) and pathophysiological processes (red). TRPV4 is expressed in many cell types in the skin including keratinocytes, melanocytes, sensory nerve endings, immune cells such a macrophages and mast cells, hair follicles and fibroblasts. Created with BioRender.

evolutionarily-refined protective mechanism that clarifies noxious cues and irritants and initiates regeneration. However, the involvement of TRPV4 in dysregulation of these processes facilitates acute and chronic diseases involving pain, itch, fibrosis, and cancer (Fig. 1).

2. TRPV4 in skin barrier function

TRPV4 has been implicated in the maintenance of skin barrier integrity. TRPV4 has been shown to contribute to inter-cellular junction formation in keratinocytes by interacting with β-catenin and enhancing cell-cell junction development and formation of tight barrier between keratinocytes (Sokabe et al., 2010). In human skin keratinocytes, activation of TRPV4 by warm temperature (33°C), or its chemical agonists lead to cell-cell junction development, augmentation of barrier integrity and acceleration of barrier recovery after disruption of the stratum corneum (Kida et al., 2012). Activation of TRPV4 was demonstrated to strengthen the tight junction associated barrier of epidermal cells, which resulted in the upregulation of occludin and claudin-4 tight junction structural proteins and tight junction regulatory factor, protein kinase C (PKC) (Akazawa et al., 2013). Additionally, knock down of *TRPV4* resulted in increased inter-cellular permeation and reduced trans-epidermal resistance (Kida et al., 2012).

Mutations in the *TRPV4* gene are known to cause autosomal dominant disorders: Charcot-Marie-Tooth type 2C (CMT2C), scapuloperoneal spinal muscular atrophy, and congenital distal spinal muscular atrophy, which are neuropathies characterized by limb, diaphragm, and laryngeal muscle weakness (Landouré et al., 2010; McCray et al., 1993). Skin biopsies from patients with CMT2C revealed reduced density of Meissner corpuscles and intra-papillary myelinated endings as well as shortened intermodal lengths (Manganelli et al., 2015). A novel heterozygous mutation identified in the *TRPV4* gene at position c.2355G>T, resulting in a tryptophan to cysteine substitution at amino acid 785 (p.Trp785Cys), was identified in family members who showed signs of congenital distal spinal muscular atrophy, skeletal abnormalities including osteonecrosis of the femoral head, and scaly skin (Liu et al., 2020a). Investigations into hereditary TRPV4 channelopathy disorders continue, and it can be considered likely that more links to epidermal signaling and function will be discovered.

3. TRPV4 in hair growth

Two recent studies revealed a contradictory role of TRPV4 in hair growth. TRPV4 was shown to be expressed on intact human hair follicles and was detected on the outer root sheath layer of the hair follicle epithelium (Szabó et al., 2019). In these human hair follicles, activation of TRPV4 via agonists disrupted hair shaft elongation, decreased intra-follicular proliferation, and induced premature catagen regression and apoptosis in culture (Szabó et al., 2019). On the other hand, a similar study in mice demonstrated that activation of TRPV4 using a small molecule agonist induced telogen to anagen transition and hair follicle regeneration by increasing the promotion of anagen promoting growth factors and downregulating the expression of anagen inhibiting factors (Yang et al., 2020). These two studies appear to reveal a contradictory role of TRPV4 in hair growth, which might be explained by the differences in the experimental set-up and species differences as Szabó et al. evaluated anagen maintenance ex vivo using cultured human hair follicles, while Yang et al examined follicle regeneration used in vivo mice studies. Further work is necessary to evaluate the role of TPRV4 in hair growth in vivo in humans.

4. TRPV4 in sensory perception

4.1 Osmo- and mechanosensation

TPRV4 was first identified as an osmosensitive ion channel and has been shown to function as an osmo-mechanosensor to maintain cellular osmotic homeostasis in many cellular systems (Liedtke et al., 2000; Liedtke & Friedman, 2003a; Moore & Liedtke, 2017). In fish, TRPV4 in skin keratinocytes senses the osmolarity of the aquatic environment and facilitates the immune effector mechanisms, as demonstrated in the fish model organism *Danio rerio* (zebrafish) (Galindo-Villegas et al., 2016). Pharmacological and genetic inhibition experiments demonstrated that skin keratinocytes through a TRPV4/Ca^{2+}/transforming growth factor (TGF)-β–activated kinase 1 (TAK1)/nuclear factor kappa B (NF-κB) signaling pathways sensed osmotic stress to induce developmental immunity (Galindo-Villegas et al., 2016). TRPV4 was implicated also in the regulation of the mRNA and protein expressions of isotocin, a protein which controls ion regulation through modulating the functions of ionocytes in zebrafish (Liu et al., 2020b). *TRPV4* gene knockdown decreased ionic contents

(Na^+, Cl^-, and Ca^{2+}) of whole larvae, the H^+-secreting function of larval skin of zebrafish, and the numbers of ionocytes and epidermal stem cells in zebrafish larval skin. These results highlight the role of TRPV4 modulating ion balance through the isotocin pathway (Liu et al., 2020b).

In the chum salmon (*Oncorhynchus keta*), TRPV4 expression pattern in the skin and gills were regulated according to changes in salinity and temperature. In particular, TRPV4 x1 and TRPV4 x2 mRNAs, which are TRPV4 variants derived through alternate splicing, were predominant in the gill and skin including at the lateral line. TRPV4 x1 expression were significantly increased with increase in salinity and temperature, whereas TRPV4 x2 mainly responded to temperature decrease. Overall, TRPV4 contributes to the regulation of hydromineral balance in the chum salmon, by varying its transcription expression and alternate splicing (Lee, Lee, & Kim, 2021).

TRPV4 has been linked to dermato-fibrotic disorders such as scleroderma where both skin keratinocytes and fibroblasts have been implicated. In terms of relevant signaling mechanisms in the skin in dermato-fibrotic disorders, a pro-fibrotic role of fibroblast-expressing TRPV4, as well as a pro-inflammatory role of TRPV4, have emerged (Abdullah et al., 2014; Akiyama et al., 2016a; Aubdool & Brain, 2011; Chen et al., 2016; Kim et al., 2016; Liu et al., 2013; Mollanazar, Smith, & Yosipovitch, 2016; Moore et al., 2013; Rahaman et al., 2014; Sharma et al., 2017). TRPV4 serves as a mechanosensitive Ca^{2+}-permeable channel and regulates matrix stiffness and TGFβ1-induced epithelial-mesenchymal transitions (EMT) (Sharma et al., 2019). TRPV4 was identified as a candidate plasma membrane mechanosensor that transmits matrix-sensing signals by translocation of yes-associated protein (YAP) and transcriptional co-activator with PDZ-binding motif (TAZ) in response to matrix stiffness and TGFβ1 in keratinocytes. TRPV4 deletion blocked both matrix stiffness-induced and TGFβ1-induced expression of YAP and TAZ proteins and was essential to long noncoding RNA expression (Sharma et al., 2019; Sharma, Ma, & Rahaman, 2020).

TRPV4 expression was found to be increased in burn scars with post-burn pruritus and the severity of the itch was correlated to the expression of TRPV4 (Yang et al., 2015). In another study, fibroblasts showed increased TRPV4 expression when stressed; their collagen synthesis was greatly attenuated after treatment with the selective TRPV4 inhibitor HC-067047 or when derived from *Trpv4* null mice (Rahman et al., 2016), The experiments suggest that this phenomenon is a more general mechanism and that TRPV4 might be critically involved in the pathogenesis of scleroderma and other fibrosing disorders (Goswami et al., 2016; Sharma et al., 2017).

4.2 Thermo-sensation

The epidermis senses a wide range of temperatures. Here, cells in the skin such as cutaneous nerve endings and keratinocytes contain temperature gated ion channels. TRPV4 was shown to be activated by warm temperatures (27–35 °C) in heterologous expression systems and in primary keratinocytes cultures (Chung, Lee, & Caterina, 2003; Guler et al., 2002; Tominaga, Liedtke, & Heller, 2007). TRPV4 is highly expressed in peripheral sensory free nerve endings and is located at cutaneous mechanosensory terminals, including Merkel cells, Meissner corpuscles, penicillate and intra-epidermal terminals, and in keratinocytes (Suzuki et al., 2003a). The role of TRPV4 in normal thermosensation in mammals is unclear as there has been contrasting results in the in vivo animal studies. Using TRPV4-deficient mice, no differences in escape latency from heat stimuli were observed in either radiant paw heating or the hot plate assays when compared to wild-type (WT) animals (Liedtke & Friedman, 2003b; Suzuki et al., 2003a). In contrast, in another study, TRPV4-deficient mice selected warmer floor temperatures in a selective gradient test, exhibited a strong preference for 34 °C, and had prolonged withdrawal latencies during acute tail heating than WT animals (Lee et al., 2005). Under hyperalgesic conditions, induced by subcutaneous injection of capsaicin or carrageenan, TRPV4-deficient mice showed longer escape latencies from a hot surface, relative to WT controls (Todaka et al., 2004). Further evidence that TRPV4 contributes to noxious heat sensation, mice with full *TrpV4* deletions or keratinocytes specific *TrpV4* deletions, were less sensitive to thermal stimuli after UVB induced hyperalgesia than control animals (Moore et al., 2013). These data suggest that TRPV4 plays a role in epidermal noxious thermosensation, a component of "forefront" signaling in the skin.

In the Japanese grass lizard, expression analysis of *Trpv4* mRNA revealed that its expression in tissues and organs is specifically controlled in cold environments and hibernation (Nagai et al., 2012). Although cold treatment and hibernation reduced TRPV4 expression in tissues, such as the brain, tongue, heart, lung, and muscle, levels of *Trpv4* mRNA in the skin remained unaffected after entering hibernation and cold treatment, suggesting that

TRPV4 in the skin may act as an environmental temperature sensor throughout the reptilian life cycle, including hibernation, perhaps maintaining critical epidermal integrity during hibernation (Nagai et al., 2012).

5. TRPV4 in cutaneous immune regulation

In addition to its role as a protective barrier, the skin is home to many immune cell types that protect against invading pathogens and tissue damage. One such cell type that expresses TRPV4 is the mast cell (Kim et al., 2010; Yang et al., 2007). Stimulation of mast cells (RBL-283 cells) either by irradiation using a low-power laser, or sheer-stress, resulted in histamine release, which was blocked by Ruthenium Red, which can also block TRPV4 (Yang et al., 2007; Yang, Chen, & Zhou, 2009). In a human mast cell (MC)-related cell line, HMC-1, the authors documented *TRPV4* mRNA expression and the response of HMC-1 cells to stimulation with the TRPV4- activator 4αPDD, which elicited a current and a Ca^{2+} signal. The role of TRPV4 in human MC was elucidated in more detail (Chen et al., 2017; Mascarenhas et al., 2017).

In rosacea, a facial skin inflammatory disorder, mast cells are exposed to high levels of the antibacterial peptide LL37, a proteolytic cleavage fragment of the innate immune precursor protein cathelicidin, which originates from neutrophils and macrophages. LL37 activates Mas-related G protein-coupled receptor member X2 (MRGPRX2) (Takahashi et al., 2018), a promiscuous G-protein-coupled receptor that can be activated by multiple pro-inflammatory mediators (Liu & Dong, 2015; McNeil et al., 2015; Meixiong et al., 2019; Navines-Ferrer et al., 2018; Tiwari et al., 2016). In a study using *TRPV4*-Crispr huMC and the TRPV4-selective inhibitor HC-067047, it was demonstrated that mast cell degranulation by either MRGPRX2-activating compound 48/80 or LL37 was TRPV4-dependent (Mascarenhas et al., 2017). Moreover, LL37 up-regulated Trpv4 mRNA, which was eliminated in *MRGPRX2*-Crispr huMC. These findings strongly suggest that MRGPRX2 signals upstream of TRPV4.

In addition to mast cells, TRPV4 is expressed in macrophages, also in the skin, and is suggested to be the link between mechanical forces and immune responses (Michalick & Kuebler, 2020). TRPV4 was found to be expressed in macrophages in various tissues such as lungs, joints, blood vessels, and skin (Hamanaka et al., 2010). In the skin, TRPV4 mediates substrate stiffness-induced macrophage polarization (Dutta, Goswami, & Rahaman, 2020). It was determined that fibrosis induced stiffer skin tissue and promoted M1 macrophage subtype in a TRPV4-dependent manner. *In vitro,* stiff matrix (50 kPa) alone increased the expression of macrophage M1 markers in a TRPV4-dependent manner, and this response was further augmented by the addition of soluble factors; which did not occur with soft matrix (1 kPa). To test the requirement for TRPV4 in M1 macrophage polarization spectrum in response to increased stiffness, gain-of-function assays were performed, where TRPV4 was reintroduced into TRPV4 knockout macrophages and this significantly upregulated the expression of M1 markers (Dutta et al., 2020).

6. TRPV4 in nociception

There are multiple lines of evidence that support a role of TRPV4 in nociception in the skin, either depending on TRPV4 in the keratinocytes, in sensory neurons and perhaps in other cell lineages. Pain behaviors linked to epidermal TRPV4 include: osmotically-evoked pain (Alessandri-Haber et al., 2003, 2005; Liedtke & Friedman, 2003a), pressure and acidic nociception (Suzuki et al., 2003b), and mechanical hyperalgesia in neuropathic (Alessandri-Haber et al., 2004; Chen, Yang, & Wang, 2011; Ding et al., 2010; Fan et al., 2021; García-Mesa et al., 2021) and inflammatory pain (Chen et al., 2013, 2014; Segond von Banchet et al., 2013). TRPV4 in sensory neurons in DRGs and TGs that innervate the skin can be sensitized by pro-inflammatory mediators, such as prostaglandin E2, activator of proteinase-receptor 2 (PAR-2), an integrator of proteolytic signaling in inflammation, leading to increased nociception to hypotonic, mild hypertonic stimuli or mechanical stimuli (Alessandri-Haber et al., 2006). It was found that TRPV4 is co-expressed with PAR2 by rat DRG neurons. Intra-plantar injection of PAR2 agonist caused mechanical hyperalgesia in mice and sensitized pain responses in a TRPV4-dependent manner (Grant et al., 2007).

TRPV4's role in facilitating and promoting inflammation and pain was also supported by observations in a mouse model of skin inflammation in which UVB radiation generated sunburn tissue damage (UV-burn) and associated pathological pain (Moore et al., 2013). Following UVB over-exposure to their hindpaws, mice with induced *Trpv4* deletions in keratinocytes and mice treated with topical TRPV4-inhibitors to their hindpaws were less sensitive to noxious thermal and mechanical stimuli versus control animals. These animals also showed dramatically reduced skin inflammation, evidenced by reduced macrophage recruitment and neutrophil infiltration and reduced secretion of inflammatory mediators such as IL6. These findings indicate that TRPV4-expressing keratinocytes of the hindpaw act as non-neural sensing and pain-generating cells, suggesting that activation of TRPV4 channels in skin keratinocytes, not in innervating sensory neurons, are sufficient to switch on neural pain networks and response-mechanisms. Exploring a possible underlying mechanism, it was determined that epidermal keratinocyte TRPV4 is essential for UVB-evoked skin tissue damage and increased expression of the pro-pain and pro-itch mediator endothelin-1 (ET-1) (Moore et al., 2013).

In regard to breast pain, TRPV4 was found be upregulated in the keratinocytes of painful breast tissue (Gopinath et al., 2005).

7. TRPV4 in itch

Itch is an irritating sensation of the skin that leads to a scratch reflex (Dhand & Aminoff, 2014). In humans, TRPV4 expression was found to be elevated in: burn scars of patients with pruritus (Yang et al., 2015); the skin biopsies of patients with chronic idiopathic pruritus(CIP) (Luo et al., 2018); and pruritic lesional skin of atopic dermatitis and psoriasis patients (Nattkemper et al., 2018; Yan et al., 2021).

The role of TRPV4 in itch has been demonstrated by a number of labs, some with conflicting results (Moore et al., 2018). Using *TRPV4* knockout mice, Akiyama et al. demonstrated the intradermal injection of serotonin, histamine, the PAR2 agonist SLIGRL, and the non-histaminergic chloroquine to assess for scratching behavior (Akiyama et al., 2016b). Serotonin induced less itching compared to the wild type mice, chloroquine enhanced scratching in the *TRPV4* knockout mice, whereas no difference was observed between wild-type and *TRPV4* knockout mice for the histamine and SLIGRL injected animals. On the other hand, other studies found that mice lacking TRPV4 had reduced histamine induced itch compared to wild type animals, whereas no difference was observed for chloroquine-induced itch (Chen et al., 2016; Kim et al., 2016).

Chen et al. took these studies further with the use of the keratinocyte specific *TRPV4* conditional knockout (cKO) mice to assess the roles of TRPV4 in the action of the histaminergic pruritogens including histamine, compound 48/40, endothelin-1, and chloroquine. Endothelin, a known pruritogen, was previously shown to be secreted in a TRPV4-dependent manner from epidermal keratinocytes (Moore et al., 2013). In the skin-specific *TRPV4* cKO animals, scratching behaviors were significantly attenuated in all histaminergic pruritogens tested but not chloroquine, suggesting a role for keratinocytic TRPV4 in histaminergic itch (Chen et al., 2016). Direct activation of TRPV4 with its agonist GSK1016790A (GSK101) induced scratching bouts and topical treatment with a TRPV4 antagonist, GSK205 was effective in attenuating histaminergic pruritogen induced scratching (Chen et al., 2016). Furthermore, it was demonstrated that histaminergic pruritus involving TRPV4 channels is signaled via the extracellular signal-regulated kinase (ERK) pathway as inhibition of mitogen-activated protein kinase (MEK), upstream of ERK, using the selective inhibitor U0126, as topical treatment reduced scratching behavior (Chen et al., 2016). In vitro, histamine, 48/80, and ET-1 produced Ca^{2+}-transients in cultured primary keratinocytes that were TRPV4-dependent, and Ca^{2+} influx via TRPV4 increased phosphorylation of extracellular signal-regulated kinase (p-ERK) in keratinocytes (Chen et al., 2016). More recently the specific contribution of TRPV4 in sensory neurons to itch was investigated by employing a sensory neuron *TRPV4* conditional knockout mouse model. It was determined that for acute itch, cKO of *TRPV4* in sensory neurons significantly attenuated scratching behaviors evoked by histamine, 48/40 and 5-HT, but not SLIGRL and chloroquine, in mice (Zhang et al., 2022). Together, these studies suggest that TRPV4 in the skin, in both keratinocytes and sensory neurons, regulates histaminergic itch.

TRPV4 has been implicated in itch caused by cholestatic liver disease. It was demonstrated that lysophosphatyl choline (LPC) is elevated in the sera of patients with cholestatic itch and in the mouse model of cholestatic itch. LPC acts as a strong pruritogen in mice and non-human primates (Chen et al., 2021). Chen et al. demonstrated that LPC caused itch by binding directly to TPRV4 channels on keratinocytes, resulting in ERK activation and the keratinocytes' vesicular extrusion of miRNA-146, which, in turn, activates TRPV1-positive proprioceptor sensory neurons (Chen et al., 2021).

Crotamiton (*N*-ethyl-o-crotonotoluidide) has been used as an anti-itch agent in humans over many decades (Kittaka, Yamanoi, & Tominaga, 2017). Using electrophysiological recordings, it was demonstrated that crotamiton inhibited TRPV4 channels, followed by large currents after crotamiton washouts. In mice, crotamiton significantly reduced the TRPV4 agonist GSK1016790A-induced scratching behaviors (Kittaka et al., 2017). The wealth of experimental evidence above highlights TRPV4 channel as an important regulator in itch and therefore should be considered as a new target for itch relief.

8. TRPV4 in skin cancer

TRPV4 channels are functionally expressed in human melanoma cell lines and pharmacological activation of TRPV4 in human melanoma cells was shown to cause severe cellular disarrangement, necrosis, and apoptosis (Olivan-Viguera et al., 2018). Another report also demonstrated that both melanocytes and melanoma cells ectopically express TRPV4. Similarly, activation of TRPV4 resulted in the decline of cell viability for melanoma A2058 and A375 cells and enhanced apoptosis of A375 cells. These studies revealed that TRPV4 mediates melanoma cell death via channel activation and calcium influx. They also characterized a new mechanism, whereby TRPV4 ion channel regulated the AKT pathway to drive the antitumor process (Zheng et al., 2019).

Recently it was reported that activation of TRPV4-induced exocytosis occurred in both heterologous expression systems and melanoma A375 cells (Li et al., 2022). Application of TRPV4 via its specific agonists resulted in vesicle priming from endoplasmic reticulum (ER) followed by morphological changes of mitochondrial crista, which lead to cell ferroptosis. A signaling mechanism underlying stimulus-triggered exocytosis was revealed, whereby when TRPV4 was activated, the resulting calcium entry enhanced protein folding and vesicle trafficking to promote exocytosis in melanoma (Li et al., 2022).

In non-melanoma skin cancer, TRPV4 was reduced in premalignant lesions when compared to the level of TRPV4 found in healthy or inflamed skin keratinocytes (Fusi et al., 2014). It was shown that TRPV4 was markedly downregulated in both the premalignant lesions of NMSC such as solar keratosis (SK) and Bowen's disease (BD), and in squamous cell carcinoma (SCC) and basal cell carcinoma (BCC). In cultured keratinocytes, cell exposure to a variety of proinflammatory mediators, including IL-8, which is released from keratinocytes upon TRPV4 stimulation, led to a reduction of TRPV4 expression (Benemei et al., 2014; Liedtke et al., 2014). It could be possible that the expression of TRPV4 differs between patients, cancer types, and sub-types and micro- and macroenvironments, and therefore larger and more targeted studies are needed to determine the role of TRPV4 in skin cancers.

9. Conclusions

In summary, TRPV4 is expressed in many cell types in the skin and plays critical roles in skin homeostasis such as barrier maintenance, hair growth, immune regulation, nociception and itch. Since TRPV4 functions in pathological states such as chronic pain, itch and skin cancer, we should not lose sight of TRPV4 as a therapeutic target for such diseases.

Acknowledgments

Research reported in this publication was supported by the National Center for Advancing Translational Sciences of the National Institutes of Health under Award Number 1KL2TR002554 and NINDS award number K01-NS121195-01A1. The content is solely the responsibility of the author and does not necessarily represent the official views of the National Institutes of Health.

I would like to thank Lily Orta for her work on Fig. 1.

References

Abdullah, H., et al. (2014). Rhinovirus upregulates transient receptor potential channels in a human neuronal cell line: Implications for respiratory virus-induced cough reflex sensitivity. *Thorax*, *69*(1), 46–54.

Akazawa, Y., et al. (2013). Activation of TRPV4 strengthens the tight-junction barrier in human epidermal keratinocytes. *Skin Pharmacology and Physiology*, *26*(1), 15–21.

Akiyama, T., et al. (2016a). Involvement of TRPV4 in Serotonin-Evoked Scratching. *The Journal of Investigative Dermatology*, *136*(1), 154–160.

Akiyama, T., et al. (2016b). Involvement of TRPV4 in serotonin-evoked scratching. *The Journal of Investigative Dermatology*, *136*(1), 154–160.

Alessandri-Haber, N., et al. (2003). Hypotonicity induces TRPV4-mediated nociception in rat. *Neuron*, *39*(3), 497–511.

Alessandri-Haber, N., et al. (2004). Transient receptor potential vanilloid 4 is essential in chemotherapy-induced neuropathic pain in the rat. *The Journal of Neuroscience*, *24*(18), 4444–4452.

Alessandri-Haber, N., et al. (2005). TRPV4 mediates pain-related behavior induced by mild hypertonic stimuli in the presence of inflammatory mediator. *Pain*, *118*(1-2), 70–79.

Alessandri-Haber, N., et al. (2006). A transient receptor potential vanilloid 4-dependent mechanism of hyperalgesia is engaged by concerted action of inflammatory mediators. *The Journal of Neuroscience*, *26*(14), 3864–3874.

Aubdool, A. A., & Brain, S. D. (2011). Neurovascular aspects of skin neurogenic inflammation. *The Journal of Investigative Dermatology. Symposium Proceedings*, *15*(1), 33–39.

Benemei, S., et al. (2014). The TRPA1 channel in migraine mechanism and treatment. *British Journal of Pharmacology*, *171*(10), 2552–2567.

Boudaka, A., Al-Yazeedi, M., & Al-Lawati, I. (2020). Role of transient receptor potential vanilloid 4 channel in skin physiology and pathology. *Sultan Qaboos University Medical Journal*, *20*(2), e138–e146.

Caterina, M. J., & Pang, Z. (2016). TRP channels in skin biology and pathophysiology. *Pharmaceuticals (Basel)*, *9*(4).

Chen, Y., Yang, C., & Wang, Z. J. (2011). Proteinase-activated receptor 2 sensitizes transient receptor potential vanilloid 1, transient receptor potential vanilloid 4, and transient receptor potential ankyrin 1 in paclitaxel-induced neuropathic pain. *Neuroscience*, *193*, 440–451.

Chen, Y., et al. (2013). Temporomandibular joint pain: A critical role for Trpv4 in the trigeminal ganglion. *Pain*, *154*(8), 1295–1304.

Chen, Y., et al. (2014). TRPV4 is necessary for trigeminal irritant pain and functions as a cellular formalin receptor. *Pain*, *155*(12), 2662–2672.

Chen, Y., et al. (2016). Transient receptor potential vanilloid 4 ion channel functions as a pruriceptor in epidermal keratinocytes to evoke histaminergic itch. *The Journal of Biological Chemistry*, *291*(19), 10252–10262.

Chen, Y., et al. (2017). TRPV4 moves toward center-fold in rosacea pathogenesis. *The Journal of Investigative Dermatology*, *137*(4), 801–804.

Chen, Y., et al. (2021). Epithelia-sensory neuron cross talk underlies cholestatic itch induced by lysophosphatidylcholine. *Gastroenterology*, *161*(1), 301–317.e16.

Chung, M. K., Lee, H., & Caterina, M. J. (2003). Warm temperatures activate TRPV4 in mouse 308 keratinocytes. *The Journal of Biological Chemistry*, *278*(34), 32037–32046.

Dhand, A., & Aminoff, M. J. (2014). The neurology of itch. *Brain*, *137*(Pt 2), 313–322.

Ding, X. L., et al. (2010). Involvement of TRPV4-NO-cGMP-PKG pathways in the development of thermal hyperalgesia following chronic compression of the dorsal root ganglion in rats. *Behavioural Brain Research*, *208*(1), 194–201.

Dutta, B., Goswami, R., & Rahaman, S. O. (2020). TRPV4 plays a role in matrix stiffness-induced macrophage polarization. *Frontiers in Immunology*, *11*, 570195.

Fan, X., et al. (2021). Role of TRPV4-P2X7 pathway in neuropathic pain in rats with chronic compression of the dorsal root ganglion. *Neurochemical Research*, *46*(8), 2143–2153.

Fuchs, E. (2009). Finding one's niche in the skin. *Cell Stem Cell*, *4*(6), 499–502.

Fusi, C., et al. (2014). Transient receptor potential vanilloid 4 (TRPV4) is downregulated in keratinocytes in human non-melanoma skin cancer. *The Journal of Investigative Dermatology*, *134*(9), 2408–2417.

Galindo-Villegas, J., et al. (2016). TRPV4-mediated detection of hyposmotic stress by skin keratinocytes activates developmental immunity. *Journal of Immunology*, *196*(2), 738–749.

García-Mesa, Y., et al. (2021). Involvement of cutaneous sensory corpuscles in non-painful and painful diabetic neuropathy. *Journal of Clinical Medicine*, *10*(19).

Gopinath, P., et al. (2005). Increased capsaicin receptor TRPV1 in skin nerve fibres and related vanilloid receptors TRPV3 and TRPV4 in keratinocytes in human breast pain. *BMC Womens Health*, *5*(1), 2.

Goswami, R., et al. (2016). TRPV4 ion channel is associated with scleroderma. *The Journal of Investigative Dermatology*.

Grant, A. D., et al. (2007). Protease-activated receptor 2 sensitizes the transient receptor potential vanilloid 4 ion channel to cause mechanical hyperalgesia in mice. *The Journal of Physiology*, *578*(Pt 3), 715–733.

Guler, A. D., et al. (2002). Heat-evoked activation of the ion channel, TRPV4. *The Journal of Neuroscience*, *22*(15), 6408–6414.

Hamanaka, K., et al. (2010). TRPV4 channels augment macrophage activation and ventilator-induced lung injury. *American Journal of Physiology. Lung Cellular and Molecular Physiology*, *299*(3), L353–L362.

Kida, N., et al. (2012). Importance of transient receptor potential vanilloid 4 (TRPV4) in epidermal barrier function in human skin keratinocytes. *Pflügers Archiv*, *463*(5), 715–725.

Kim, K. S., et al. (2010). Functional expression of TRPV4 cation channels in human mast cell line (HMC-1). *Korean Journal of Physiology & Pharmacology*, *14*(6), 419–425.

Kim, S., et al. (2016). Facilitation of TRPV4 by TRPV1 is required for itch transmission in some sensory neuron populations. *Science Signaling*, *9*(437), ra71.
Kittaka, H., Yamanoi, Y., & Tominaga, M. (2017). Transient receptor potential vanilloid 4 (TRPV4) channel as a target of crotamiton and its bimodal effects. *Pflügers Archiv*, *469*(10), 1313–1323.
Landouré, G., et al. (2010). Mutations in TRPV4 cause charcot-marie-tooth disease type 2C. *Nature Genetics*, *42*(2), 170–174.
Lee, H. J., Lee, S. Y., & Kim, Y. K. (2021). Molecular characterization of transient receptor potential vanilloid 4 (TRPV4) gene transcript variant mRNA of chum salmon Oncorhynchus keta in response to salinity or temperature changes. *Gene*, *795*, 145779.
Lee, H., et al. (2005). Altered thermal selection behavior in mice lacking transient receptor potential vanilloid 4. *The Journal of Neuroscience*, *25*(5), 1304–1310.
Li, M., et al. (2022). Activation of TRPV4 induces exocytosis and ferroptosis in human melanoma cells. *International Journal of Molecular Sciences*, *23*(8).
Liedtke, W., & Friedman, J. M. (2003a). Abnormal osmotic regulation in trpv4-/- mice. *Proceedings of the National Academy of Sciences of the United States of America*, *100*(23), 13698–13703.
Liedtke, W., & Friedman, J. M. (2003b). Abnormal osmotic regulation in trpv4-/- mice. *Proceedings of the National Academy of Sciences of the United States of America*, *100*, 13698–13703.
Liedtke, W., et al. (2000). Vanilloid receptor-related osmotically activated channel (VR-OAC), a candidate vertebrate osmoreceptor. *Cell*, *103*(3), 525–535.
Liedtke, W., et al. (2003). Mammalian TRPV4 (VR-OAC) directs behavioral responses to osmotic and mechanical stimuli in C. elegans. *Proceedings of the National Academy of Sciences of the United States of America*, *100*, 14531–14536.
Liedtke, W., et al. (2014). Keratinocyte growth regulation TRP-ed up over downregulated TRPV4? *The Journal of Investigative Dermatology*, *134*(9), 2310–2312.
Liu, Q., & Dong, X. (2015). The role of the Mrgpr receptor family in itch. *Handbook of Experimental Pharmacology*, *226*, 71–88.
Liu, B., et al. (2013). TRPA1 controls inflammation and pruritogen responses in allergic contact dermatitis. *The FASEB Journal*, *27*(9), 3549–3563.
Liu, Y., et al. (2020a). Novel TRPV4 mutation in a large Chinese family with congenital distal spinal muscular atrophy, skeletal dysplasia and scaly skin. *Journal of the Neurological Sciences*, *419*, 117153.
Liu, S. T., et al. (2020b). Transient receptor potential vanilloid 4 modulates ion balance through the isotocin pathway in zebrafish (Danio rerio). *American Journal of Physiology. Regulatory, Integrative and Comparative Physiology*, *318*(4), R751–r759.
Luo, J., et al. (2018). Transient receptor potential vanilloid 4-expressing macrophages and keratinocytes contribute differentially to allergic and nonallergic chronic itch. *The Journal of Allergy and Clinical Immunology*, *141*(2), 608–619.e7.
Manganelli, F., et al. (2015). Charcot-marie-tooth disease: New insights from skin biopsy. *Neurology*, *85*(14), 1202–1208.
Mascarenhas, N. L., et al. (2017). TRPV4 mediates mast cell activation in cathelicidin-induced rosacea inflammation. *The Journal of Investigative Dermatology*, *137*(4), 972–975.
McCray, B. A., et al. (1993). Autosomal dominant TRPV4 disorders. In M. P. Adam, et al. (Eds.), *GeneReviews(®)*. Seattle, WA: University of Washington, Seattle Copyright © 1993-2022, University of Washington, Seattle. GeneReviews is a registered trademark of the University of Washington, Seattle. All rights reserved.
McNeil, B. D., et al. (2015). Identification of a mast-cell-specific receptor crucial for pseudo-allergic drug reactions. *Nature*, *519*(7542), 237–241.

Meixiong, J., et al. (2019). Activation of mast-cell-expressed mas-related G-protein-coupled receptors drives non-histaminergic itch. *Immunity*.

Michalick, L., & Kuebler, W. M. (2020). TRPV4-A missing link between mechanosensation and immunity. *Frontiers in Immunology, 11*, 413.

Mollanazar, N. K., Smith, P. K., & Yosipovitch, G. (2016). Mediators of chronic pruritus in atopic dermatitis: Getting the itch out? *Clinical Reviews in Allergy and Immunology, 51*(3), 263–292.

Moore, C., & Liedtke, W. B. (2017). Frontiers in neuroscience osmomechanical-sensitive TRPV channels in mammals. In T. L. R. Emir (Ed.), *Neurobiology of TRP channels* (pp. 85–94). Boca Raton, FL: CRC Press/Taylor & Francis (c) 2018 by Taylor & Francis Group, LLC.

Moore, C., et al. (2013). UVB radiation generates sunburn pain and affects skin by activating epidermal TRPV4 ion channels and triggering endothelin-1 signaling. *Proceedings of the National Academy of Sciences of the United States of America, 110*(34), E3225–E3234.

Moore, C., et al. (2018). Regulation of pain and itch by TRP channels. *Neuroscience Bulletin, 34*(1), 120–142.

Nagai, K., et al. (2012). Structure and hibernation-associated expression of the transient receptor potential vanilloid 4 channel (TRPV4) mRNA in the Japanese grass lizard (Takydromus tachydromoides). *Zoological Science, 29*(3), 185–190.

Nattkemper, L. A., et al. (2018). The genetics of chronic itch: Gene expression in the skin of patients with atopic dermatitis and psoriasis with severe itch. *The Journal of Investigative Dermatology, 138*(6), 1311–1317.

Navines-Ferrer, A., et al. (2018). MRGPRX2-mediated mast cell response to drugs used in perioperative procedures and anaesthesia. *Scientific Reports, 8*(1), 11628.

Olivan-Viguera, A., et al. (2018). Pharmacological activation of TRPV4 produces immediate cell damage and induction of apoptosis in human melanoma cells and HaCaT keratinocytes. *PLoS One, 13*(1), e0190307.

Patapoutian, A. (2005). TRP channels and thermosensation. *Chemical Senses, 30*(Suppl. 1), i193–i194.

Rahaman, S. O., et al. (2014). TRPV4 mediates myofibroblast differentiation and pulmonary fibrosis in mice. *The Journal of Clinical Investigation*.

Rahman, M., et al. (2016). Electrophysiological characterization of voltage-dependent calcium currents and TRPV4 currents in human pulmonary fibroblasts. *American Journal of Physiology. Lung Cellular and Molecular Physiology, 310*(7), L603–L614.

Roosterman, D., et al. (2006). Neuronal control of skin function: The skin as a neuroimmunoendocrine organ. *Physiological Reviews, 86*(4), 1309–1379.

Segond von Banchet, G., et al. (2013). Neuronal IL-17 receptor upregulates TRPV4 but not TRPV1 receptors in DRG neurons and mediates mechanical but not thermal hyperalgesia. *Molecular and Cellular Neurosciences, 52*, 152–160.

Sharma, S., Ma, L., & Rahaman, S. O. (2020). Role of TRPV4 in matrix stiffness-induced expression of EMT-specific LncRNA. *Molecular and Cellular Biochemistry, 474*(1-2), 189–197.

Sharma, S., et al. (2017). TRPV4 ion channel is a novel regulator of dermal myofibroblast differentiation. *American Journal of Physiology-Cell Physiology*. ajpcell 00187 2016.

Sharma, S., et al. (2019). TRPV4 regulates matrix stiffness and TGFβ1-induced epithelial-mesenchymal transition. *Journal of Cellular and Molecular Medicine, 23*(2), 761–774.

Sokabe, T., et al. (2010). The TRPV4 channel contributes to intercellular junction formation in keratinocytes. *The Journal of Biological Chemistry, 285*(24), 18749–18758.

Suzuki, M., et al. (2003a). Localization of mechanosensitive channel TRPV4 in mouse skin. *Neuroscience Letters, 353*(3), 189–192.

Suzuki, M., et al. (2003b). Impaired pressure sensation in mice lacking TRPV4. *The Journal of Biological Chemistry*, *278*(25), 22664–22668.
Szabó, I. L., et al. (2019). TRPV4 is expressed in human hair follicles and inhibits hair growth in vitro. *The Journal of Investigative Dermatology*, *139*(6), 1385–1388.
Takahashi, T., et al. (2018). Cathelicidin promotes inflammation by enabling binding of self-RNA to cell surface scavenger receptors. *Scientific Reports*, *8*(1), 4032.
Tiwari, V., et al. (2016). Mas-related G protein-coupled receptors offer potential new targets for pain therapy. *Advances in Experimental Medicine and Biology*, *904*, 87–103.
Todaka, H., et al. (2004). Warm temperature-sensitive transient receptor potential vanilloid 4 (TRPV4) plays an essential role in thermal hyperalgesia. *The Journal of Biological Chemistry*, *279*(34), 35133–35138.
Tominaga, M. (2007). Frontiers in neuroscience the role of TRP channels in thermosensation. In W. B. Liedtke, & S. Heller (Eds.), *TRP ion channel function in sensory transduction and cellular signaling cascades*. Boca Raton, FL: CRC Press/Taylor & Francis Copyright © 2007, Taylor & Francis Group, LLC.
Tominaga, M., & Caterina, M. J. (2004). Thermosensation and pain. *Journal of Neurobiology*, *61*(1), 3–12.
Yan, J., et al. (2021). Cimifugin relieves pruritus in psoriasis by inhibiting TRPV4. *Cell Calcium*, *97*, 102429.
Yang, W., Chen, J., & Zhou, L. (2009). Effects of shear stress on intracellular calcium change and histamine release in rat basophilic leukemia (RBL-2H3) cells. *Journal of Environmental Pathology, Toxicology and Oncology*, *28*(3), 223–230.
Yang, W. Z., et al. (2007). Effects of low power laser irradiation on intracellular calcium and histamine release in RBL-2H3 mast cells. *Photochemistry and Photobiology*, *83*(4), 979–984.
Yang, Y. S., et al. (2015). Increased expression of three types of transient receptor potential channels (TRPA1, TRPV4 and TRPV3) in burn scars with post-burn pruritus. *Acta Dermato-Venereologica*, *95*(1), 20–24.
Yang, P., et al. (2020). Transient stimulation of TRPV4-expressing keratinocytes promotes hair follicle regeneration in mice. *British Journal of Pharmacology*, *177*(18), 4181–4192.
Zheng, J., et al. (2019). Mechanism for regulation of melanoma cell death via activation of thermo-TRPV4 and TRPV2. *Journal of Oncology*, *2019*, 7362875.
Zhang Q, Dias F, Fang Q, Henry G, Wang Z, Suttle A, et al. (2022). Involvement of sensory neurone-TRPV4 in acute and chronic itch behaviours. Acta Dermato-Venereologica 102: adv00651. doi:10.2340/actadv.v102.1621. PMID: 35146528; PMCID: PMC9121103.

CHAPTER SEVEN

Function and regulation of thermosensitive ion channel TRPV4 in the immune system

Tusar Kanta Acharya[a,b], Ram Prasad Sahu[a,b], Satish Kumar[a], Shamit Kumar[a,b], Tejas Pravin Rokade[a,b], Ranabir Chakraborty[a], Nishant Kumar Dubey[a,b], Deep Shikha[a,b], Saurabh Chawla[a], and Chandan Goswami[a,b,*]

[a]School of Biological Sciences, National Institute of Science Education and Research (NISER)-Bhubaneswar, Khurda, Odisha, India
[b]Homi Bhabha National Institute, Mumbai, India
*Corresponding author: e-mail address: chandan@niser.ac.in

Contents

Abstract

Transient receptor potential vanilloid sub-type 4 (TRPV4) is a six transmembrane protein that acts as a non-selective Ca^{2+} channel. Notably, TRPV4 is present in almost all animals, from lower eukaryotes to humans and is expressed in diverse tissue and cell

ISSN 1063-5823
https://doi.org/10.1016/bs.ctm.2022.07.002

types. Accordingly, TRPV4 is endogenously expressed in several types of immune cells that represent both innate and adaptive immune systems of higher organism. TRPV4 is known to be activated by physiological temperature, suggesting that it acts as a molecular temperature sensor and thus plays a key role in temperature-dependent immune activation. It is also activated by diverse endogenous ligands, lipid metabolites, physical and mechanical stimuli. Both expression and function of TRPV4 in various immune cells, including T cells and macrophages, are also modulated by multiple pro- and anti-inflammatory compounds. The results from several laboratories suggest that TRPV4 is involved in the immune activation, a phenomenon with evolutionary significance. Because of its diverse engagement in the neuronal and immune systems, TRPV4 is a potential therapeutic target for several immune-related disorders.

Abbreviations

ConA concanavalin-A
IFN interferon
IL interleukin
LPS lipopolysaccharide
MS multiple sclerosis
SOCE store operated calcium entry
TRPV transient receptor potential vanilloid receptor
TNF tumor necrosis factor
4α-PDD 4α-phorbol 12, 13-didecanoate

1. Introduction

Cellular functions are intimately linked with the ionic balance and homeostasis of different ions within the cytosol and other subcellular organelles. Ion channels, pumps, carrier proteins and several metal-binding proteins typically regulate this ionic homeostasis. In this context, ion channels are specialized proteins present on the cell surface that allow influx of charged ions across the plasma membrane. Ion movement through the ion channel occurs through the diffusion of ions along the concentration gradient across the membrane. Most of the ion channels form a specific "pore" structure in their open conformation, which is formed by the alpha-helices present in the lipid bilayer (Fischbarg et al., 1994). Ion channels are typically classified based on their gating properties and the ions passing through the channel pore, the number of pores they possess (one-pore or two-pore channels), number of transmembrane segments (includes, 2TM, 6TM, 12TM, 24TM ion channels) they have and their preferred location where they become active. For example, "voltage-gated" ion channels are typically activated by changes in the voltage across the cell membrane. Similarly, "ligand-gated" ion channels are activated by their interaction with the ligands. However, in recent times, a third group of ion channels has been identified which is fundamentally different from classical ligand-gated or voltage-gated channels. These members belong to Transient Receptor Potential (TRP) Channels family.

Recent reports endorse TRP channels as very important for physiology and disease context. TRP channels are a special group of ion channels present on the plasma membrane of animal cells. TRP channels are present in lower eukaryotes too, including fungi, but are generally missing in plants. The first TRP gene was identified in a *Drosophila* mutant with visual abnormalities (Minke, 2010). Subsequently, many TRP channels have been identified and each of these channels displays unique behaviors. TRP channels are "non-selective" in nature as activation of these channels causes an influx of Ca^{2+}, K^{+}, Mg^{+}, Zn^{2+}, Co^{2+}, Sr^{2+}, Fe^{+}, etc. TRP channels are divided into distinct groups or families according to their characteristics and behavior- TRPC (canonical), TRPP (polycystin), TRPM (melastatin), TRPML (mucolipin), TRPA (ankyrin), TRPV (vanilloid), and TRPN (NOMPC-like) (Blair et al., 2019). The molecular uniqueness, expression, function, and regulation of these channels vary in a species-specific and also tissue-dependent manner. One key feature of some TRP channels is their ability to be activated by temperature. Among all the thermo-gated TRP channels, TRPV4 is important as it gets activated at physiological temperatures ($\geq$27 °C). Other TRPV channels, namely TRPV1 ($\geq$42 °C) and TRPV2 ($\geq$52 °C), are activated at higher than physiological temperatures. As temperature increase is a key factor for triggering immune activation, only a few thermosensitive ion channels are important in this context. TRPV4 is one of these few channels that are important for temperature-dependent immune activation, especially at the physiological temperature.

2. Ion channels in immune cells

The function of immune cells is intimately related to the presence and function of diverse ion channels and levels of cations and anions in the cytosol. This is also supported by the fact that specific point mutations in key ion channels (alternatively termed as channelopathies) lead to the development of chronic immune disorders and/or immune deficiencies

(Maus et al., 2015; Vaeth & Feske, 2018). Immune disorders due to mutations in ion channels also cause disorders in other related tissues causing pathophysiological problems, like abnormal neuro-immune cross-talks (Vincent, Lang, & Kleopa, 2006). In multiple organ systems, the roles of ion channels in regulating key immune and inflammatory responses have been elucidated. Store-operated Ca^{2+}-entry (SOCE) into immune cells is a major pathway for the generation of an immune response (Hogan, Lewis, & Rao, 2009). Mutations in STIM1 (Ca^{2+}-sensor) and/or ORA1 (Ca^{2+} channel) causes severe immunodeficiency phenotypes in humans (Feske, 2011). In general, a higher level of intracellular Ca^{2+} is required for proper immune functions, including T cell activation, suggesting the importance of Ca^{2+}-channels in the immune system. Other than Ca^{2+}, divalent cations, namely Mg^{2+} and Zn^{2+} are also crucial as these divalent ions play important roles as secondary messengers to regulate intracellular signaling. Ca^{2+} and K^{+} ions also regulate the membrane potential of various immune cells. Mg^{2+} is crucial for immune cells, especially in the context of proliferation and viability. In immune cells, most of the Mg^{2+} ions are conducted by TRPM7. Zn^{2+} transportation is also controlled by transporter protein, i.e. by ZIP-transporters and Zir-like proteins. These ZIP proteins regulate Zn^{2+}-influx, relevant in T cell and mast cells. Lymphocytes (T cells and B cells) regulate cell proliferation, cell motility, gene expression, and other homeostatic functions through Ca^{2+}, Mg^{2+}, and Zn^{2+}. Other than these divalent cations, immune cells also have HVCH (hydrogen voltage-gated channel), H_v1 (voltage-gated proton channel), K_{2P} (potassium channel), and $GABA_A$ receptors (ligand-gated chloride channel), all of which are important for ion homeostasis (Feske, Wulff, & Skolnik, 2015). Ionotropic P_2X receptors are another group of ion channels (ligand-gated cation channels) in immune cells that enable the response to extracellular ATP. Accordingly, ion channel members, namely P_2RX_1, P_2RX_2, P_2RX_4, and P_2RX_7 are expressed in immune cells involved in innate and adaptive immune response (Feske et al., 2015).

3. TRPV4 gene, variants and mutations

In humans, the TRPV4 gene is composed of 15 exons and is confined on chromosomes 12q23 to q24.1 (ENSEMBL: ENSG00000111199). So far, five splice variants of TRPV4 have been reported, named TRPV4A-E.

Based on structural similarity of the variants, TRPV4-A and TRPV4-D belong to group-I, while TRPV4-B, TRPV4-C and TRPV4-E belong to group II. Group I TRPV4 variants contain canonical sequence, TRPV4-A lacks exon 7, and TRPV4-D contains short deletion in exon 2 (residues 27-61). Both group-I variants TRPV4-A and TRPV4-D are present on the cell membrane and perform identical channel and regulatory roles. Group-II variants (TRPV4-B, C and E) show deletion of N-terminal, TRPV4-B & C lack exon 7, 5 respectively, while TRPV4-E lacks exon 5 and 7. These deletions in the Group-II variants result in debilitated oligomerization and intracellular retention of channel products, which can be found in the ER (Endoplasmic reticulum) (Arniges, Fernández-Fernández, Albrecht, Schaefer, & Valverde, 2006; Everaerts, Nilius, & Owsianik, 2010). At the genome level, TRPV4 has numerous missense variations and not all missense variations cause drastic phenotypic changes or lethality (Ghosh, Kaur, Kumar, & Goswami, 2016). However, there are several "gain-of-function" point mutations of TRPV4 that induce severe pathophysiological disorders, causing neuronal, and muscular-skeletal diseases (Das & Goswami, 2019; Verma, Kumar, & Goswami, 2010).

At the protein level, all TRPV4 mammalian homologues show significant sequence identity (~98%) and almost similar length. TRPV4 contains ~871 amino acids with amino and carboxyl termini facing the cytoplasm and a core region comprising six membrane-spanning α-helices (TM1-6) and a linking putative cationic pore-loop located between TM5 and TM6. This topography matches with voltage-gated ion channels (Liedtke et al., 2000; Strotmann, Harteneck, Nunnenmacher, Schultz, & Plant, 2000). Crystal structure with full-length TRPV4 sequence is still awaited, although a recent study used cryo-EM (Electron microscopy) at a resolution of 3.5 nm to investigate TRPV4 structure. At the N-terminal cytoplasmic domain, a stretch of six Ankyrin-repeat domains (ANK, approximately 33-amino acid motif) is present, which is mainly involved in protein-protein interactions. At the amino terminus, PRD (Proline-rich domain) near the ankyrin repeat is also present, which is thought to help the mechanosensitive characteristics of the TRPV4 channel (Shigematsu, Sokabe, Danev, Tominaga, & Nagayama, 2010). In *C elegans*, the behavioral defects due to mutation in *osm-9* gene can be rescued by hTRPV4, suggesting that in spite of the differences in sequence between invertebrates to vertebrates, TRPV4 channels share structure-function relationships across species to a large extent (Liedtke, Tobin, Bargmann, & Friedman, 2003).

4. TRPV4 as a functional tetrameric protein

At the protein level, TRPV4 forms a functional tetramer (Deng et al., 2018). This TRPV4 tetramer acts as a non-selective polymodal cation channel, which was initially discovered as an osmo-mechanosensitive channel (Liedtke et al., 2000; Strotmann et al., 2000). It can also be activated by an array of stimuli like cell swelling, temperature, endogenous ligands and a large number of synthetic/exogenous activators. Endogenous activators include specific products of enzymatic reactions, namely PLA2 (Phospholipase A2), arachidonic acid and its non-oxidative metabolite anandamide, as these compounds are known to stimulate TRPV4 activity in HEK293 and retinal Müller cells (Watanabe et al., 2003). So far, many TRPV4 exogenous ligands have been discovered as activators (Table 1). For example, bis-andrographolide A (BAA, extracted from plant *Andrographis paniculate*); other natural compounds include flavone apigenin (4',5,7-trihydroxyflavone), and citric acid, phorbol esters, i.e. 4α-phorbol 12,13-didecanoate (4α-PDD), and phorbol 12-myristate 13-acetate and 4α-phorbol 12,13-dihexanoate (4α-PDH), all act as activators (Watanabe et al., 2002). Other than natural compounds, some synthetic ligands have also been discovered as potent TRPV4 modulators, including RN-1747 (acts as inhibitor) and GSK1016790A (acts as activator) (Ma et al., 2012; Thorneloe et al., 2008; Vincent et al., 2009; Vriens, Owsianik, Janssens, Voets, & Nilius, 2007). For blocking the activity of TRPV4, ruthenium red is frequently used as a non-specific TRPV4 inhibiting agent (Voets et al., 2002). TRPV4 is also activated by an array of biochemical stimuli that are exogenous, or endogenous in nature. TRPV4 also interacts with lipid and arachidonic acid metabolites generated endogenously (Watanabe et al., 2003). TRPV4 is activated by biomechanical and physical stimuli, like shear stress, stretch, and cell membrane swelling, mechanical deformation, osmotic stimuli, pH and heat. Due to these properties, TRPV4 is associated with numerous physiological functions i.e. with osmolarity-sensing in kidneys, sheer-stress sensing in arteries, neurological responses, and the regulation of osteogenesis. In mice, the absence of TRPV4 is linked with altered pressure/vasodilatory responses, osmosensing, sensory and motor neuropathies, and development of fibrosis in the lung, skin, and cornea (Hamanaka et al., 2010; Liedtke et al., 2000; Masuyama et al., 2008; Thodeti et al., 2009; Thorneloe et al., 2012; Vriens et al., 2004).

Table 1 TRPV4 expression and functions in various immune cells.

Cell-type	Molecular mediator	Mechanism	Effect	References
Macrophage	4-α-Phorbol didecanoate (4αPDD)		Activates TRPV4 ion channel and significantly elevates intracellular calcium, superoxide, and nitric oxide production	Hamanaka et al. (2010)
	LPS	Induction of anti-inflammatory/ pro-resolution cytokines	Stimulates the phagocytosis process	Scheraga et al. (2016)
	GSK1016790A	Induces PI_3K, PKC, and RhoA signaling pathways	Increases cytosolic intracellular calcium and sodium ions	Baratchi et al. (2019)
	HC-067047		Inhibits TRPV4 channel and has shown antinociceptive properties in different pain models, including inflammatory, orofacial, visceral and neuropathic pain	Dias et al. (2019)
T-cell	DMAPP		Elevated intracellular Ca^{2+} levels in TRPV4-transfected HEK293T cells.	(Bang et al., 2012)
	4α-PDD and RN1734	Immune activation, differentiation, host–pathogen interaction cellular response pathways	4α-PDD increases the influx of Ca^{2+}. Inhibition of TRPV4 (RN1734) inhibits calcium influx into soluble CD3 stimulated T cells	Majhi et al. (2015)
	Epoxyeicos- trienoic acids (EETs)		Mediate TRPV4 activation in response to either hypoosmotic shock or mechanical stimulation	Berna-Erro et al. (2017)
	Gadolinium		An inhibitor of stretch-activated ion channels and nonselective inhibitor of extracellular Ca^{2+}-entry. May act on other channels as well	Goldenberg, Ravindran, and Kuebler (2015)
	RN-1734		Completely inhibits both ligand- and hypotonicity-activated TRPV4	Vincent et al. (2009)

Continued

Table 1 TRPV4 expression and functions in various immune cells.—cont'd

Cell-type	Molecular mediator	Mechanism	Effect	References
Microglia	4α-PDD	TRPV4 regulates cellular Ca^{2+} level, migration, motility	TRPV4 is endogenously expressed in rat primary microglia and microglia cell line N9	Chakraborty and Goswami (2022)
	RN1734	TRPV4 inhibition decreases calcium influx and decreases the levels of IL-1β and TNF-α	TRPV4 inhibition alleviated demyelination and inhibits glia activation	Liu et al. (2018)
	GSK1016790A	TRPV4 activation increases cellular Ca^{2+} level and increases migration	TRPV4 is expressed in primary mouse microglia	Nishimoto et al. (2021)
Dendritic cell	GSK1016790A	Uptake of IgG-coated beads through receptor-mediated mechanisms operating in dendritic cell	Increase in intracellular Ca^{2+} concentration in dendritic cells	Naert et al. (2019)
Mast cell	LL37 and HC 067047	A potential mechanism by which LL37 can promote TRPV4 upregulation and Mas-related G-protein coupled receptors may play a role in LL37's induction of TRPV4 expression	TRPV4 expression appears to play a role in mast cell activation	Mascarenhas et al. (2016)
Neutrophils	GSK2193874, HC-067047	Putative mechanism of platelet activating factor (PAF)-induced lung vascular barrier	In neutrophils, TRPV4 inhibition or deficiency attenuated the PAF-induced increase in intracellular calcium	Yin et al. (2016)

5. Temperature effect and presence of TRPV4 in immune cells: Physiological and pathophysiological contexts

Extreme sensitivity of immune functions towards a slight shift in body temperature makes thermosensitive ion channels as potential candidates for immune regulation. TRP channels are activated at varying temperatures, and participate in immune regulation in various ways, among which temperature response remains a significant mechanism (Güler et al., 2002; Watanabe et al., 2002). Elevation of basal body temperature to encounter infections mediated by various pathogens is an essential activator of the innate immune system. It has been reported that the cytotoxic activities of T cells from human umbilical cord blood and adult blood enhanced at slightly increased temperature, i.e. at $\leq$40 °C, but the activities decreased if exposed to 42 °C for 1 h (Shen et al., 1994). Furthermore, CD8+ T cells, when exposed to mild hyperthermia, significantly enhanced the production of antigen-specific IFN-γ (interferon-γ) (Mace, Zhong, Kokolus, & Repasky, 2012). Fever-like whole-body hyperthermia also increases the morphology and distribution of T cells in tissues. Subsequently, it increases the level and distribution of Protein Kinase C (PKC) within T cells, which is typically a Ca^{2+}-dependent process (Wang, Ostberg, & Repasky, 1999). These results suggest the involvement of ion channels in the T cells that are Ca^{2+-}-permeable and/or thermosensitive. Apart from T cells, hyperthermia also regulates the activities of Dendritic cells (DC) and Natural killer cells (NK cells). An increase in temperature enhances the phagocytic activities of DCs by augmenting IFN-γ and TNF-α (Evans, Repasky, & Fisher, 2015; Knippertz et al., 2011). It has also been demonstrated that an increase in temperature results in increased cytotoxic activity and recruitment of the NK cells to tumor sites (Zänker & Lange, 1982).

Since the discovery of the TRPV4 channel, it has been widely explored in diverse and unique biological systems. Due to its thermosensitivity at the physiological temperature range, the functionality and potential of TRPV4 in immune regulation were recognized decades ago. TRPV4 expression was first discovered in human leukocytes in 2008 (Spinsanti et al., 2008). After that, multiple groups investigated the functions of TRPV4 in various immune cells. Yin et al. in 2016 established that TRPV4 is indeed present in both murine and human neutrophils and TRPV4 deficiency reduces the pro-inflammatory response, neutrophil adhesion, reactive oxygen species generation, and chemotaxis (Michalick & Kuebler, 2020; Yin et al., 2016). TRPV4 activation by pharmacological activator 4α-PDD in keratinocyte cell line increases pro-inflammatory responses and induces the release of IL-8 level, which, in turn, indicates the role of TRPV4 in neutrophil recruitment (Fusi et al., 2014). Inhibition of TRPV4 suppresses acid-induced pulmonary inflammation by decreasing the neutrophils, macrophages, chemokines, and cytokines associated with these cell types. Notably, TRPV4 is expressed in a wide range of immune cells. Importantly, multiple endogenous inflammatory compounds that sensitize and activate TRPV4 (e.g., Arachidonic acid, metabolites, Prostaglandin (PGE_2) and histamine) have been identified (Alessandri-Haber et al., 2003; Cenac et al., 2010; Sulk & Steinhoff, 2015).

6. TRPV4 expression and function in T cell

Intracellular Ca^{2+} plays a crucial role in T cell function. T cell activation and regulation are very heterogeneous. Under resting conditions, T cells, B cells and macrophages contain a basal amount of Ca^{2+}, which increases with immune activation following the exposure to B cell receptor, T cell receptor, cytokines, and chemokines. It has also been reported that the chelation of intracellular Ca^{2+} by BAPTA-AM or reduction of intracellular Ca^{2+} by extracellular EGTA causes a decrease in IL-2 production, IL-2 receptor expression and proliferation of naïve T cells in response to an activation stimulus (Vig & Kinet, 2009; Watman, Crespo, Davis, Poiesz, & Zamkoff, 1988). T cells can also be activated by a diverse range of temperatures, which paves the path to understanding several thermosensitive TRP channels in T cell biology. Although there are a few studies on the role of TRPV4 in T cells, TRPV4 emerges as a potential candidate for T cell regulation. It has been reported that TRPV4 is endogenously expressed in human T cell lines, primary human T cells and primary murine splenic T cells (Fig. 1). TRPV4-specific agonist, 4α-PDD, increases Ca^{2+}-levels in T cells soon after application (Fig. 1). Moreover, T cell receptor activation by concanavalin-A (ConA) and CD3/CD28 elevates the expression levels of TRPV4. In contrast, blocking the activity of TRPV4 decreases Ca^{2+}-levels and regulates effector cytokine production by suppressing tumor necrosis factor, IL-2 and IFN γ release. Simultaneous inhibition of TRPV1 and TRPV4 by specific inhibitors also synergistically reduces the process of ConA or T cell receptor-mediated T cell activation

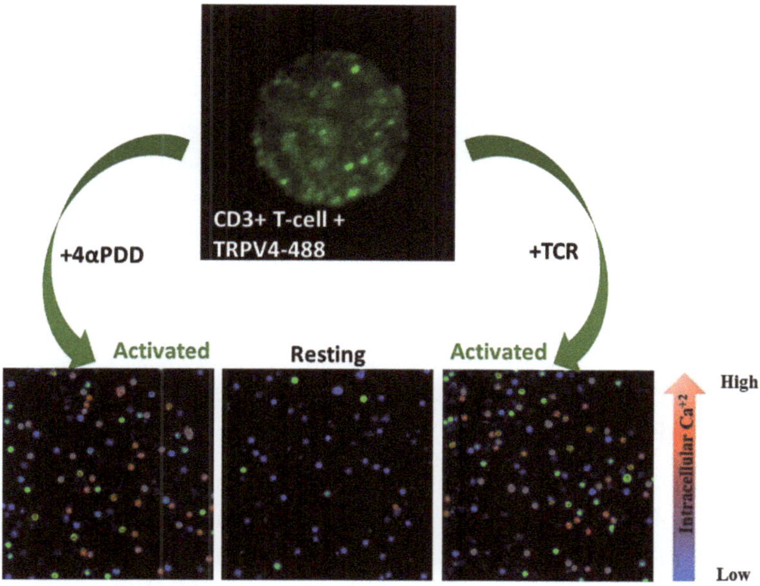

Fig. 1 Endogenous TRPV4 in resting T cell regulates Ca^{2+}-influx required for T cell activation. The upper panel indicates a single murine T cell immunostained for TRPV4 (green) and subsequently subject to confocal imaging. The lower panel images indicate the relative levels of Ca^{2+} in a population of T cells. These live cells were loaded with Ca^{2+}-sensitive dye Fluo4 and treated with TRPV4-specific agonist 4αPDD or TCR. More number of cells show a high level of intracellular Ca^{2+}. All the original images were pseudocolored (red fluorescence indicates more Ca^{2+} while blue indicates a low level of intracellular Ca^{2+}).

process (Majhi et al., 2015). The expression of TRPV4 is reported in human primary $CD4^+$ T cells (Ghoneum et al., 2018). Stimulating $CD4^+$ T cells by DPV576, an aqueous mixture of nanodiamond (ND) and nanoplatinum (NP) along with CD3/CD28, decreases the expression of TRPV1 but did not affect TRPV4 expression. Taken together, TRPV4 expression and function are highly dependent on the nature of the stimuli and the duration of exposure of the compound for T cell receptor activation. Fever mediates immunological consequences by modulating innate immune cells, like T cells. When the $CD4^+$ T cells were stimulated by anti-CD3 and anti-CD28 at 39°C (a slightly higher temperature than physiological, but in the range of fever temperatures), the T cell responses were Th2-mediated, as there was an enhancement of IL4/5/13. Accordingly, the level of Th2–relevant transcription factor GATA3 is upregulated and the expression of Th1-relevant transcription factor Tbet is reduced.

Further, the enhanced expression of GATA3 relied on the activation of notch signaling pathway. However, TRPV4 and TRPV1 antagonists inhibited the switching of Th2 and decreased the amount of IL13. These data suggest that TRPV-mediated skewing of Th2 for the effector T cell response is primarily regulated by Notch activation (Umar et al., 2020).

The evidence in the literature suggests diversified expression and immunomodulation of TRPV4 in T cells and their importance in Ca^{2+}-homeostasis (Fig. 1). Overall, there are very few findings on the mechanistic illustration of TRPV4 in T cell function, which opens a new area of research in the application of TRPV4 in T cell biology.

7. TRPV4 expression and function in mast cell

TRPV4 is functionally expressed in human mast cells. Activation of TRPV4 by pharmacological compound 4α-PDD increases the intracellular Ca^{2+} level and hypersensitivity induced by histamine. The hypersensitivity reaction can be inhibited significantly by silencing TRPV4 with specific siRNA (Cenac et al., 2010; Kim et al., 2010; Stokes, Shimoda, Koblan-Huberson, Adra, & Turner, 2004). The mast cell is the key regulator of IgE-dependent bronchoconstriction in airway smooth muscle. Recently it has been shown that TRPV4 mediates IgE-independent activation of mast cells to induce bronchoconstriction during the onset of asthma. In this condition, activation of TRPV4 increases Ca^{2+} and ATP release, which triggers the P_2X_4-dependent release of cysteinyl leukotrienes that activates mast cells resulting in airway smooth muscle contraction. The bronchoconstriction can be rescued by TRPV4-specific antagonist GSK2193874 (Bonvini et al., 2020). Activation of TRPV4 induces substantial Ca^{2+}-rise in primary mast cells isolated from mice. FcεRI (high-affinity IgE receptor), Mrgprb2 (Mas-related G-protein coupled receptor member B2) and ET_A (endothelin receptors$_A$) are well known surface receptors, which upon stimulation, activate mast cells by elevating intracellular Ca^{2+} and degranulation of mast cells respectively. However, in TRPV4-deficient mice, Ca^{2+} levels did not increase and degranulation mediated by FcεRI, Mrgprb2, and ET_A receptors was visible (Solis-Lopez et al., 2017). Chronic inflammatory skin condition Rosacea is a mast cell-mediated disease. Increased serine protease activity and high levels of cathelicidin fragments like LL-37 mediate Rosacea. TRPV4 expression increases almost 20-fold in mast cells treated with LL-37 (cathelicidin-derived antimicrobial peptide) compared to normal mouse skin. In contrast, the degranulation activity of mast cells was reduced by approximately 50% due to the knockdown of TRPV4 (Di Nardo, Mascarenhas, & Wang, 2017; Mascarenhas, Wang, Chang, & Di Nardo, 2016). Taken together, the data suggest the involvement of TRPV4 in complex immune disorders.

8. TRPV4 in dendritic cells

Innate immunity is a well-orchestrated process regulated by a plethora of immune cells. Among all types of immune cells, Dendritic cells are also known as the sentinels of the immune system. DCs are professional antigen-presenting cells that have the ability to educate T, B and NK cells at specific infections (Stockwin, McGonagle, Martin, & Blair, 2000). DCs are bone marrow-derived cells that have a remarkable potential toward recognition of different pathogen-associated molecular patterns (PAMPs) due to the specialized distribution of pattern recognition receptors (PRR) like Toll-like receptors, C-type lectins and intracellular nucleic acid sensors (Martin-Gayo & Yu, 2019). These events are regulated by specific cytosolic Ca^{2+}-levels. Cellular Ca^{2+} metabolism is tightly regulated by a series of Ca^{2+}-permeable ion channels, including CRAC channels, L-type Ca^{2+} channels, purinergic receptors and TRP channels (Shumilina, Huber, & Lang, 2011). So far, multiple TRP channels have been identified in DCs. There are very few reports available depicting the function of TRPV4 in Dendritic cells. Szollosi et al. have shown the expression of TRPV4 in human monocyte-derived DCs. The authors found that an increase in temperature to 43 °C reduces the endocytic activity of DCs and silencing of TRPV2 prevents the reduction of endocytic activity. In contrast, TRPV1 and TRPV4 were also expressed in the DCs, but pharmacological (antagonist) and genetic inhibition (RNAi) of TRPV4 and TRPV1 did not prevent the reduction of endocytic activity in DCs (Szöllősi et al., 2013). TRPV4 is also expressed in mouse CD11c^{+} bone marrow-derived DCs (BMDCs). In CD11c^{+} BMDCs, TRPV4 is predominantly expressed in the plasma membrane and pharmacological activation by GSK1016790A (TRPV4-specific agonist) increases intracellular Ca^{2+}-levels. The cells were unresponsive to the agonist in the presence of TRPV4 antagonist HC067047. Similarly, in TRPV4 knockout cells, GSK1016790A could not increase the Ca^{2+}-levels confirming the functional importance of TRPV4. Moreover, TRPV4 channels are downregulated in mature DCs. During the differentiation process of immature to mature DCs, the activity and expression of TRPV4 were downregulated in the presence of lipopolysaccharide (LPS). TRPV4 deficient cells did not lower the nuclear translocation of NF-κB in response to LPS. Accordingly, the lack of TRPV4 did not prevent the upregulation of pro-inflammatory cytokines, IL-6, IL-12 or maturation markers like CD40, CD80, and CD86 (Naert, López-Requena, Voets, Talavera, & Alpizar, 2019). In summary, TRPV4 is expressed in DCs. Modulation of TRPV4 channel activity regulates the various functions of DCs. However, the mechanistic involvement of TRPV4 in DCs is not clearly understood and further investigation is needed.

9. TRPV4 in macrophage and relevant functions

The macrophage functions include pro-inflammatory responses like phagocytosis, inflammation, migration, expression of the inflammatory protein, and proliferation. These functions are sensitive to changes in stiffness of their surrounding matrix, which possibly depends on mechano-activated ion channels, including TRPV4 (Blakney, Swartzlander, & Bryant, 2012; Féréol et al., 2006; McWhorter, Wang, Nguyen, Chung, & Liu, 2013; Scheraga et al., 2016). Thus, mechanosensitive and/or mechanically-activated ion channels may transduce various biomechanical signals into specific cellular responses and inflammation. Since TRPV4 is a mechanically-activated channel, it can be activated by both biochemical and biomechanical stimuli (Güler et al., 2002; Loukin, Zhou, Su, Saimi, & Kung, 2010; Mochizuki et al., 2009; Strotmann et al., 2000).

TRPV4 has both pro-inflammatory functions (like phagocytosis, adhesion, and reactive oxygen species production) as well as anti-inflammatory effects (like secretion of pro-resolution cytokines and bacterial clearance) in macrophages (Hamanaka et al., 2010; Scheraga et al., 2016). In macrophages, TRPV4 may exert both protective and detrimental effects on the host tissue by facilitating bacterial clearance in infection while promoting parenchymal injury in sterile inflammation (Balakrishna et al., 2014; Michalick et al., 2017; Yin et al., 2016). TRPV4 can play a role in force-dependent cytoskeletal changes in other systems/cell types (Goswami, Kuhn, Heppenstall, & Hucho, 2010). Thus, it was hypothesized that TRPV4 might play a role in macrophage phagocytosis (Hamanaka et al., 2010; Rahaman et al., 2014). Bacterial LPS-stimulates phagocytosis of *E. coli* particles in the presence of Ca^{2+} by 151%±3% in murine BMDMs. For LPS to induce macrophage phagocytosis of *E. coli* particles, extracellular Ca^{2+} is required. LPS-stimulation of phagocytosis is completely abolished in BMDMs from TRPV4-KO mice (*Trpv4*$^{-/-}$) and decreased by 67% ± 9% upon downregulation of TRPV4 with siRNA. The ability of LPS to induce phagocytosis through TRPV4 is dependent on matrix stiffness in a varied range comparable to that seen in the inflamed or fibrotic lung. Finally, TRPV4 mediates the LPS signal to release anti-inflammatory/pro-resolution cytokines in macrophages. Collectively, these data implicate TRPV4 in the pathogenesis of lung infection/bacterial pneumonia by integrating the LPS and the matrix stiffness signals for macrophage phagocytosis and pro-resolution cytokine release (Scheraga et al., 2016). One of the studies identified the anti-inflammatory role of PAR2 (protease activated receptor 2) in alveolar macrophages by suppression of toll-like receptor 4 (TLR4)-induced inflammation. The PAR2-mediated cAMP generation inhibited TRPV4-dependent Ca^{2+}-signaling in alveolar macrophages to resolve TLR-elicited inflammation. The reduction of TRPV4 levels using siRNA or other antagonists in PAR2 null mice attenuated Ca^{2+}-entry, reduced the levels of proinflammatory cytokines, phosphorylation of NFκB and NFAT. The blockade of TRPV4 in PAR2 null mouse alveolar macrophages (after the LPS challenge) promoted the resolution of inflammation and reversed lung injury (Rayees et al., 2019). This study suggests that TRPV4 plays an essential role in mediating inflammation in TLR4-induced responses in alveolar macrophages. There are reports that TRPV4 is expressed in wound tissue macrophages (tissue-resident macrophages). A corneal alkali burn wound healing model was used to determine the role of TRPV4 in corneal fibrosis. Immunohistochemistry data showed that TRPV4 knockout mice failed to exhibit the expected fibrosis-associated increase in numbers of polymorphonuclear leukocytes and accumulation of macrophages. It was concluded that alkali-induced corneal fibrosis and inflammation were, in part, dependent on TRPV4 present in macrophage (Okada et al., 2016).

Biochemical factors including cytokines, chemokines, modified low-density lipoprotein (LDL), and bacterial lipopolysaccharide (LPS) help in initiating macrophage-mediated inflammatory responses (Goswami et al., 2017; Mosser & Edwards, 2008; Rahaman et al., 2006). Plasma membrane of macrophages has mechanosensitive/thermosensitive ion channel TRPV4, which may be acting as the receptor for these chemical and mechanical stimuli. Interestingly emerging data have shown that the macrophage TRPV4 is involved in a variety of inflammatory diseases, including acute lung injury/acute respiratory disease syndrome, atherosclerosis, foreign body response (FBR), skin disease, and fibrosis (Goswami et al., 2017;

Hamanaka et al., 2007; Okada et al., 2016). These studies suggest that TRPV4 may have a critical role in regulating inflammatory responses like cytokine production, foam cell formation, giant cell formation, and phagocytosis. Ca^{2+}-dependent signaling is associated with various macrophage-specific inflammatory responses, including atherogenesis, migration, phagocytosis, and foam cell formation (Melendez & Tay, 2008; Nunes & Demaurex, 2010). In this context, TRPV4 along with other ion pumps, maintain macrophage Ca^{2+}-homeostasis, and its dysregulation can lead to variable pathophysiological conditions (Blakney et al., 2012; Gupta, Goswami, Alharbi, Biswas, & Rahaman, 2019; Hansen & Taylor, 2016). TRPV4 is also required for oxidized low-density lipoprotein (oxLDL) internalization and for subsequent formation of macrophage foam cells, and the loss of TRPV4 function (either genetic or pharmacological) abrogates foam cell formation (Goswami et al., 2017). Hence, TRPV4-dependent Ca^{2+}-influx has multiple roles in macrophages.

10. Microglia

Our present understanding of the effects of TRPV4 modulation on microglial physiology is at its infancy. Only a decade ago Konno and colleagues (Konno et al., 2012) reported a reduction in the release of pro-inflammatory cytokine tumor necrosis factor α (TNF-α) and upregulation of ischemia-related β-galactoside binding lectin (galectin3) upon pharmacological activation of TRPV4 in lipopolysaccharide (LPS)-treated microglia. In a cuprizone-induced demyelinating mouse model, TRPV4 expression was found to be elevated in the corpus callosum, besides the neuroinflammatory signs of microglial activation (Liu et al., 2018). Pharmacological inhibition of TRPV4 not only reduced reactivity of microglia, but also promoted remyelination. Other than anti-inflammatory properties of TRPV4 activation, recent studies have also reported a pro-inflammatory response of the same. Pharmacological activation of TRPV4 significantly upregulates an inflammatory response in a pilocarpine model of temporal lobe epilepsy in mice by causing microglial activation, elevation in the level of NLRP3 inflammasome, and pro-inflammatory cytokines IL-1β, IL-6, and TNF-α. Inhibition of the channel ameliorated these changes and improved survival of neurons, thereby offering a neuroprotective effect (Wang et al., 2019). In an experimental system of spinal cord injury (SCI), phase of inflammation coincided with increased expression of TRPV4 in cells at the endothelial-microglial junctions, activation of which caused degradation of tight junctions, increased scarring of the injury and endothelial cell damage. All these detrimental manifestations were suppressed upon channel inhibition (Kumar et al., 2020). Recently we have shown that TRPV4 and TRPM8 are expressed in rat primary microglia and microglial cell line N9 and regulates cellular Ca^{2+}-level, cellular morphology, migration and motility (Chakraborty & Goswami, 2022).

11. Astrocytes

Because of the non-excitable nature of astrocytes, these cells communicate with each other via Ca^{2+} ions. As such, activation of TRPV4 in astrocytes can initiate gap junction-mediated astrocyte communication and eventual response to neurotransmission by the release of gliotransmitters (Shibasaki, Ikenaka, Tamalu, Tominaga, & Ishizaki, 2014). Upon cerebral ischemia, astrocytic TRPV4 level increases concomitantly with increased reactivity of these cells in the hippocampus, suggesting its role in astrocyte-regulated neuroinflammatory response following ischemic insults (Butenko et al., 2012). Like in microglia, astrocytes also acquire a reactive state upon TRPV4 activation in demyelinating mice and epilepsy models, suggestive of an induction of pro-inflammatory response (Liu et al., 2018; Wang et al., 2019). Future research is required to understand the mechanisms of TRPV4 regulated generation of inflammatory responses via microglia (innate immune cells of the brain) and astrocytes (non-immune cells with significant roles in neuroinflammation).

12. Roles of TRPV4 in neuroimmune interactions

Over several years of neuroimmunology studies, the brain has been considered to have a functional disconnect with the immune system. As such, inflammation of the central nervous system (CNS) was considered to happen only if peripheral immune cells could surpass the blood-brain barrier (BBB) (Engelhardt, Vajkoczy, & Weller, 2017). However, recent advancements in our understanding not only describe the uniqueness of CNS immune system along the meninges of the brain (Louveau et al., 2015) but also ascribe a multitude of functions to the *bona fide* macrophages of the brain, i.e. microglia (Li & Barres, 2018). Under healthy and diseased conditions, microglia function to maintain and restore homeostasis in the brain, thereby generating pro- and/or anti-inflammatory responses upon CNS insults. Microglia express pattern recognition receptors (PRRs) that recognize pathogen-associated molecular patterns (PAMPs) upon microbial infections in the brain, allowing for the generation of an innate immune response (Gadani, Walsh, Lukens, & Kipnis, 2015; Kigerl, de Rivero Vaccari, Dietrich, Popovich, & Keane, 2014). Pro-inflammatory responses can also be generated in the absence of PAMPs, mediated via tissue-endogenous damage-associated molecular patterns (DAMPs) that are released upon CNS damage, as in cases of neurodegenerative diseases, thereby causing sustained neuroinflammation (Glass, Saijo, Winner, Marchetto, & Gage, 2010). In the brain, microglial response to injuries is also supported by structural and functional changes in astrocytes, which also respond to CNS insults by acquiring a reactive state. Similar to microglia, astrocytes can also be neuroprotective (A2 astrocytes; anti-inflammatory) or neurotoxic (A1 astrocytes; pro-inflammatory) in nature, depending on the intracellular signaling pathways involved (Ding et al., 2021). Consequently, secretion of pro-inflammatory cytokines in the brain is also contributed by astrocytes. The pro-inflammatory transcription factor NFκB in astrocytes is sufficient to generate a global neuroinflammatory response in several experimental models of neurodegenerative diseases (Brambilla et al., 2005, 2009, 2012, 2014; Fu et al., 2010; Khorooshi, Babcock, & Owens, 2008). Being important regulators of structural integrity of the blood-brain barrier (BBB), astrocytes dysfunctions could lead to infiltration of peripheral immune cells into the brain parenchyma that enhances neuroinflammation (Argaw et al., 2012).

In the brain, the involvement of Ca^{2+} channels in neuroinflammation caused upon aging has also been reported (Navakkode, Liu, & Soong, 2018). Besides immune receptors present on microglia, several other ion channels also mediate such inflammatory responses (Kettenmann, Hanisch, Noda, & Verkhratsky, 2011). Pro-inflammatory cytokines cause an elevation in intracellular Ca^{2+}-level in microglia, leading to their activation (Goghari, Franciosi, Kim, Lee, & McLarnon, 2000). Ion channels on astrocytes also regulate inflammatory responses generated in the brain. Reduced expression or mis-localization of the water channel aquaporin 4 (AQP4, which is also functionally linked with TRPV4 in few cases) limits the clearance of protein aggregates from the brain parenchyma, which in turn triggers neuroinflammation (Zou et al., 2019). The generation of neuroinflammatory responses requires the participation of several Ca^{2+} channels. Greater pharmacological interest lies in channels that can be targeted for therapeutic interventions, among which are members of the TRP superfamily of non-selective cation channels. Both microglia and astrocytes express TRP channels that regulate intracellular Ca^{2+}-dynamics and versatile functions like cytokine production, osmotic sensing, motility, migration, cytokine production, and gene expression among others (Echeverry, Rodriguez, & Torres, 2016; Verkhratsky, Reyes, & Parpura, 2013). Microglia differentially express members of the melastatin (TRPM), canonical (TRPC), and vanilloid (TRPV) sub-families of TRP channels (Kettenmann et al., 2011; Shirakawa & Kaneko, 2018). On the other hand, members of TRPC, TRPV, and ankyrin (TRPA) sub-families have been reported in astrocytes (Verkhratsky et al., 2013). Recent studies have elucidated the implications of heat-sensitive TRPV4 in microglial activity regulation (Nishimoto et al., 2021).

13. TRPV4 in auto-immune disorders

Autoimmune disorders in simple terms are the condition where the immune cells start attacking the self-cells causing tissue damages that lead to severe consequences. Since it is already established that various ion channels, including the TRP channels are expressed in immune cells, these channels are also potentially involved in autoimmune diseases and can affect its occurrence and severity. Intra-cellular Ca^{2+}-homeostasis is critical for normal immune functions. Reports suggest differential expression of TRP channels in diverse immune cells in case of autoimmune conditions, which may be linked with the abnormal Ca^{2+}-levels within these cells that induce further downstream signaling cascades.

Multiple Sclerosis (MS) is a chronic autoimmune disease wherein immune cells degrade the myelin sheath of neurons. Patient samples from relapsing-remitting multiple sclerosis reveal an altered expression of the TRPV4 in peripheral blood mononuclear cells (PBMCs). The expression of TRPV4 significantly decreases in the patients compared to the healthy individuals, which indicates a potential role of these channels in the development of MS (Çakır, Saçmacı, & Sabah-Özcan, 2021). Similarly, in case of Psoriasis, which is another autoimmune disorder (where the PBMC's are involved in pathological processes), the expression of TRPV4 decreases in the PBMCs of the patient (Özcan, Gürel, & Çakır, 2021). In contrast, expression of TRPV4 is upregulated in Ulcerative Colitis (UC), another autoimmune disorder (Rizopoulos, Papadaki-Petrou, & Assimakopoulou,

2018). UC comes under the umbrella of Inflammatory Bowel Disorders (IBDs) and TRPV4 plays a crucial role in these disorders. Since the expression of TRPV4 increases in UC, pharmacological inhibition of TRPV4 by its specific antagonist RN1734 resulted in decreased inflammation and pain associated with UC (Fichna et al., 2012). Therefore, targeting TRPV4 may be a potential strategy to reduce the inflammatory responses in such cases. In the case of Rheumatoid Arthritis, functional expression of TRPV4 is detected in synoviocytes, which are involved in joint destruction by chronic inflammation. Moreover, reports suggest that TRPV4 may act downstream to immunological signaling as its expression and function enhances upon release of TNF-α cytokine and can potentially activate the synoviocytes to attack the joints (Kochukov et al., 2009). TRPV4 is expressed in healthy skin tissues. Autoimmune bullous skin disorders are the conditions where large blisters are formed on the skin due to invasion by the immune cells. These autoimmune disorders are of various types, namely pemphigus Vulgaris (PV), bullous pemphigoid (BP), dermatitis herpetiformis (DH), and epidermolysis bullosa acquisita (EBA). TRPV4 expression is altered in such chronic conditions and decreases significantly to normal skin tissues in these types of bullous disorders (Bai, Zou, Jiang, & Wang, 2015). The molecular mechanisms of how TRPV4 modulates the intricacies of autoimmune disorders is yet to be explored in detail. Nevertheless, the involvement of TRPV4 in various autoimmune conditions is being established and TRPV4 stands out as a potential target in curing autoimmune disorders.

14. Steroids in TRPV4-mediated immunomodulation

It is not a mere coincidence that the evolution of adaptive immunity at the cell and tissue level, and steroidogenesis from cholesterol at the level of molecular evolution, started during the vertebrate evolution (Baker, 2011; Danilova, 2006). Additionally, the TRPV4 sequence appears to co-evolve with the cholesterol biosynthesis pathway. Several steroids interact with a fragment of TRPV4, also suggesting the interplay among them, which most likely has significant importance in bone mineralization and membrane fluidity regulation, which shaped the vertebrate evolution to what it is today (Kumari et al., 2015). The vertebrate-specific TRPV4 ion channels appeared circa 440 million years before. TRPV4 seem to be regulated by a multitude of endogenously produced substances that are also immune-modulatory, namely arachidonic acid (AA) and its metabolites epoxyeicosatrienoic acid (EET) (Zheng et al., 2013), serotonin, histamine

(Cenac et al., 2010), and sex hormones like progesterone (Jung et al., 2009). Many immunological disorders are also sex-dependent to some extent. For example, autoimmune diseases like Systemic Lupus Erythematosus (Weckerle & Niewold, 2011), rheumatoid arthritis (Zink, Listing, Klindworth, & Zeidler, 2001) and chronic autoimmune thyroiditis (Mariotti et al., 2009) are predominant in females. Although the female hormones estrogen and progesterone are said to have an immunosuppressant role in mothers during pregnancy, their mechanism of action is still not elucidated (Munoz-Suano, Hamilton, & Betz, 2011). The intracellular progesterone receptor-mediated immunosuppression in humans and mice in various cell types including natural killer cells, macrophages, Dendritic cells, T cells (Gilliver, 2010), eosinophils (Hamano, Terada, Maesako, Numata, & Konno, 1998), and mast cells (Padilla et al., 1990) are reported. But recent reports suggest that membranous receptors present in immune cells cause rapid stimulation (often in seconds or faster) in response to some steroids (Thomas, 2008). This indicates that several Ca^{2+} channels, including TRP channel members, might be involved in such immune modulation. Recently, a rapid Ca^{2+}-influx via TRPV4 ion channels in the oviduct was also reported (Jung et al., 2018). Also, Progesterone has a direct effect on TRPV4 ion channels by downregulating its expression via a genomic pathway (Jung et al., 2009). Thus, the correlation of progesterone and TRPV4 channel function may influence the immune response, which needs further investigation. It is worth noting that Dimethylallyl pyrophosphate (DMAPP), a mevalonate pathway (Cholesterol biosynthesis) metabolite, has very recently been noted to play an important role in the stimulation of T cells (Okuno et al., 2020). As a proof-of-principle, this compound induced acute inflammation in neurons via TRPV4 activation (Bang, Yoo, Yang, Cho, & Hwang, 2012). Altogether, results suggest that endogenous steroids, their metabolites and precursors can modulate TRPV4 and cause immune modulation.

15. Temperature, immunity, and importance of TRPV4 in the evolution of the immune system

The requirement of immune response starts from discrimination of self from non-self, very early in evolution. For example, restriction enzymes in prokaryotes, immunoglobulin and lectins in Porifera and Cnidaria, phagocytosis in Amoeba, etc., are examples of "self vs non-self-discrimination" (Buchmann, 2014; Cooper, 2010). The primitive immune response remained conserved throughout evolution and is termed as "innate immunity," though the nature of such response varies. As for adaptive immunity, it is believed that it developed only from the vertebrate evolution (Cooper, 2010) or, a sudden functional and structural change of the metazoan innate immune system resulted in the vertebrate evolution with the adaptive immune system (Boehm, Iwanami, & Hess, 2012). Nevertheless, the temperature has a profound effect on entire biological processes, including the immune system, which can "sense" temperature shift due to various mechanisms. Notably, during the evolution process from lower organism to higher organism, changes in both environmental temperature as well as body temperature were obvious and the immune systems had to adapt to these temperature changes. One of these sensory inputs is due to the presence of temperature-sensitive proteins, like TRP channels, HSF (heat shock factors) etc., within the immune cells, which, when activated, allows an influx of Ca^{2+} and activation of the immune response. Thus, the "shift-in-temperature" has a profound effect on immune system regulation, which somehow seems to be conserved in the evolutionary context. Primitive forms of life, such as insects and other primitive multicellular organisms, have a predominantly innate immune response. In contrast, more complex organisms, and most vertebrates, possess both innate as well adaptive immune responses (Murdock et al., 2012). Both innate as well as acquired immune responses are sensitive to temperature variations. However, the acquired immune response is considered to be more temperature-dependent. Most of the recent studies show temperature-dependent innate immune response. For example, it has been demonstrated that temperature-dependent innate immune response limits the replication of viruses at warm temperatures (Foxman et al., 2015).

Though temperature's role in the immune system's regulation in endotherms is well established, the effect of temperature on immune defense in ectothermic animals where immune cells are directly exposed to external temperature is not very well documented (Boltana et al., 2013). Experimental results confirm that *Gasterosteus aculeatus,* a fish, shows temperature-dependent immune activation (Dittmar, Janssen, Kuske, Kurtz, & Scharsack, 2014). Poikilothermic animals like fish are complex model organisms to study temperature-dependent immune regulation as their body temperature fluctuates with the surrounding and at the same time, they possess complete vertebrate immune system with innate and adaptive immunity. Nevertheless, experimental evidence suggests that temperature affects not only immune response in vertebrates but also influences invertebrate immune traits and responses as well. Insect larvae, when injected with

LPS, exhibit increases in antibacterial responses when maintained at a high temperature as compared with when maintained at lower temperatures (Catalán, Wozniak, Niemeyer, Kalergis, & Bozinovic, 2012).

Studies have demonstrated that both endotherms and ectotherms develop a febrile response to infection. Endotherms respond to infection with a rapid increase in body temperature known as fever. In contrast, ectotherms lack intrinsic thermogenesis but employ "behavioral fever" by either moving to warmer places or increasing their physical activity. This phenomenon helps to improve survival in ectotherms (fish, amphibians, and several invertebrates) (Boltana et al., 2013). The body temperature of Ectotherms is subjected to seasonal changes in temperature, a significant factor that further has a profound effect on immune responses and thereby survivability. The numbers of circulating lymphocytes which are also the most abundant white blood cell in amphibians, reptiles, and fish, show higher cell count during warmer temperatures (Rollins-Smith & Woodhams, 2012). Thus, the animal's ability to elicit an active immune response decreases during winters in ectotherms.

The role of ion channels has long been known in the immune cells for essential physiological functions. Conservation of thermosensitive TRP channel sequences suggests its possible role in vertebrate as well as invertebrate evolution (Saha et al., 2017, 2019). TRPV4 is one such thermosensitive ion channel that responds to a warmer yet physiological temperature. The role of TRPV4 has been studied in various innate immune cell functions and activation (Majhi et al., 2015; Scheraga, Southern, Grove, & Olman, 2020). Most studies related to TRPV4 immune-related functions have already been done on mammalian immune cells and tissues. Recently, it has been found that TRPV4 also is expressed in avian immune cells. Tissues like "Bursa-of-fabricius," Spleen, Thymus, Pancreas and Caecum are the hub of immune cells and are involved in their maturation and differentiation. The study has shown the expression of TRPV4 in immune tissues in the avian system *Anas platyrhynchos* (Majhi et al., 2018). Several studies point to a potential role of TRPV4 ion channel in the immune adaptation and animal evolution as a response to environmental temperature (Fig. 2). TRPV4 might also have an impact in overall aging system, a process that has not been explored in higher organisms. For example, OSM-9 protein in *Caenorhabditis elegans* has been identified as a functional homologue of mammalian TRPV4. OSM-9 deficient *Caenorhabditis elegans* live longer, likely due to reduced Ca^{2+}-flux and related inflammatory response (Liedtke et al., 2003). The suppression of Ca^{2+}-flux

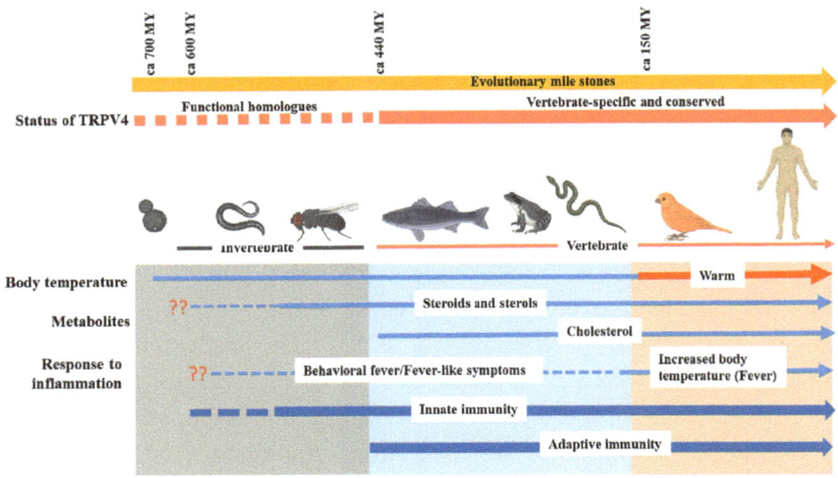

Fig. 2 Schematic representation of the importance of TRPV4 in the evolution of the animal kingdom and overall immunity. The hTRPV4 represents a highly efficient thermosensitive ion channel that remains conserved in all vertebrates. However, functional homologues are present in invertebrates as well as in yeast. TRPV4 function has been highly influenced by physiological temperature and specific metabolites that affect the biophysical properties of lipid bilayer. These aspects might have a strong influence on the development of innate as well as adaptive immunity. *This figure was prepared using https://biorender.com.*

affects phosphatase calcineurin orthologue *tax-6* in worms, which is known to have a role in the aging process (Mair et al., 2011).

16. Conclusion

Though initially perceived as a peripheral neuron-specific molecule engaged in neuronal sensory functions, endogenous expression of TRPV4 and many other TRP channels in diverse immune cells has been reported by several groups (Fig. 3). The expression of TRPV4 in immune cells seems to have an evolutionary relationship but this theory is not conclusively proven yet. Notably, altered expression of TRPV4 causes multiple immunological disorders and related complications such as neuro-immune or immune-endocrine abnormalities. Mutations in TRPV4 often cause complex diseases that are also transmissible to the next generation. Activation of TRPV4 channel by physical stimuli, i.e. by changes in pressure, osmolarity and temperature, makes it a potential therapeutic target that can be regulated by altering these stimuli in a chemical-drug-free manner.

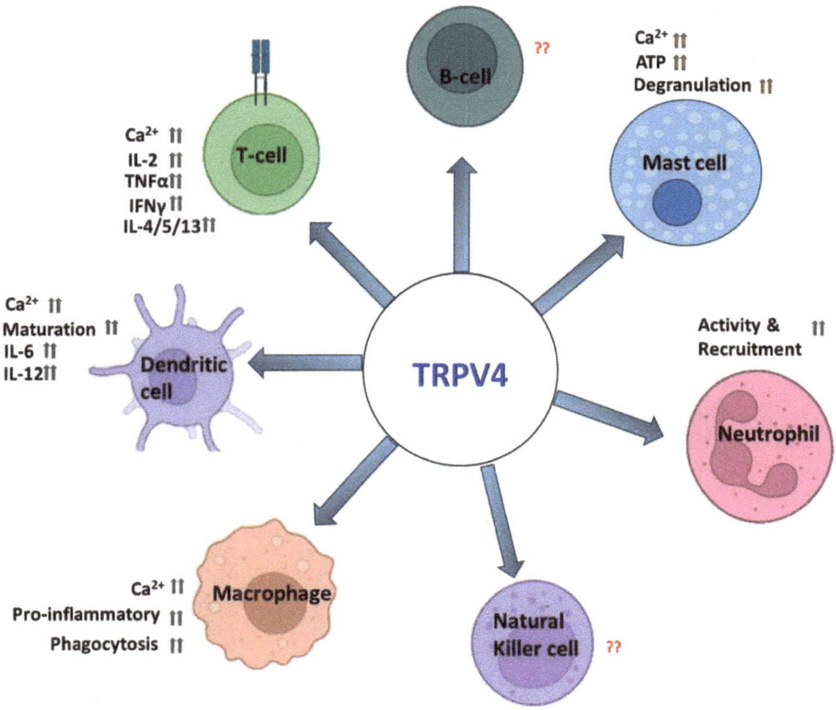

Fig. 3 Schematic diagram of presence and function of TRPV4 in different immune cells. *This figure was prepared using https://biorender.com.*

Alteration of TRPV4 activity by specific endogenous or exogenous lipids and other ligands is another alternative for therapeutic targeting. However, further detailed studies are needed to verify these possibilities.

Acknowledgments

All the authors acknowledge the intellectual input and scientific discussion on this topic. All authors acknowledge the intramural support from NISER, DAE. Some of the figures are prepared using https://biorender.com.

References

Alessandri-Haber, N., Yeh, J. J., Boyd, A. E., Parada, C. A., Chen, X., Reichling, D. B., et al. (2003). Hypotonicity induces TRPV4-mediated nociception in rat. *Neuron*, *39*, 497–511.

Argaw, A. T., Asp, L., Zhang, J., Navrazhina, K., Pham, T., Mariani, J. N., et al. (2012). Astrocyte-derived VEGF-A drives blood-brain barrier disruption in CNS inflammatory disease. *The Journal of Clinical Investigation*, *122*, 2454–2468.

Arniges, M., Fernández-Fernández, J. M., Albrecht, N., Schaefer, M., & Valverde, M. A. (2006). Human TRPV4 channel splice variants revealed a key role of ankyrin domains in multimerization and trafficking. *Journal of Biological Chemistry, 281*, 1580–1586.

Bai, J., Zou, X., Jiang, Y., & Wang, Q. (2015). Expression of transient receptor potential lvanilloidreceptor 4 protein in autoimmune bullous skin disorders. *Journal of Southern Medical University, 35*, 1349–1351.

Baker, M. E. (2011). Origin and diversification of steroids: Co-evolution of enzymes and nuclear receptors. *Molecular and Cellular Endocrinology, 334*, 14–20.

Balakrishna, S., Song, W., Achanta, S., Doran, S. F., Liu, B., Kaelberer, M. M., et al. (2014). TRPV4 inhibition counteracts edema and inflammation and improves pulmonary function and oxygen saturation in chemically induced acute lung injury. *American Journal of Physiology-Lung Cellular and Molecular Physiology, 307*, L158–L172.

Bang, S., Yoo, S., Yang, T., Cho, H., & Hwang, S. W. (2012). Nociceptive and pro-inflammatory effects of dimethylallyl pyrophosphate via TRPV4 activation. *British Journal of Pharmacology, 166*, 1433–1443.

Baratchi, S., Keov, P., Darby, W. G., Lai, A., Khoshmanesh, K., Thurgood, P., et al. (2019). The TRPV4 agonist GSK1016790A regulates the membrane expression of TRPV4 channels. *Frontiers in Pharmacology, 10*, 6.

Berna-Erro, A., Izquierdo-Serra, M., Sepúlveda, R. V., Rubio-Moscardo, F., Doñate-Macián, P., Serra, S. A., et al. (2017). Structural determinants of 5′, 6′-epoxyeicosatrienoic acid binding to and activation of TRPV4 channel. *Scientific Reports, 7*, 1–11.

Blair, N. T., Carvacho, I., Chaudhuri, D., Clapham, D. E., DeCaen, P., Delling, M., et al. (2019). In *Transient receptor potential channels (version 2019.4) in the IUPHAR/BPS Guide to Pharmacology Database IUPHAR/BPS Guide to Pharmacology CITE 2019.*

Blakney, A. K., Swartzlander, M. D., & Bryant, S. J. (2012). The effects of substrate stiffness on the in vitro activation of macrophages and in vivo host response to poly (ethylene glycol)-based hydrogels. *Journal of Biomedical Materials Research Part A, 100*, 1375.

Boehm, T., Iwanami, N., & Hess, I. (2012). Evolution of the immune system in the lower vertebrates. *Annual Review of Genomics and Human Genetics, 13*, 127–149.

Boltana, S., Rey, S., Roher, N., Vargas, R., Huerta, M., Huntingford, F. A., et al. (2013). Behavioural fever is a synergic signal amplifying the innate immune response. *Proceedings of the Royal Society B: Biological Sciences, 280*, 20131381.

Bonvini, S. J., Birrell, M. A., Dubuis, E., Adcock, J. J., Wortley, M. A., Flajolet, P., et al. (2020). Novel airway smooth muscle–mast cell interactions and a role for the TRPV4-ATP axis in non-atopic asthma. *European Respiratory Journal, 56.*

Brambilla, R., Bracchi-Ricard, V., Hu, W.-H., Frydel, B., Bramwell, A., Karmally, S., et al. (2005). Inhibition of astroglial nuclear factor κB reduces inflammation and improves functional recovery after spinal cord injury. *The Journal of Experimental Medicine, 202*, 145–156.

Brambilla, R., Dvoriantchikova, G., Barakat, D., Ivanov, D., Bethea, J. R., & Shestopalov, V. I. (2012). Transgenic inhibition of astroglial NF-κB protects from optic nerve damage and retinal ganglion cell loss in experimental optic neuritis. *Journal of Neuroinflammation, 9*, 1–12.

Brambilla, R., Morton, P. D., Ashbaugh, J. J., Karmally, S., Lambertsen, K. L., & Bethea, J. R. (2014). Astrocytes play a key role in EAE pathophysiology by orchestrating in the CNS the inflammatory response of resident and peripheral immune cells and by suppressing remyelination. *Glia, 62*, 452–467.

Brambilla, R., Persaud, T., Hu, X., Karmally, S., Shestopalov, V. I., Dvoriantchikova, G., et al. (2009). Transgenic inhibition of astroglial NF-κB improves functional outcome in experimental autoimmune encephalomyelitis by suppressing chronic central nervous system inflammation. *The Journal of Immunology, 182*, 2628–2640.

Buchmann, K. (2014). Evolution of innate immunity: clues from invertebrates via fish to mammals. *Frontiers in Immunology, 5*, 459.
Butenko, O., Dzamba, D., Benesova, J., Honsa, P., Benfenati, V., Rusnakova, V., et al. (2012). The increased activity of TRPV4 channel in the astrocytes of the adult rat hippocampus after cerebral hypoxia/ischemia. *PLoS One, 7*, e39959.
Çakır, M., Saçmacı, H., & Sabah-Özcan, S. (2021). Selected transient receptor potential channel genes' expression in peripheral blood mononuclear cells of multiple sclerosis. *Human & Experimental Toxicology, 40*, S406–S413.
Catalán, T. P., Wozniak, A., Niemeyer, H. M., Kalergis, A. M., & Bozinovic, F. (2012). Interplay between thermal and immune ecology: Effect of environmental temperature on insect immune response and energetic costs after an immune challenge. *Journal of Insect Physiology, 58*, 310–317.
Cenac, N., Altier, C., Motta, J.-P., d'Aldebert, E., Galeano, S., Zamponi, G. W., et al. (2010). Potentiation of TRPV4 signalling by histamine and serotonin: An important mechanism for visceral hypersensitivity. *Gut, 59*, 481–488.
Chakraborty, R., & Goswami, C. (2022). Both heat-sensitive TRPV4 and cold-sensitive TRPM8 ion channels regulate microglial activity. *Biochemical and Biophysical Research Communications, 611*, 132–139.
Cooper, E. L. (2010). Evolution of immune systems from self/not self to danger to artificial immune systems (AIS). *Physics of Life Reviews, 7*, 55–78.
Danilova, N. (2006). The evolution of immune mechanisms. *Journal of Experimental Zoology Part B: Molecular and Developmental Evolution, 306*, 496–520.
Das, R., & Goswami, C. (2019). TRPV4 expresses in bone cell lineages and TRPV4-R616Q mutant causing Brachyolmia in human reveals "loss-of-interaction" with cholesterol. *Biochemical and Biophysical Research Communications, 517*, 566–574.
Deng, Z., Paknejad, N., Maksaev, G., Sala-Rabanal, M., Nichols, C. G., Hite, R. K., et al. (2018). Cryo-EM and X-ray structures of TRPV4 reveal insight into ion permeation and gating mechanisms. *Nature Structural & Molecular Biology, 25*, 252–260.
Di Nardo, A., Mascarenhas, N., & Wang, Z. (2017). TRPV4 is a key driver for mast cell activation in LL37mediated skin inflammation. *Journal of Dermatological Science, 86*, e69.
Dias, F. C., Alves, V. S., Matias, D. O., Figueiredo, C. P., Miranda, A. L. P., Passos, G. F., et al. (2019). The selective TRPV4 channel antagonist HC-067047 attenuates mechanical allodynia in diabetic mice. *European Journal of Pharmacology, 856*, 172408.
Ding, Z.-B., Song, L.-J., Wang, Q., Kumar, G., Yan, Y.-Q., & Ma, C.-G. (2021). Astrocytes: A double-edged sword in neurodegenerative diseases. *Neural Regeneration Research, 16*, 1702.
Dittmar, J., Janssen, H., Kuske, A., Kurtz, J., & Scharsack, J. P. (2014). Heat and immunity: An experimental heat wave alters immune functions in three-spined sticklebacks (G asterosteus aculeatus). *Journal of Animal Ecology, 83*, 744–757.
Echeverry, S., Rodriguez, M. J., & Torres, Y. P. (2016). Transient receptor potential channels in microglia: Roles in physiology and disease. *Neurotoxicity Research, 30*, 467–478.
Engelhardt, B., Vajkoczy, P., & Weller, R. O. (2017). The movers and shapers in immune privilege of the CNS. *Nature Immunology, 18*, 123–131.
Evans, S. S., Repasky, E. A., & Fisher, D. T. (2015). Fever and the thermal regulation of immunity: The immune system feels the heat. *Nature Reviews Immunology, 15*, 335–349.
Everaerts, W., Nilius, B., & Owsianik, G. (2010). The vanilloid transient receptor potential channel TRPV4: From structure to disease. *Progress in Biophysics and Molecular Biology, 103*, 2–17.
Féréol, S., Fodil, R., Labat, B., Galiacy, S., Laurent, V. M., Louis, B., et al. (2006). Sensitivity of alveolar macrophages to substrate mechanical and adhesive properties. *Cell Motility and the Cytoskeleton, 63*, 321–340.

Feske, S. (2011). Immunodeficiency due to defects in store-operated calcium entry. *Annals of the New York Academy of Sciences*, *1238*, 74–90.

Feske, S., Wulff, H., & Skolnik, E. Y. (2015). Ion channels in innate and adaptive immunity. *Annual Review of Immunology*, *33*, 291–353.

Fichna, J., Mokrowiecka, A., Cygankiewicz, A., Zakrzewski, P., Małecka-Panas, E., Janecka, A., et al. (2012). Transient receptor potential vanilloid 4 blockade protects against experimental colitis in mice: A new strategy for inflammatory bowel diseases treatment? *Neurogastroenterology & Motility*, *24*, e557–e560.

Fischbarg, J., Cheung, M., Li, J., Iserovich, P., Czegledy, F., Kuang, K., et al. (1994). Are most transporters and channels beta barrels? *Molecular and Cellular Biochemistry*, *140*, 147–162.

Foxman, E. F., Storer, J. A., Fitzgerald, M. E., Wasik, B. R., Hou, L., Zhao, H., et al. (2015). Temperature-dependent innate defense against the common cold virus limits viral replication at warm temperature in mouse airway cells. *Proceedings of the National Academy of Sciences of the United States of America*, *112*, 827–832.

Fu, E. S., Zhang, Y. P., Sagen, J., Candiotti, K. A., Morton, P. D., Liebl, D. J., et al. (2010). Transgenic inhibition of glial NF-kappa B reduces pain behavior and inflammation after peripheral nerve injury. *PAIN®*, *148*, 509–518.

Fusi, C., Materazzi, S., Minocci, D., Maio, V., Oranges, T., Massi, D., et al. (2014). Transient receptor potential vanilloid 4 (TRPV4) is downregulated in keratinocytes in human non-melanoma skin cancer. *Journal of Investigative Dermatology*, *134*, 2408–2417.

Gadani, S. P., Walsh, J. T., Lukens, J. R., & Kipnis, J. (2015). Dealing with danger in the CNS: The response of the immune system to injury. *Neuron*, *87*, 47–62.

Ghoneum, M. H., Gimzewski, J. K., Ghoneum, A., Katano, H., Paw, U. C. N., & Agrawal, A. (2018). Inhibition of TRPV1 channel activity in human CD4+ T cells by nanodiamond and nanoplatinum liquid, DPV576. *Nanomaterials*, *8*, 770.

Ghosh, A., Kaur, N., Kumar, A., & Goswami, C. (2016). Why individual thermo sensation and pain perception varies? Clue of disruptive mutations in TRPVs from 2504 human genome data. *Channels*, *10*, 339–345.

Gilliver, S. C. (2010). Sex steroids as inflammatory regulators. *The Journal of Steroid Biochemistry and Molecular Biology*, *120*, 105–115.

Glass, C. K., Saijo, K., Winner, B., Marchetto, M. C., & Gage, F. H. (2010). Mechanisms underlying inflammation in neurodegeneration. *Cell*, *140*, 918–934.

Goghari, V., Franciosi, S., Kim, S. U., Lee, Y. B., & McLarnon, J. G. (2000). Acute application of interleukin-1β induces Ca2+ responses in human microglia. *Neuroscience Letters*, *281*, 83–86.

Goldenberg, N. M., Ravindran, K., & Kuebler, W. M. (2015). TRPV4: Physiological role and therapeutic potential in respiratory diseases. *Naunyn-Schmiedeberg's Archives of Pharmacology*, *388*, 421–436.

Goswami, C., Kuhn, J., Heppenstall, P. A., & Hucho, T. (2010). Importance of non-selective cation channel TRPV4 interaction with cytoskeleton and their reciprocal regulations in cultured cells. *PLoS One*, *5*, e11654.

Goswami, R., Merth, M., Sharma, S., Alharbi, M. O., Aranda-Espinoza, H., Zhu, X., et al. (2017). TRPV4 calcium-permeable channel is a novel regulator of oxidized LDL-induced macrophage foam cell formation. *Free Radical Biology and Medicine*, *110*, 142–150.

Güler, A. D., Lee, H., Iida, T., Shimizu, I., Tominaga, M., & Caterina, M. (2002). Heat-evoked activation of the ion channel, TRPV4. *Journal of Neuroscience*, *22*, 6408–6414.

Gupta, N., Goswami, R., Alharbi, M. O., Biswas, D., & Rahaman, S. O. (2019). TRPV 4 is a regulator in *P. gingivalis* lipopolysaccharide-induced exacerbation of macrophage foam cell formation. *Physiological Reports, 7*, e14069.

Hamanaka, K., Jian, M.-Y., Townsley, M. I., King, J. A., Liedtke, W., Weber, D. S., et al. (2010). TRPV4 channels augment macrophage activation and ventilator-induced lung injury. *American Journal of Physiology-Lung Cellular and Molecular Physiology, 299*, L353–L362.

Hamanaka, K., Jian, M.-Y., Weber, D. S., Alvarez, D. F., Townsley, M. I., Al-Mehdi, A. B., et al. (2007). TRPV4 initiates the acute calcium-dependent permeability increase during ventilator-induced lung injury in isolated mouse lungs. *American Journal of Physiology-Lung Cellular and Molecular Physiology, 293*, L923–L932.

Hamano, N., Terada, N., Maesako, K.-I., Numata, T., & Konno, A. (1998). In *Effect of sex hormones on eosinophilic inflammation in nasal mucosa Allergy and Asthma Proceedings* (p. 263). OceanSide Publications.

Hansen, L., & Taylor, W. R. (2016). Is increased arterial stiffness a cause or consequence of atherosclerosis? *Atherosclerosis, 249*, 226–227.

Hogan, P. G., Lewis, R. S., & Rao, A. (2009). Molecular basis of calcium signaling in lymphocytes: STIM and ORAI. *Annual Review of Immunology, 28*, 491–533.

Jung, C., Fandos, C., Lorenzo, I. M., Plata, C., Fernandes, J., Gené, G. G., et al. (2009). The progesterone receptor regulates the expression of TRPV4 channel. *Pflügers Archiv-European Journal of Physiology, 459*, 105–113.

Jung, C., Fernández-Dueñas, V., Plata, C., Garcia-Elias, A., Ciruela, F., Fernández-Fernández, J. M., et al. (2018). Functional coupling of GABAA/B receptors and the channel TRPV4 mediates rapid progesterone signaling in the oviduct. *Science Signaling, 11*, eaam6558.

Kettenmann, H., Hanisch, U.-K., Noda, M., & Verkhratsky, A. (2011). Physiology of microglia. *Physiological Reviews, 91*, 461–553.

Khorooshi, R., Babcock, A. A., & Owens, T. (2008). NF-κB-driven STAT2 and CCL2 expression in astrocytes in response to brain injury. *The Journal of Immunology, 181*, 7284–7291.

Kigerl, K. A., de Rivero Vaccari, J. P., Dietrich, W. D., Popovich, P. G., & Keane, R. W. (2014). Pattern recognition receptors and central nervous system repair. *Experimental Neurology, 258*, 5–16.

Kim, K. S., Shin, D. H., Nam, J. H., Park, K. S., Zhang, Y. H., Kim, W. K., et al. (2010). Functional expression of TRPV4 cation channels in human mast cell line (HMC-1). *The Korean Journal of Physiology & Pharmacology, 14*, 419–425.

Knippertz, I., Stein, M. F., Dörrie, J., Schaft, N., Müller, I., Deinzer, A., et al. (2011). Mild hyperthermia enhances human monocyte-derived dendritic cell functions and offers potential for applications in vaccination strategies. *International Journal of Hyperthermia, 27*, 591–603.

Kochukov, M. Y., McNearney, T. A., Yin, H., Zhang, L., Ma, F., Ponomareva, L., et al. (2009). Tumor necrosis factor-alpha (TNF-α) enhances functional thermal and chemical responses of TRP cation channels in human synoviocytes. *Molecular Pain, 5*. 1744-8069-1745-1749.

Konno, M., Shirakawa, H., Iida, S., Sakimoto, S., Matsutani, I., Miyake, T., et al. (2012). Stimulation of transient receptor potential vanilloid 4 channel suppresses abnormal activation of microglia induced by lipopolysaccharide. *Glia, 60*, 761–770.

Kumar, H., Lim, C. S., Choi, H., Joshi, H. P., Kim, K.-T., Kim, Y. H., et al. (2020). Elevated TRPV4 levels contribute to endothelial damage and scarring in experimental spinal cord injury. *Journal of Neuroscience, 40*, 1943–1955.

Kumari, S., Kumar, A., Sardar, P., Yadav, M., Majhi, R. K., Kumar, A., et al. (2015). Influence of membrane cholesterol in the molecular evolution and functional regulation of TRPV4. *Biochemical and Biophysical Research Communications*, *456*, 312–319.

Li, Q., & Barres, B. A. (2018). Microglia and macrophages in brain homeostasis and disease. *Nature Reviews Immunology*, *18*, 225–242.

Liedtke, W., Choe, Y., Martí-Renom, M. A., Bell, A. M., Denis, C. S., Hudspeth, A., et al. (2000). Vanilloid receptor–related osmotically activated channel (VR-OAC), a candidate vertebrate osmoreceptor. *Cell*, *103*, 525–535.

Liedtke, W., Tobin, D. M., Bargmann, C. I., & Friedman, J. M. (2003). Mammalian TRPV4 (VR-OAC) directs behavioral responses to osmotic and mechanical stimuli in Caenorhabditis elegans. *Proceedings of the National Academy of Sciences of the United States of America*, *100*, 14531–14536.

Liu, M., Liu, X., Wang, L., Wang, Y., Dong, F., Wu, J., et al. (2018). TRPV4 inhibition improved myelination and reduced glia reactivity and inflammation in a cuprizone-induced mouse model of demyelination. *Frontiers in Cellular Neuroscience*, *392*.

Loukin, S., Zhou, X., Su, Z., Saimi, Y., & Kung, C. (2010). Wild-type and brachyolmia-causing mutant TRPV4 channels respond directly to stretch force. *Journal of Biological Chemistry*, *285*, 27176–27181.

Louveau, A., Smirnov, I., Keyes, T. J., Eccles, J. D., Rouhani, S. J., Peske, J. D., et al. (2015). Structural and functional features of central nervous system lymphatic vessels. *Nature*, *523*, 337–341.

Ma, X., He, D., Ru, X., Chen, Y., Cai, Y., Bruce, I. C., et al. (2012). Apigenin, a plant-derived flavone, activates transient receptor potential vanilloid 4 cation channel. *British Journal of Pharmacology*, *166*, 349–358.

Mace, T. A., Zhong, L., Kokolus, K. M., & Repasky, E. A. (2012). Effector CD8+ T cell IFN-γ production and cytotoxicity are enhanced by mild hyperthermia. *International Journal of Hyperthermia*, *28*, 9–18.

Mair, W., Morantte, I., Rodrigues, A. P., Manning, G., Montminy, M., Shaw, R. J., et al. (2011). Lifespan extension induced by AMPK and calcineurin is mediated by CRTC-1 and CREB. *Nature*, *470*, 404–408.

Majhi, R. K., Maity, A., Senapati, M. R., Behera, P. C., Mandal, A. K., & Giri, S. C. (2018). *Endogenous expression of thermo-sensitive ion channels TRPV1 and TRPV4 in immune tissues of avian species (duck, Anas platyrhynchos)*. PeerJ Preprints.

Majhi, R. K., Sahoo, S. S., Yadav, M., Pratheek, B. M., Chattopadhyay, S., & Goswami, C. (2015). Functional expression of TRPV channels in T cells and their implications in immune regulation. *The FEBS Journal*, *282*, 2661–2681.

Mariotti, S., Prinzis, A., Ghiani, M., Cambuli, V. M., Pilia, S., Marras, V., et al. (2009). Puberty is associated with a marked increase of the female sex predominance in chronic autoimmune thyroiditis. *Hormone Research in Paediatrics*, *72*, 52–56.

Martin-Gayo, E., & Yu, X. G. (2019). Role of dendritic cells in natural immune control of HIV-1 infection. *Frontiers in Immunology*, *10*, 1306.

Mascarenhas, N. L., Wang, Z., Chang, Y.-L., & Di Nardo, A. (2016). TRPV4 mediates mast cell activation in cathelicidin-induced rosacea inflammation. *The Journal of Investigative Dermatology*, *137*, 972–975.

Masuyama, R., Vriens, J., Voets, T., Karashima, Y., Owsianik, G., Vennekens, R., et al. (2008). TRPV4-mediated calcium influx regulates terminal differentiation of osteoclasts. *Cell Metabolism*, *8*, 257–265.

Maus, M., Jairaman, A., Stathopulos, P. B., Muik, M., Fahrner, M., Weidinger, C., et al. (2015). Missense mutation in immunodeficient patients shows the multifunctional roles of coiled-coil domain 3 (CC3) in STIM1 activation. *Proceedings of the National Academy of Sciences of the United States of America*, *112*, 6206–6211.

McWhorter, F. Y., Wang, T., Nguyen, P., Chung, T., & Liu, W. F. (2013). Modulation of macrophage phenotype by cell shape. *Proceedings of the National Academy of Sciences of the United States of America, 110*, 17253–17258.

Melendez, A. J., & Tay, H. K. (2008). Phagocytosis: A repertoire of receptors and Ca2+ as a key second messenger. *Bioscience Reports, 28*, 287–298.

Michalick, L., Erfinanda, L., Weichelt, U., van der Giet, M., Liedtke, W., & Kuebler, W. M. (2017). Transient receptor potential vanilloid 4 and serum glucocorticoid–regulated kinase 1 are critical mediators of lung injury in overventilated mice in vivo. *Anesthesiology, 126*, 300–311.

Michalick, L., & Kuebler, W. M. (2020). TRPV4—A missing link between mechanosensation and immunity. *Frontiers in Immunology, 413*.

Minke, B. (2010). The history of the Drosophila TRP channel: The birth of a new channel superfamily. *Journal of Neurogenetics, 24*, 216–233.

Mochizuki, T., Sokabe, T., Araki, I., Fujishita, K., Shibasaki, K., Uchida, K., et al. (2009). The TRPV4 cation channel mediates stretch-evoked Ca^{2+} influx and ATP release in primary urothelial cell cultures. *Journal of Biological Chemistry, 284*, 21257–21264.

Mosser, D. M., & Edwards, J. P. (2008). Exploring the full spectrum of macrophage activation. *Nature Reviews Immunology, 8*, 958–969.

Munoz-Suano, A., Hamilton, A. B., & Betz, A. G. (2011). Gimme shelter: The immune system during pregnancy. *Immunological Reviews, 241*, 20–38.

Murdock, C., Paaijmans, K. P., Bell, A. S., King, J. G., Hillyer, J. F., Read, A. F., et al. (2012). Complex effects of temperature on mosquito immune function. *Proceedings of the Royal Society B: Biological Sciences, 279*, 3357–3366.

Naert, R., López-Requena, A., Voets, T., Talavera, K., & Alpizar, Y. A. (2019). Expression and functional role of TRPV4 in bone marrow-derived CD11C+ cells. *International Journal of Molecular Sciences, 20*, 3378.

Navakkode, S., Liu, C., & Soong, T. W. (2018). Altered function of neuronal L-type calcium channels in ageing and neuroinflammation: Implications in age-related synaptic dysfunction and cognitive decline. *Ageing Research Reviews, 42*, 86–99.

Nishimoto, R., Derouiche, S., Eto, K., Deveci, A., Kashio, M., Kimori, Y., et al. (2021). Thermosensitive TRPV4 channels mediate temperature-dependent microglia movement. *Proceedings of the National Academy of Sciences of the United States of America, 118*.

Nunes, P., & Demaurex, N. (2010). The role of calcium signaling in phagocytosis. *Journal of Leukocyte Biology, 88*, 57–68.

Okada, Y., Shirai, K., Miyajima, M., Reinach, P. S., Yamanaka, O., Sumioka, T., et al. (2016). Loss of TRPV4 function suppresses inflammatory fibrosis induced by alkali-burning mouse corneas. *PLoS One, 11*, e0167200.

Okuno, D., Sugiura, Y., Sakamoto, N., Tagod, M. S., Iwasaki, M., Noda, S., et al. (2020). Comparison of a novel bisphosphonate prodrug and zoledronic acid in the induction of cytotoxicity in human Vγ2Vδ2 T cells. *Frontiers in Immunology, 1405*.

Özcan, S., Gürel, G., & Çakır, M. (2021). Gene expression profiles of transient receptor potential (TRP) channels in the peripheral blood mononuclear cells of psoriasis patients. *Human & Experimental Toxicology, 40*, 1234–1240.

Padilla, L., Reinicke, K., Montesino, H., Villena, F., Asencio, H., Cruz, M., et al. (1990). Histamine content and mast cells distribution in mouse uterus: The effect of sexual hormones, gestation and labor. *Cellular and Molecular Biology, 36*, 93–100.

Rahaman, S. O., Grove, L. M., Paruchuri, S., Southern, B. D., Abraham, S., Niese, K. A., et al. (2014). TRPV4 mediates myofibroblast differentiation and pulmonary fibrosis in mice. *The Journal of Clinical Investigation, 124*, 5225–5238.

Rahaman, S. O., Lennon, D. J., Febbraio, M., Podrez, E. A., Hazen, S. L., & Silverstein, R. L. (2006). A CD36-dependent signaling cascade is necessary for macrophage foam cell formation. *Cell Metabolism, 4*, 211–221.

Rayees, S., Joshi, J. C., Tauseef, M., Anwar, M., Baweja, S., Rochford, I., et al. (2019). PAR2-mediated cAMP generation suppresses TRPV4-dependent Ca^{2+} signaling in alveolar macrophages to resolve TLR4-induced inflammation. *Cell Reports*, *27*(793–805), e794.

Rizopoulos, T., Papadaki-Petrou, H., & Assimakopoulou, M. (2018). Expression profiling of the transient receptor potential vanilloid (TRPV) channels 1, 2, 3 and 4 in mucosal epithelium of human ulcerative colitis. *Cells*, 7, 61.

Rollins-Smith, L. A., & Woodhams, D. C. (2012). *Amphibian immunity*. New York: Oxford University Press.

Saha, S., Ghosh, A., Tiwari, N., Kumar, A., Kumar, A., & Goswami, C. (2017). Preferential selection of Arginine at the lipid-water-interface of TRPV1 during vertebrate evolution correlates with its snorkeling behaviour and cholesterol interaction. *Scientific Reports*, 7, 1–21.

Saha, S., Sucharita, S., Majhi, R. K., Tiwari, A., Ghosh, A., Pradhan, S. K., et al. (2019). TRPA1 is selected as a semi-conserved channel during vertebrate evolution due to its involvement in spermatogenesis. *Biochemical and Biophysical Research Communications*, *512*, 295–302.

Scheraga, R. G., Abraham, S., Niese, K. A., Southern, B. D., Grove, L. M., Hite, R. D., et al. (2016). TRPV4 mechanosensitive ion channel regulates lipopolysaccharide-stimulated macrophage phagocytosis. *The Journal of Immunology*, *196*, 428–436.

Scheraga, R. G., Southern, B. D., Grove, L. M., & Olman, M. A. (2020). The role of TRPV4 in regulating innate immune cell function in lung inflammation. *Frontiers in Immunology*, *11*, 1211.

Shen, R.-N., Lu, L., Young, P., Shidnia, H., Hornback, N. B., & Broxmeyer, H. E. (1994). Influence of elevated temperature on natural killer cell activity, lymphokine-activated killer cell activity and lectin-dependent cytotoxicity of human umbilical cord blood and adult blood cells. *International Journal of Radiation Oncology Biology Physics*, *29*, 821–826.

Shibasaki, K., Ikenaka, K., Tamalu, F., Tominaga, M., & Ishizaki, Y. (2014). A novel subtype of astrocytes expressing TRPV4 (transient receptor potential vanilloid 4) regulates neuronal excitability via release of gliotransmitters. *Journal of Biological Chemistry*, *289*, 14470–14480.

Shigematsu, H., Sokabe, T., Danev, R., Tominaga, M., & Nagayama, K. (2010). A 3.5-nm structure of rat TRPV4 cation channel revealed by Zernike phase-contrast cryoelectron microscopy. *Journal of Biological Chemistry*, *285*, 11210–11218.

Shirakawa, H., & Kaneko, S. (2018). Physiological and pathophysiological roles of transient receptor potential channels in microglia-related CNS inflammatory diseases. *Biological and Pharmaceutical Bulletin*, *41*, 1152–1157.

Shumilina, E., Huber, S. M., & Lang, F. (2011). Ca^{2+} signaling in the regulation of dendritic cell functions. *American Journal of Physiology-Cell Physiology*, *300*, C1205–C1214.

Solis-Lopez, A., Kriebs, U., Marx, A., Mannebach, S., Liedtke, W., Caterina, M., et al. (2017). Analysis of TRPV channel activation by stimulation of FCεRI and MRGPR receptors in mouse peritoneal mast cells. *PLoS One*, *12*, e0171366.

Spinsanti, G., Zannolli, R., Panti, C., Ceccarelli, I., Marsili, L., Bachiocco, V., et al. (2008). Quantitative real-time PCR detection of TRPV1–4 gene expression in human leukocytes from healthy and hyposensitive subjects. *Molecular Pain*, *4*. 1744-8069-1744-1751.

Stockwin, L. H., McGonagle, D., Martin, I. G., & Blair, G. E. (2000). Dendritic cells: Immunological sentinels with a central role in health and disease. *Immunology and Cell Biology*, *78*, 91–102.

Stokes, A. J., Shimoda, L. M., Koblan-Huberson, M., Adra, C. N., & Turner, H. (2004). A trpv2–pka signaling module for transduction of physical stimuli in mast cells. *The Journal of Experimental Medicine*, *200*, 137–147.

Strotmann, R., Harteneck, C., Nunnenmacher, K., Schultz, G., & Plant, T. D. (2000). OTRPC4, a nonselective cation channel that confers sensitivity to extracellular osmolarity. *Nature Cell Biology*, *2*, 695–702.

Sulk, M., & Steinhoff, M. (2015). Role of TRP channels in skin diseases. In *TRP channels as therapeutic targets* (pp. 293–323). Elsevier.

Szöllősi, A. G., Oláh, A., Tóth, I. B., Papp, F., Czifra, G., Panyi, G., et al. (2013). Transient receptor potential vanilloid-2 mediates the effects of transient heat shock on endocytosis of human monocyte-derived dendritic cells. *FEBS Letters*, *587*, 1440–1445.

Thodeti, C. K., Matthews, B., Ravi, A., Mammoto, A., Ghosh, K., Bracha, A. L., et al. (2009). TRPV4 channels mediate cyclic strain-induced endothelial cell reorientation through integrin-to-integrin signaling. *Circulation Research*, *104*, 1123–1130.

Thomas, P. (2008). Characteristics of membrane progestin receptor alpha (mPRα) and progesterone membrane receptor component 1 (PGMRC1) and their roles in mediating rapid progestin actions. *Frontiers in Neuroendocrinology*, *29*, 292–312.

Thorneloe, K. S., Cheung, M., Bao, W., Alsaid, H., Lenhard, S., Jian, M.-Y., et al. (2012). An orally active TRPV4 channel blocker prevents and resolves pulmonary edema induced by heart failure. *Science Translational Medicine*, *4*. 159ra148-159ra148.

Thorneloe, K. S., Sulpizio, A. C., Lin, Z., Figueroa, D. J., Clouse, A. K., McCafferty, G. P., et al. (2008). N-((1S)-1-{[4-((2S)-2-{[(2, 4-dichlorophenyl) sulfonyl] amino}-3-hydroxypropanoyl)-1-piperazinyl] carbonyl}-3-methylbutyl)-1-benzothiophene-2-carboxamide (GSK1016790A), a novel and potent transient receptor potential vanilloid 4 channel agonist induces urinary bladder contraction and hyperactivity: Part I. *Journal of Pharmacology and Experimental Therapeutics*, *326*, 432–442.

Umar, D., Das, A., Gupta, S., Chattopadhyay, S., Sarkar, D., Mirji, G., et al. (2020). Febrile temperature change modulates CD4 T cell differentiation via a TRPV channel-regulated Notch-dependent pathway. *Proceedings of the National Academy of Sciences of the United States of America*, *117*, 22357–22366.

Vaeth, M., & Feske, S. (2018). Ion channelopathies of the immune system. *Current Opinion in Immunology*, *52*, 39–50.

Verkhratsky, A., Reyes, R. C., & Parpura, V. (2013). TRP channels coordinate ion signalling in astroglia. *Reviews of Physiology, Biochemistry and Pharmacology*, *166*, 1–22.

Verma, P., Kumar, A., & Goswami, C. (2010). TRPV4-mediated channelopathies. *Channels*, *4*, 319–328.

Vig, M., & Kinet, J.-P. (2009). Calcium signaling in immune cells. *Nature Immunology*, *10*, 21–27.

Vincent, F., Acevedo, A., Nguyen, M. T., Dourado, M., DeFalco, J., Gustafson, A., et al. (2009). Identification and characterization of novel TRPV4 modulators. *Biochemical and Biophysical Research Communications*, *389*, 490–494.

Vincent, A., Lang, B., & Kleopa, K. A. (2006). Autoimmune channelopathies and related neurological disorders. *Neuron*, *52*, 123–138.

Voets, T., Prenen, J., Vriens, J., Watanabe, H., Janssens, A., Wissenbach, U., et al. (2002). Molecular determinants of permeation through the cation channel TRPV4. *Journal of Biological Chemistry*, *277*, 33704–33710.

Vriens, J., Owsianik, G., Janssens, A., Voets, T., & Nilius, B. (2007). Determinants of 4α-phorbol sensitivity in transmembrane domains 3 and 4 of the cation channel TRPV4. *Journal of Biological Chemistry*, *282*, 12796–12803.

Vriens, J., Watanabe, H., Janssens, A., Droogmans, G., Voets, T., & Nilius, B. (2004). Cell swelling, heat, and chemical agonists use distinct pathways for the activation of the cation channel TRPV4. *Proceedings of the National Academy of Sciences of the United States of America*, *101*, 396–401.

Wang, X.-Y., Ostberg, J. R., & Repasky, E. A. (1999). Effect of fever-like whole-body hyperthermia on lymphocyte spectrin distribution, protein kinase C activity, and uropod formation. *The Journal of Immunology*, *162*, 3378–3387.

Wang, Z., Zhou, L., An, D., Xu, W., Wu, C., Sha, S., et al. (2019). TRPV4-induced inflammatory response is involved in neuronal death in pilocarpine model of temporal lobe epilepsy in mice. *Cell Death & Disease*, *10*, 1–10.
Watanabe, H., Davis, J. B., Smart, D., Jerman, J. C., Smith, G. D., Hayes, P., et al. (2002). Activation of TRPV4 channels (hVRL-2/mTRP12) by phorbol derivatives. *Journal of Biological Chemistry*, *277*, 13569–13577.
Watanabe, H., Vriens, J., Prenen, J., Droogmans, G., Voets, T., & Nilius, B. (2003). Anandamide and arachidonic acid use epoxyeicosatrienoic acids to activate TRPV4 channels. *Nature*, *424*, 434–438.
Watanabe, H., Vriens, J., Suh, S. H., Benham, C. D., Droogmans, G., & Nilius, B. (2002). Heat-evoked activation of TRPV4 channels in a HEK293 cell expression system and in native mouse aorta endothelial cells. *Journal of Biological Chemistry*, *277*, 47044–47051.
Watman, N. P., Crespo, L., Davis, B., Poiesz, B. J., & Zamkoff, K. W. (1988). Differential effect on fresh and cultured T cells of PHA-induced changes in free cytoplasmic calcium: Relation to IL-2 receptor expression, IL-2 production, and proliferation. *Cellular Immunology*, *111*, 158–166.
Weckerle, C. E., & Niewold, T. B. (2011). The unexplained female predominance of systemic lupus erythematosus: Clues from genetic and cytokine studies. *Clinical Reviews in Allergy & Immunology*, *40*, 42–49.
Yin, J., Michalick, L., Tang, C., Tabuchi, A., Goldenberg, N., Dan, Q., et al. (2016). Role of transient receptor potential vanilloid 4 in neutrophil activation and acute lung injury. *American Journal of Respiratory Cell and Molecular Biology*, *54*, 370–383.
Zänker, K., & Lange, J. (1982). Whole body hyperthermia and natural killer cell activity. *The Lancet*, *319*, 1079–1080.
Zheng, X., Zinkevich, N. S., Gebremedhin, D., Gauthier, K. M., Nishijima, Y., Fang, J., et al. (2013). Arachidonic acid–induced dilation in human coronary arterioles: Convergence of signaling mechanisms on endothelial TRPV 4-mediated Ca^{2+} entry. *Journal of the American Heart Association*, *2*, e000080.
Zink, A., Listing, J., Klindworth, C., & Zeidler, H. (2001). The national database of the German collaborative arthritis centres: I. Structure, aims, and patients. *Annals of the Rheumatic Diseases*, *60*, 199–206.
Zou, W., Pu, T., Feng, W., Lu, M., Zheng, Y., Du, R., et al. (2019). Blocking meningeal lymphatic drainage aggravates Parkinson's disease-like pathology in mice overexpressing mutated α-synuclein. *Translational Neurodegeneration*, *8*, 1–17.

CHAPTER EIGHT

TRPV4: Cell type-specific activation, regulation and function in the vertebrate eye

Luka Lapajne[a,b,†], Christopher N. Rudzitis[a,†], Brenan Cullimore[a], Daniel Ryskamp[a], Monika Lakk[a], Sarah N. Redmon[a], Oleg Yarishkin[a], and David Krizaj[a,c,d,*]

[a]Department of Ophthalmology & Visual Sciences, University of Utah School of Medicine, Salt Lake City, UT, United States
[b]Department of Ophthalmology, University Medical Centre, University of Ljubljana, Ljubljana, Slovenia
[c]Department of Neurobiology, University of Utah, Salt Lake City, UT, United States
[d]Department of Bioengineering, University of Utah, Salt Lake City, UT, United States
*Corresponding author: e-mail address: david.krizaj@hsc.utah.edu

Contents

† Equal contribution.

Current Topics in Membranes, Volume 89
ISSN 1063-5823
https://doi.org/10.1016/bs.ctm.2022.07.005

Abstract

The architecture of the vertebrate eye is optimized for efficient delivery and transduction of photons and processing of signaling cascades downstream from phototransduction. The cornea, lens, retina, vasculature, ciliary body, ciliary muscle, iris and sclera have specialized functions in ocular protection, transparency, accommodation, fluid regulation, metabolism and inflammatory signaling, which are required to enable function of the retina—light sensitive tissue in the posterior eye that transmits visual signals to relay centers in the midbrain. This process can be profoundly impacted by non-visual stimuli such as mechanical (tension, compression, shear), thermal, nociceptive, immune and chemical stimuli, which target these eye regions to induce pain and precipitate vision loss in glaucoma, diabetic retinopathy, retinal dystrophies, retinal detachment, cataract, corneal dysfunction, ocular trauma and dry eye disease. TRPV4, a polymodal nonselective cation channel, integrate non-visual inputs with homeostatic and signaling functions of the eye. The TRPV4 gene is expressed in most if not all ocular tissues, which vary widely with respect to the mechanisms of TRPV4 channel activation, modulation, oligomerization, and participation in protein- and lipid interactions. Under- and overactivation of TRPV4 may affect intraocular pressure, maintenance of blood–retina barriers, lens accommodation, neuronal function and neuroinflammation. Because TRPV4 dysregulation precipitates many pathologies across the anterior and posterior eye, the channel could be targeted to mitigate vision loss.

1. Introduction

The vertebrate eye has remained remarkably similar over 500 million years, from our early lamprey-like ancestors to humans (Lamb, Collin, & Pugh, 2007). The eye is enclosed in bone and perforated to pass the optic nerve and vasculature, with photons entering across the cornea and refracted

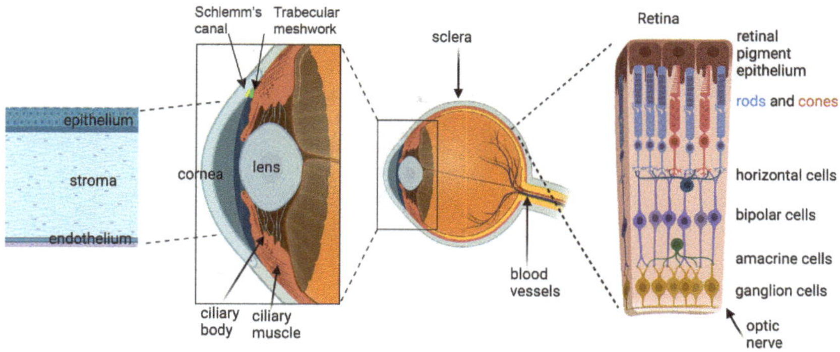

Fig. 1 Schematic representation of the vertebrate eye (modified after Biorender.com). The multicellular structures in the anterior eye (cornea, lens, ciliary body & muscle, trabecular meshwork) control photon access to the eye, intraocular pressure and light refraction onto the light-sensitive tissue (retina) in the posterior eye. The retina is composed of many cell types; phototransduction takes place in rod and cone photoreceptors, which transmit signals down an excitatory bipolar-RGC pathway that is modulated by horizontal and amacrine cells; only the main neuronal classes and the RPE are shown; glia, endothelial cells, pericytes and exotic subtypes of neuron not depicted).

by a contractile crystalline lens onto the retina, a photosensitive tissue (Fig. 1) that transmits the visual signal to the brain. The anterior segment of the eye includes the ciliary body (which produces aqueous humor), trabecular meshwork (the primary drainage mechanism for the aqueous humor), ciliary muscle (secondary drainage mechanism, and regulator of lens contraction) and iris, which determines the color of the eye, and is anatomically divided into two compartments (anterior and posterior chamber) important for aqueous humor circulation. The cornea transitions into an opaque sclera (connective tissue that protects the eye), is innervated by the trigeminal nerve, and consists of multiple layers with different functions (Fig. 2). The retina is supplied by two vascular beds (choroidal and posterior) and functions as a feature extractor that distills 3D information about the shape, color, motion information of the visual stimulus from the 2D pattern of light refracted onto photoreceptors. This information is transmitted by the optic nerve to visual centers in the midbrain.

The Transient Receptor Potential Vanilloid Isoform 4 (TRPV4) channel is expressed in most, if not all, ocular structures as a plasma membrane channel, with possible auxiliary functions in cilia and intracellular store release (Krizaj, Cordeiro, & Strauss, 2022; Luo et al., 2014; Ryskamp et al., 2016, 2011). While the channel pore is likely homotetrameric, heteromerization with TRPV1 and TRPC1 subunits has been suggested

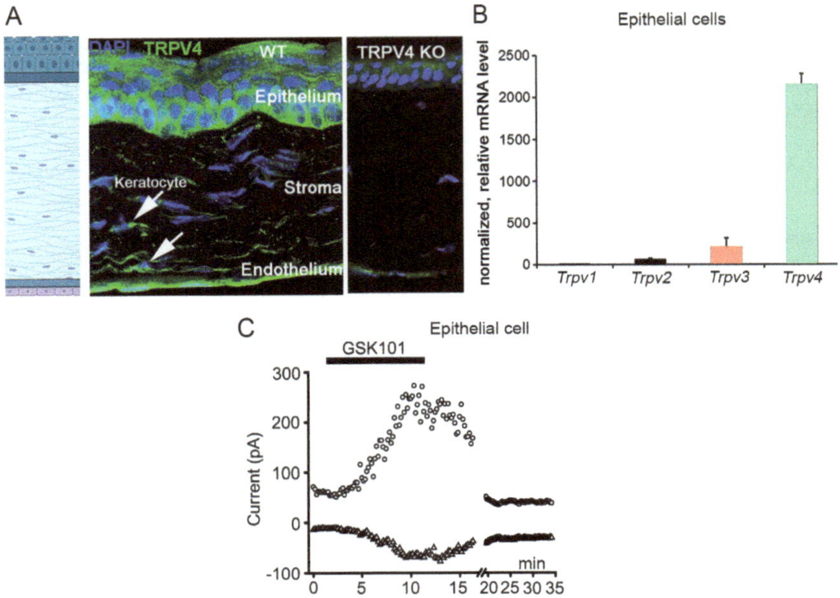

Fig. 2 (A) Antibody labeling. TRPV4 expression is observed in all cellular layers of the cornea, including epithelial, stromal (keratocyte) and endothelial cells. The epithelial layer shows graded basal-to-apical expression. Expression specificity validated in corneas from TRPV4$^{-/-}$ mice developed by W. Liedtke. (B) Relative transcript levels, mouse corneal epithelial cells. TRPV4 expression is strong. (C) Corneal epithelial cell, voltage-clamped at Vm = −100 and +100 mV. At room temperature, the agonist GSK1016790A induces a slow, reversible current. Note that the amplitude of the current will be much smaller at the resting potential (~ −30 mV).

for retinal endothelial and glial cells, respectively (Jo, Lakk, Rudzitis, & Križaj, 2022; O'Leary et al., 2019). Among its likely ocular functions are: cell and tissue volume regulation (Jo et al., 2016, 2015), wound healing and cell proliferation (Lapajne et al., 2020; Okada, Sumioka, Reinach, Miyajima, & Saika, 2022), lens contractility and accommodation (Chen et al., 2019; Shahidullah & Delamere, 2022), production and secretion of aqueous humor (Jo et al., 2016; Ryskamp et al., 2016), corneal and retinal inflammation (Okada et al., 2022; Ryskamp et al., 2014), regulation of epithelial and endothelial barriers (Lapajne et al., 2020; Phuong et al., 2017; Ríos et al., 2019), neuronal excitability (Gao, Yang, Jacoby, Wu, & Pang, 2019; Ryskamp et al., 2011) and apoptosis (Li, Cheng, Zhang, Sun, & Wu, 2021; Ryskamp et al., 2011). Global TRPV4$^{-/-}$ mice don't exhibit obvious morphological ocular abnormalities: knockout animals develop normal eye size and intraocular pressure, and their light-evoked neuronal responses and visual behavior are indistinguishable from those recorded in wild type mice (Lakk et al., 2018; Phuong et al., 2017; Ryskamp et al., 2016; Yarishkin, Phuong, Bretz, et al., 2018; Yarishkin, Phuong, Lakk, & Križaj, 2018). The absence of obvious ocular phenotypes is consistent with the normal development, appearance, fertility and behavior noted in original KO studies by Suzuki, Mizuno, Kodaira, and Imai (2003) and Liedtke and Friedman (2003). Pronounced TRPV4-dependent phenotypes emerge, however, in pathophysiological states such as ocular hypertension, blood–retina barrier breakdown, cholesterolemia and ischemia/hyperoxia (Cappelli et al., 2021; Križaj, 2019; Križaj et al., 2014). In addition, TRPV4 channelopathies perturb eye development and function (Thibodeau et al., 2017) and may be associated with juvenile glaucoma (Young, Whisenhunt, Martin, & Tompson, 2022). In addition to recent reviews of TRPV4 signaling within the eye (Guarino, Paruchuri, & Thodeti, 2020; Thébault, 2021), this review briefly summarizes the ocular landscape of TRPV4 expression and concludes with a reminder TRPV4 function is ultimately defined by its expressing cell type.

2. Cornea

The cornea maintains optical transparency and protects the eye from mechanical, chemical and microbial damage. All cellular layers of the cornea—the epithelium, stroma and endothelium (Fig. 2) prominently express TRPV4.

2.1 Corneal epithelial cells

The corneal epithelium consists of multiple cellular layers forming a barrier that resists the flow of tear fluid and protects the eye from mechanical and chemical injury, desiccation and temperature changes. The cells continually experience intraocular pressure, tear tonicity and $\sim$14 dyn/cm^2 shear from osmotically driven transepithelial ion and water gradients. Chemical injury and mechanical damage from contact lens abrasion, keratoablative interventions, acute mechanical trauma and eyelid sliding over corneal surfaces with reduced moistness may result in loss of corneal transparency. Moreover, epithelial cells function as surrogate Schwann cells to corneal afferents and thus their injury can induce debilitating pain due to overexcitation of mechanonociceptors in the trigeminal nerve.

Transcript expression of the vanilloid TRP subfamily in mouse corneal epithelial cells is dominated by *Trpv4* (Lapajne et al., 2020) (Fig. 2).

Immunohistochemistry and analysis of TRPV4^{eGFP} corneas show robust expression in central and limbal regions. A well-defined basal-to-squamous gradient of TRPV4-ir suggests potential functions in differentiation/stemness and/or barrier permeability. The selective agonist TRPV4 GSK1016790A (GSK101) evokes outwardly rectifying whole cell currents and $[Ca^{2+}]_i$ elevations in dissociated, cultured and wholemount corneal epithelial preparations (Fig. 2C). The channel is activated by moderate temperature and swelling, and may mediate regulatory volume decrease (RVD) in the presence of large hypotonic gradients (Lapajne et al., 2020; Pan et al., 2008). TRPV4 activation in intact corneal epithelial sheets manifests as massive propagation of transepithelial calcium waves that are attenuated by the selective antagonist HC067047, probenecid, suramin and BAPTA-AM. This suggests a model whereby TRPV4 is upstream from Pnx1 hemichannel-mediated ATP release and P2 purinergic receptors (Lapajne et al., 2020), as proposed previously for human pulmonary fibroblasts (Rahman, Sun, Mukherjee, Nilius, & Janssen, 2018), cholangiocyte cilia (Gradilone et al., 2007), renal epithelia (Silva & Garvin, 2008) and the urothelium (Roberts et al., 2020). We suggest that the TRPV4-Pnx1-P2 receptor axis modulates corneal sensitivity and nociception during mechanical injury, acid/alkaline pH exposure and wound healing.

2.2 Corneal stroma

The stroma makes up to ~90% of the thickness of the cornea populated by keratocytes, myofibroblasts and stromal stem cells. Stromal keratocytes label strongly with TRPV4 antibodies and express the TRPV4^{eGFP} marker (Lapajne et al., 2020). TRPV4 may contribute to scar formation resulting from mechanical/chemical injury by promoting TGFβ-dependent transdifferentiation of keratinocytes into myofibroblasts (Okada et al., 2016). Overactivation of keratocyte TRPV4 channels could contribute to cell swelling, calcium overloads and purinergic excitation that underlie corneal edema and transparency loss.

2.3 Corneal endothelial cells

The corneal endothelium maintains corneal transparency by transporting fluid from the stroma into the anterior chamber. Human and mouse corneal endothelial cells express TRPV4 (Lapajne et al., 2020; Mergler et al., 2011) (Fig. 2B). Immortalized cells respond to the selective agonist 4α-PDD, hypotonicity and moderate heat (<38 °C) with inward currents and calcium signals (Mergler et al., 2011). Overexpression in an immortalized human corneal endothelial cell line suppressed metabolism and reduced viability but surprisingly did not impact monolayer permeability (Donau et al., 2022).

2.4 Corneal nerves

As the most densely innervated surface epithelium in our body, the cornea is exquisitely sensitive to pain mediated by Aδ-mechanoreceptors and polymodal C-type nociceptors from the ophthalmic branch of the fifth (trigeminal) nerve. Subsets of afferents express TRPV1, Piezo2 and TRPM8 (Fernández-Trillo et al., 2020) but insertion of the TRPV4 gene into trigeminal neurons promotes NGF-mediated healing of the corneal epithelium (Okada et al., 2019, 2022).

3. Sclera

The sclera, a connective tissue that envelops much of the eye (Fig. 1), is composed of collagen-bound ECM with embedded scleral fibroblasts. Stretch is a major regulator of scleral ECM deposition and presumably regulates fibroblast responses to myopia and glaucoma. In contrast to a vast literature on TRPV4 expression in generic fibroblasts (Adapala et al., 2021), not much is known about TRPV4 expression in scleral cells. The tissue is TRPV4-ir, and it might be interesting to examine whether scleral deformation affects a TRPV4-dependent myopic loop with the retina.

4. Ciliary body

The ciliary body consists of nonpigmented epithelial cells (NPE), pigmented epithelial cells (PE) and the ciliary muscle, which together produce aqueous humor and stabilize the lens. TRPV4 expression within the pig and mouse CB is largely confined to NPE cells (Delamere, Mandal, & Shahidullah, 2016; Jo et al., 2016), which respond to the agonist GSK101 and hypotonicity with $[Ca^{2+}]_i$ elevations. Analogy with the choroid plexus (Bothwell, Omileke, Patabendige, & Spratt, 2021; Preston et al., 2018) suggests TRPV4 might regulate transepithelial ion fluxes and secretion of aqueous humor, potentially together with modulation of neurohormone secretion (Alkozi, de Lara, Sánchez-Naves, & Pintor, 2017; Alkozi & Pintor, 2015). If, similar to retinal glia and overexpressing oocytes (Toft-Bertelsen et al., 2019), NPE TRPV4 channels are capable of sensing small osmotic gradients, they could play a role in steady-state release of the aqueous fluid. Another potential function is regulation of RVD, which was observed in NPE, but not PE, cells (Edelman, Sachs, & Adorante, 1994).

5. Trabecular meshwork

The trabecular meshwork (TM) is a key outflow pathway for the aqueous humor and as such a primary determinant of IOP and hypertensive glaucoma (Križaj, 2019), located just underneath the sclera (Fig. 1). TM cells are subjected to various forms of mechanical stress that include IOP-induced stretching of ECM beams, compression and aqueous humor shear. The pathways by which biomechanical stimuli impact homeostatic and pathological function of TM cells are not completely understood, but TRPV4 is suspected to play an important role due to strong expression and regulation of multiple signaling mechanisms in mouse and human TM. Both rodent and mammalian tissues express TRPV4 mRNA and protein; TRPV4 antibodies colocalize with the TM marker αSMA in situ (Fig. 3A) and label the cell membrane, primary cilia and intracellular compartments in cultured cells (Luo et al., 2014; Ryskamp et al., 2016). The TM TRPV4 channel is activated by GSK101, moderate temperature, swelling, ECM strain, lipid stretch (Fig. 4B) and partially by low ($0.5\,\text{dyn/cm}^2$) levels of shear (Ryskamp et al., 2016; Yarishkin et al., 2021; Yarishkin, Phuong, Bretz, et al., 2018; Yarishkin, Phuong, Lakk, & Križaj, 2018). Its activation by pressure is delayed relative to the rapid onset of the mechanosensitive Piezo1

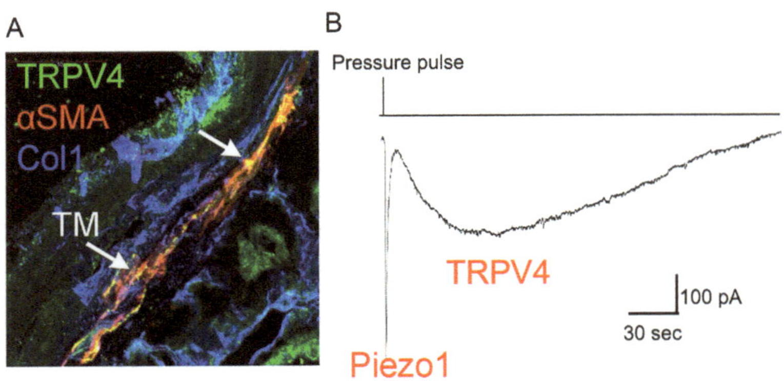

Fig. 3 Trabecular meshwork in a vertical section from a human eye, labeled for TRPV4, the tissue marker αSMA (alpha smooth muscle actin) and the ECM marker Collagen 1. (B) Pressure clamp, whole cell recording, cell held at −100 mV. 25 mm Hg pressure pulse induces a rapid inward current mediated by Piezo1, followed by a delayed TRPV4 activation.

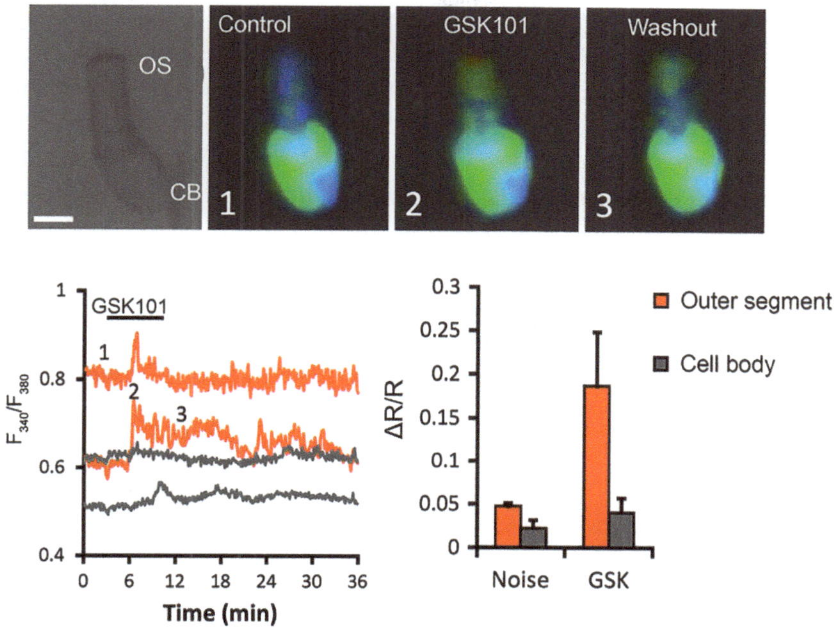

Fig. 4 Tiger salamander rods. The photoreceptor cells were loaded with the calcium indicator Fura-2AM and exposed to GSK1016790A (50nM). $[Ca^{2+}]i$ elevations are observed at the tip of the outer segment (OS) in contrast to the weak signals observed in the cell bodies. Scale bar = 10 µm.

current (Fig. 3B), presumably due to obligatory activation of the PLA2-eicosanoid cascade (Ryskamp et al., 2016). Strain-induced TRPV4 activation regulates gene expression, promotes actin polymerization, increases the density and size of focal adhesions and facilitates the secretion of extracellular matrix (Lakk & Križaj, 2021; Ryskamp et al., 2016). Studies implicated ciliary and membrane TRPV4 in hypotension (Luo et al., 2014; Patel et al., 2021), hypertension (Ryskamp et al., 2016) and in myofibroblast differentiation that maintains elevated IOP (Lakk & Križaj, 2021). Topical application of the agonist GSK1016790A reduced IOP in a nocturnal model of ocular hypertension (Patel et al., 2021), yet intracameral injections and eye drop application of the antagonist HC067047 also lower IOP (Redmon et al., 2021; Ryskamp et al., 2016). The hypotensive effect was ascribed to stimulation of TM-intrinsic eNOS synthase (Patel et al., 2021) but this is under debate (Overby & Stamer, 2021). Indeed, the agonist GSK1016790A increased outflow resistance and the antagonist HC067047 increased outflow facility in the Glauconix model of conventional outflow (Ryskamp et al., 2016).

An important aspect of TRPV4 signaling pertains to its role in lipid metabolism. Radial stretch significantly increases the membrane phosphatidycholine:cholesterol ratio (Lakk et al., 2021). Because TRPV4 functions as a primary sensor of TM strain (Lakk & Križaj, 2021; Ryskamp et al., 2016), it could regulate the membrane through its lipid composition. Conversely, TRPV4 activity is itself impacted by membrane cholesterol, a planar polycyclic lipid that promotes membrane stiffness while keeping the membrane fluid. Cholesterol depletion facilitates agonist- and mechanically induced TRPV4 activation in trabecular meshwork cells (Lakk et al., 2021), suppresses it in glia (Lakk, Yarishkin, Baumann, Iuso, & Križaj, 2017) and does not affect RGC TRPV4 activation (Lakk and Križaj, unpublished). Unlike endothelial cells (Saliez et al., 2008), TRPV4 in trabecular cells does not partition into lipid rafts or caveolar domains, with the exception of ~75 kDa variant of the protein, which co-immunoprecipitates with Caveolin-1. Lakk et al. (2021) suggested that TRPV4 variants might be differentially susceptible to caveolar interactions, with loss of cholesterol interactions outside of lipid rafts resulting in gain-of-function for the channel.

6. Ciliary muscle

The ciliary muscle connects the ciliary body the lens through the zonules of Zinn (Fig. 1). Its contraction and relaxation modulate hydrostatic pressure in the lens by altering water content. Mouse and human ciliary muscle immunolabels with TRPV4 antibodies (Ryskamp et al., 2016).

7. Schlemm's canal

This lymphatic-vascular structure is in series with juxtacanalicular TM and functions as a drain for aqueous humor outflow. Immunohistochemistry shows strong TRPV4 expression in the canal of Schlemm (Ryskamp et al., 2016).

8. Lens

The crystalline lens consists of lens capsule, epithelial cells, and cortical and nuclear fiber cells that comprise lens cortex and nucleus, respectively. TRPV4 was implicated in maintaining lens transparency, survival and accommodation by regulating force-dependent ion fluxes between lens epithelial and lens fiber cells. In mice, TRPV4 was localized to epithelial and lens fiber cells whereas pig TRPV4 is mainly epithelial (Nakazawa, Donaldson, & Petrova, 2019; Shahidullah & Delamere, 2022). Delamere's group showed that swelling-induced Ca^{2+} influx opens connexin and possibly pannexin hemichannels to mediate ATP release, which in turn stimulates Src family kinases (SFKs) and downstream Na/K-ATPase activity (Shahidullah, Mandal, & Delamere, 2012). TRPV4 channels in lens epithelial cells may be able to detect remote damage in lens nucleus (Shahidullah & Delamere, 2022; Shahidullah, Mandal, & Delamere, 2015). Lens fiber cells in adults may lose virtually all organelles yet appear to retain functional TRPV4 channels (Nakazawa et al., 2019). TRPV4 activity seems to have little impact on quality of vision (Chen, Petrova, Qiu, & Donaldson, 2022) and the retinal light response (Yarishkin, Phuong, Lakk, et al., 2018). Its function as a transducer of hydrostatic pressure within the lens may involve a negative feedback loop with TRPV1-dependent signaling. TRPV4 was suggested to sense positive changes in hydrostatic pressure (i.e., tracking cell swelling), while its cognate TRPV1 may be activated by cell shrinkage (Gao, Sun, White, Delamere, & Mathias, 2015).

9. Retina

The vertebrate retina is a complex tissue composed of more than 100 cell types that include many types of neurons and glia, endothelial, smooth muscle-like and immune cells. The flow of visual signal consists of a vertical glutamatergic pathway mediated by excitatory photoreceptor, bipolar and ganglion cell synapses and inhibitory/modulatory contributions from horizontal networks mediated by horizontal and amacrine cells, respectively. Retinal glia consist of Müller cells which provide critical ionic, metabolic and modulatory of support to neurons, astrocytes that line the inner limiting membrane (ILM) and (in vascularized retinas) control permeability of the blood–retina barrier, and microglia as the resident immune cells. In mammals, TRPV4 channels are largely expressed in the inner retina, with by far the strongest expression in RGCs and Müller cells (Lakk et al., 2018). TRPV4 has also been localized to rat but not mouse, bipolar cells and endothelial cells (Gao et al., 2019; Phuong et al., 2017; Ryskamp et al., 2011). The absence of light-evoked photoreceptor, bipolar and amacrine phenotypes in mammalian retinas (Yarishkin, Phuong, Bretz, et al., 2018) suggests its functions may be confined to excitatory signaling in the inner plexiform layer. Accordingly, $TRPV4^{-/-}$ mice do not exhibit changes in retinal b-wave (outer retinal function) or amacrine cell-mediated oscillatory potentials (Yarishkin, Phuong, Lakk, et al., 2018) while TRPV4 agonists promote RGC excitability and reactive gliosis (Ryskamp et al., 2014). TRPV4 overactivation induces RGC apoptosis (Ryskamp et al., 2011) whereas inhibition improves neuronal survival following retinal detachment (Matsumoto et al., 2018; Taylor, Arnér, & Ghosh, 2017).

9.1 RPE

The RPE recycles visual photopigment and acts as a restraining barrier to between the choroid, a non-fenestrated blood vessel, and the retina. TRPV4 inhibition may help maintain the integrity of the outer retina–blood barrier in diabetic retinas (Arredondo Zamarripa et al., 2017).

9.2 Photoreceptors

Mammalian photoreceptors do not express TRPV4 channels and photoreceptor signaling is unaffected by deletion of the TRPV4 gene (Yarishkin, Phuong, Lakk, et al., 2018) or exposure to GSK101 (Ryskamp et al., 2011). In contrast, pilot experiments suggest that the agonist GSK101 may evoke Ca^{2+} signals in rod photoreceptors from the tiger salamander retina. These responses were detected in the outer segment, the modified ciliary compartment that hosts the phototransduction machinery (Fig. 4) rather than the cell body/inner segment region, which hosts the majority of non-cyclic nucleotide-gated calcium channels (Krizaj & Copenhagen, 2002). A more detailed recent study reports that pressure steps evoke cationic currents within the cell body that are sensitive to TRPV antagonists and may include a TRPV4 component (Pang, Gao, & Wu, 2021). In contrast to mammals, many amphibian photoreceptors undergo large diurnal changes in shape and photosensitivity (Krizaj & Copenhagen, 1998). It is possible that TRPV4, potentially in conjunction with other mechanotransducers (Pang et al., 2021), regulates these "retinomotor" movements. Consistent with this conjecture, amphibian and teleost retinomotor movements require calcium and cytoskeletal signaling (Nagle & Burnside, 1984) and we and others showed that TRPV4-mediated Ca^{2+} influx can be upstream from actomyosin and microtubular remodeling (Goswami, Kuhn, Heppenstall, & Hucho, 2010; Lakk & Križaj, 2021).

9.3 Retinal ganglion cells

Retinal ganglion cells (RGCs) are projection neurons that transfer visual information from the retina to ~50 midbrain areas (Martersteck et al., 2017).

The >30 types of RGCs have distinct patterns of dendritic stratification, excitatory/inhibitory input, rod/cone input, and axonal projections. Comparison of *Trpv4* gene expression between purified retinal cell populations revealed it to be the strongest in RGCs, with the following vanilloid gene expression: *Trpv4* > *Trpv2* ≫ *Trpv1* = *Trpv3* (Lakk et al., 2018) (Fig. 4B). TRPV4-immunoractivity has been reported in teleost (salmon; Nisembaum et al., 2022), tiger salamander (Ryskamp & Križaj, unpublished observations), rat (Li et al., 2021; Sappington et al., 2015), pig (Taylor et al., 2017), non-human primate (Gao et al., 2019) and human (Ryskamp et al., 2011) RGCs, with expression generally most pronounced in cells with large somata. GSK101 and phorbol esters evoke a nonselective inward cationic current and elevate $[Ca^{2+}]_i$ in rodent and primate RGCs (Gao et al., 2019; Lakk et al., 2018; Ryskamp et al., 2014, 2011) (Fig. 6A). TRPV4 agonists induce RGC apoptosis (Ryskamp et al., 2011) whereas TRPV4 inhibition and gene deletion improve RGC survival in pig (Taylor et al., 2017) and mouse (Matsumoto et al., 2018) models of retinal detachment.

9.4 Müller glia and retinal homeostasis

Müller cells constitute ~90% of the retinal glial population with crucial functions in water transport, recycling of neurotransmitters, synaptogenesis, immune signaling, vascular function, and K^+ siphoning (Musada et al., 2020; Reichenbach & Bringmann, 2010; Vecino, Rodriguez, Ruzafa, Pereiro, & Sharma, 2016). TRPV4 expression was reported for mouse (Jo et al., 2015; Ryskamp et al., 2014), rat (Li et al., 2021; Sappington et al., 2015), pig (Taylor et al., 2017), non-human primate (Gao et al., 2019) and human (Ryskamp et al., 2011) Müller glia, as well as an immortalized human cell line (Netti et al., 2017). In contrast to sparse TRPV4 expression in brain astrocytes (20–30%; Pivonkova et al., 2018; Shibasaki, Ikenaka, Tamalu, Tominaga, & Ishizaki, 2014), every Müller cell manifests robust TRPV4-immunoreactivity and responsiveness to GSK101 (Jo et al., 2015; Ryskamp et al., 2014). Retinal detachment, intraocular pressure, swelling and indentation induce inward currents and $[Ca^{2+}]_i$ increases that are associated with TRPV4 activation (Jo et al., 2022; Ryskamp et al., 2014). Moreover, TRPV4 agonists mimic, and antagonists inhibit, mechanically evoked calcium signals, and TRPV4$^{-/-}$ cells exhibit reduced sensitivity to stretch, swelling, and retinal detachment (Jo et al., 2022; Ryskamp et al., 2014). Additivity of polymodal TRPV4 signaling has been unmasked by combing pressure and moderate temperature stimuli (Matsumoto et al., 2018) or swelling with GSK101 (Toft-Bertelsen, Križaj, & MacAulay, 2017). The increase in cytosolic [Ca^{2+}] evoked by swelling of native Müller cells is suppressed by TRPV4 antagonists and *Trpv4* gene deletion while hypotonicity-induced Ca^{2+} increases in immortalized Müller cells do not appear to involve TRPV4 (Netti et al., 2017). In primary glia, TRPV4 inhibition obliterates calcium signals induced by weak (~5–15 mOsm) stimuli and suppresses 40–70% of the calcium response evoked by large (120–190 mOsm) hypotonic gradients (Ryskamp et al., 2014; Toft-Bertelsen et al., 2019). The activating stimulus is the rate of water influx (swelling), mediated by the AQP4 water channel that constitutes the principal driver of membrane expansion and TRPV4 activation (Jo et al., 2015; Toft-Bertelsen et al., 2017). Interestingly, TRPV4 activity itself facilitates calcium-dependent increase in cell volume through mechanisms that remain to be studied (Jo et al., 2015). Müller cell volume regulation is Ca^{2+}-dependent (Jo et al., 2015), yet the osmosensitivity of RVD may be too low for it to play significant physiological roles.

Overactivation or underactivation of Müller glial TRPV4 signaling induce reactive gliosis, suggesting that the channel participates in both homeostatic maintenance and inflammatory activation (Ryskamp et al., 2014). Activation of the channel does not stimulate the NLRP3 inflammasome (Li et al., 2021) yet conditional ablation of the channel protects the retina from detachment-induced inflammation (Matsumoto et al., 2018), indicating a link to proinflammatory pathways. Analogy with CNS astroglia (Diaz-Otero et al., 2019; Pivonkova et al., 2018; Turovsky et al., 2020) suggests potential functions for TRPV4 in Müller glial endfeet in neurovascular coupling, functional hyperemia, intravascular pressure, ischemic edema and barrier permeability. The channel may also affect RGC viability via release of cytokines MCP-1 and TNFα from Müller cells (Li et al., 2021; Matsumoto et al., 2018).

9.5 Microglia and immune signaling

In contrast to cortical microglia in which *Trpv4* dominates thermoTRP expression, its expression in mouse retinal microglia is weak and was therefore missed in an early immunohistochemical study (Redmon et al., 2021; Ryskamp et al., 2011). GSK101 and hypotonicity evoke small (10–20 pA) currents but substantial $[Ca^{2+}]_i$ increases and hyperpolarizations (~10 mV) that are presumably mediated by a Ca^{2+}-activated K^+ conductance. Microglial TRPV4 activation in intact retinas is associated with Ca^{2+}-dependent retraction of higher-order branches. Perfusion of intact retinas with the TRPV4 antagonist HC06047 did not affect microglial $[Ca^{2+}]_i$, resting membrane potential or branching. This does not preclude TRPV4 involvement in dendritic surveillance in "resting" cells under *in vivo* conditions. Indeed, Tominaga's group recently showed that TRPV4 modulates microglial motility in response to moderate temperature (Nishimoto et al., 2021).

9.6 Astrocytes

Astrocytes in intact retina are immuno-negative for TRPV4 (Ryskamp et al., 2011) and analysis of *Trpv4* mRNA in astrocytes purified from mouse retinas suggests it to be low relative to RGCs and Müller cells. However, modest *Trpv4* expression was reported in astrocytes from the optic nerve head of wild type and DBA/2J mice (Choi, Sun, & Jakobs, 2015), suggesting it may be region-, strain- and context-specific.

9.7 Endothelial cells and blood–retina barrier regulation

The retina is associated with two different classes of endothelial cell (EC): choroidal in the outer retina and microvascular (capillary) in the inner retina. TRPV4 antibodies weakly label capillaries in the intact mouse retina (Phuong et al., 2017) whereas transgenic retinas that express GFP downstream from the TRPV4 promotor show strong fluorescence (Redmon et al., 2021). Among the many physiological processes in retinal microvascular ECs that are sensitive TRPV4 are Ca^{2+} homeostasis, angiogenesis, cytoskeletal remodeling and junctional permeability (Arredondo Zamarripa et al., 2017; Cappelli et al., 2021; Phuong et al., 2017). TRPV4 agonists increased in vitro and in vivo permeability of endothelial barriers, increased $[Ca^{2+}]_i$, and promoted calcium-dependent disassembly of VE-cadherin:β-catenin-containing adherens/occludens junctions and reorganization of the actin cytoskeleton. The agonist GSK101 evoked the "canonical" response signature typical PLA2-dependent TRPV4 responses (White et al., 2016): outward rectification, reversal at ~0 mV and delayed onset-to-peak and a sustained plateau (Ryskamp et al., 2014). The potency of GSK101 in elevating $[Ca^{2+}]_i$ and modulating barrier permeability in human ECs (EC_{50} ~ 1.6 nM) is similar to affinities reported for mouse and human TRPV4 channels (EC_{50} 3–20 nM; Thorneloe et al., 2008) and ~10-fold higher compared to TRPV4-expressing retinal glia, epithelial and trabecular meshwork cells. TRPV4 activation is coupled to Calcium-Induced Calcium release, with SERCA blockers reducing the amplitude of GSK101-evoked response by ~31% in human

(Phuong et al., 2017) and ~75% in bovine ECs (Monaghan et al., 2015). The large reversible decrease in the microvascular transendothelial resistance induced by GSK101 correlates with remodeling of cortical F-actin, redistribution of VE-cadherin from interdigitated processes between adjacent cells to the cell membrane, dislocation of β-catenin and decreased expression of the tight junction protein occludin and is mirrored in vivo by ~30% increase in capillary Evans Blue extravasation (Phuong et al., 2017). The time course of recovery from increased transendothelial permeability is slower (~hours) compared to the relatively rapid response to the agonist (~minutes), presumably due to the compensatory activation of downstream signaling pathways (MAPKs/Src/Raf/Ras, RhoA and myosin-light chain kinase pathways). We hypothesize that Ca^{2+} influx controls local actin remodeling and bonding between VE-cadherin and β-catenin, with relocation of TRPV4 channels from adherens junctions possibly controlled by the focal adhesion kinase, integrins and the Rho/Rho kinase pathway (Bagnell, Sumner, & McCray, 2022). Consistent with this, TRPV4-mediated calcium influx impaired filopodial ruffling and stress fiber organization in cells transfected with F-*actin:mApple* constructs (Phuong et al., 2017). Another important process may be vascular tone, with TRPV4 as a coordinator of signaling between endothelial cells, pericytes, and glia that maintains the balance between vasodilation and vasoconstriction (Sonkusare et al., 2012; White et al., 2016).

Shear flow, endothelial stretch and eicosanoids could maintain vascular autoregulation and regulate redistribution of blood flow from superficial to deep vascular plexi by stimulating TRPV4 channels through local elevations of EETs and 20-HETE. This process is almost certainly disturbed under pathological circumstances such as edema, blast trauma, transendothelial invasion of monocytes and ocular hypertension (Križaj, 2019). Studies in cultured cells implicated TRPV4 in regulation of retinal endothelial cell migration, sprouting and tube formation. TRPV4 inhibition promotes developmental angiogenesis (O'Leary et al., 2019) yet deletion of the channel does not appear to affect retinal blood vessel development in mice (Cappelli et al., 2021). Cultured human retinal endothelial cells respond to unidirectional strain by reorienting their long axis orthogonally in a TRPV4-dependent manner (Cappelli et al., 2021). Sensitivity to swelling and cholesterol implicates TRPV4 in diabetic retinal vascular phenotypes. Consistent with this, TPRV4 blockers suppress BRB breakdown in a rodent model of diabetes (Arredondo Zamarripa et al., 2017) and TRPV4 knockdown may reduce oxygen-induced neovascularization and VEGF-VEGFR2 signaling (Cappelli et al., 2021; O'Leary et al., 2019) yet hyperglycemia also downregulates TRPV4 expression (Monaghan et al., 2015).

An extensive body of work has documented a central role for TRPV4 in retina, brain and body edema formation, often in partnership with aquaporins (Kitchen et al., 2020; Michinaga, Onishi, Shimizu, Mizuguchi, & Hishinuma, 2021; Ríos et al., 2019; Weber et al., 2020). Studies on the way are exploring whether cytotoxic and vascular edema in the retina might be controlled with therapeutic targeting of TRPV4 and/or AQP4 proteins (Iuso & Križaj, 2016). Given that suppression of TRPV4 channels promotes physiological revascularization of ischemic retinas (O'Leary et al., 2019), pharmacological interventions should carefully consider the biological context.

10. One protein to confound them all

The ubiquitous expression of the TRPV4 channels across ocular tissues mirrors its expression across the CNS, PNS and body (Krizaj et al., 2022; White et al., 2016). The uniquely divergent patterns of TRPV4 expression, activation and modulation within the eye, however, believe simplistic evaluations of its functional importance.

10.1 TRPV4 protein

Different ocular cells appear to express different versions of the TRPV4 protein. Western blots precipitate two molecular weight bands from protein lysates of human retinal endothelial cells and mouse RGCs (Phuong et al., 2017; Ryskamp et al., 2011) but show four bands in TM cells (Lakk et al., 2021), mirroring kidney, ventricular myocyte and choroid epithelial expression (Hochstetler et al., 2020; Lakk et al., 2021; Zhao et al., 2012). The bands could reflect presence of an unglycosylated ~75 kDa splice variant and multiple glycosylated variants, membrane vs intracellular pools, splice variants, long vs short forms of the protein or differential association with lipid rafts (Lakk et al., 2021).

10.2 Same cell type in retina vs other parts of CNS

TRPV4 expression in retinal microglia is lower compared to their cortical counterparts, which furthermore show a different sequence of relative thermoTRP expression (Redmon et al., 2021). Thus, microglia may express region-specific sensomes that equip them for differential responsiveness to the external milieu.

10.3 Same cell type in different regions of the eye

The choroid is a large blood vessel that perfuses the anterior part of the retina. TRPV4 expression in choroidal ECs is weaker relative to microvascular (inner retinal) ECs (Phuong et al., 2017), indicating heterogeneity of TRPV4 signaling across dissimilar vascular beds within the eye.

10.4 Expression across different cell types

Relative expression of TRPV4 differs across retinal neurons vs glial and endothelial cells (Phuong et al., 2017; Redmon et al., 2021). The same stimulus may therefore produce different response magnitudes across cell types.

10.5 Biophysical properties of the TRPV4 current

Fig. 5 shows the I–V relationships for cell types from different ocular tissues. The currents were acquired under voltage clamp by the same investigator on the same experimental setup under identical conditions (except for the Müller glia, which were recorded under K^+-free conditions). The voltage-dependence of the GSK1016790A-evoked current ranges from minimally to moderately strong, showing outward rectification at positive membrane potentials in RGCs and microglia but a more linear relationship in Müller glia and epithelial cells.

10.6 Time course and mechanism of activation

Fig. 6 shows the time course of GSK1016790A-activated current simultaneously recorded in representative RGC and Müller cells. The former shows an inactivating Ca^{2+} response with a rapid onset, in contrast to the slow onset and sustained elevation in the glia. The responses in trabecular meshwork cells are nested in-between the extremes shown by neurons and glia by show a peak $[Ca^{2+}]_i$ response that relaxes to a steady plateau that can be maintained for hours (Yarishkin et al., 2022). Sustained activation in Müller cells, microglia, TM cells, corneal epithelial cells, nonpigmentary epithelial cells of the ciliary body and microvascular endothelial cells requires an intermediary PLA2-eicosanoid step while RGC signals appear to be independent of eicosanoid signaling (Jo et al., 2016; Lapajne et al., 2020; Phuong et al., 2017; Ryskamp et al., 2014, 2011). The difference can be unmasked with the PLA2 activator Sc-3034, which has no effect on RGC $[Ca^{2+}]_i$, but elevates $[Ca^{2+}]_i$ in glia in a TRPV4-dependent manner (Fig. 6B). Interestingly, overexpression of TRPV4 in an oocyte system produces a PLA2-insensitive TRPV4 current (Toft-Bertelsen et al., 2019) and TRPV4 is functional in yeast, which lacks PLA2 and polyunsaturated fatty acids (Loukin, Su, & Kung, 2009). In some cells, PLA2 might be downstream from a GPCR

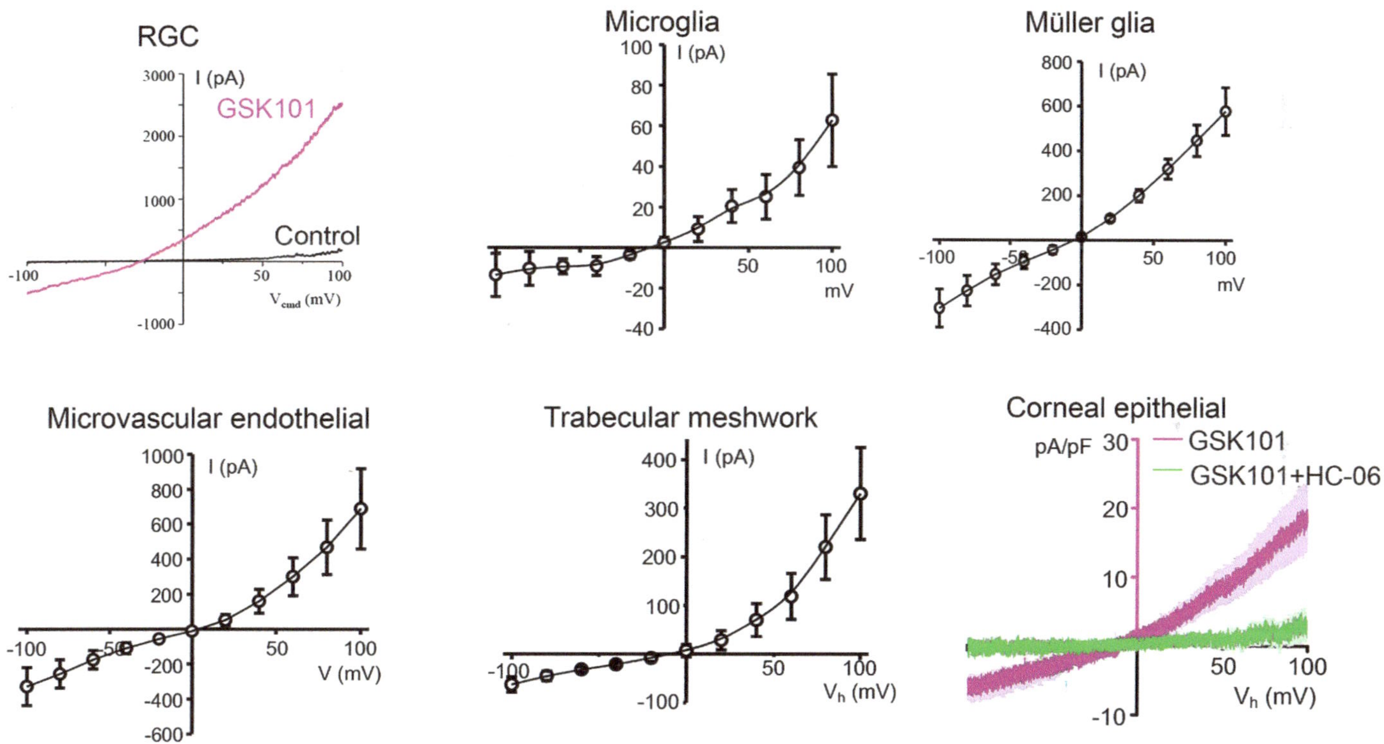

Fig. 5 Whole cell voltage clamp in ocular cells stimulated with 25 nM GSK1016790A (with exception of the endothelial cell, exposed to 1 nM of the agonist). The agonist-evoked response varies in terms of the voltage-dependence of the evoked current, as in the shape of the I-V relationship, and current amplitudes/densities. All responses were blocked by blocker HC067047 (2–5 μM).

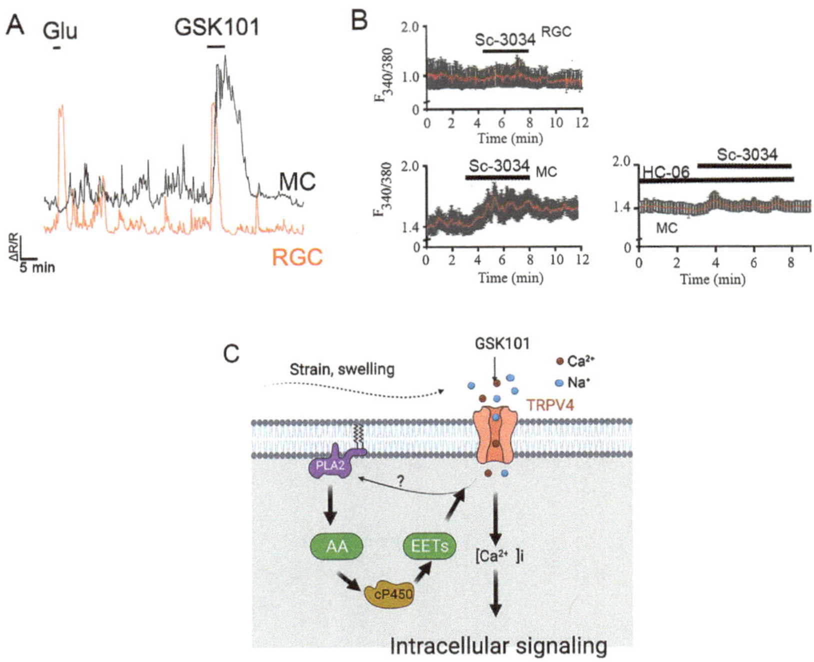

Fig. 6 The PLA2-dependence of TRPV4 signaling is cell type-specific. (A) Calcium imaging. Simultaneous recording from a mouse RGC and Müller cell stimulated with GSK1016790A shows a rapid inactivating response in the neuron and a slow-onset sustained response in the glial cell. Compare the amplitude of TRPV4-mediated response to the calcium signal evoked by saturating concentration of glutamate (100 μM). (B) Calcium imaging. Müller cells (MC) but not RGCs respond to the PLA2 activator Sc-3034. The TRPV4 antagonist HC067047 blocks the Sc-3034-induced $[Ca^{2+}]_i$ signal in glia. (C) Model of PLA2-dependent TRPV4 signaling in retinal glia, endothelial and epithelial cells.

(Mamenko, Zaika, Jin, O'Neil, & Pochynyuk, 2011) or activated by an initial Ca^{2+} influx across the TRP channel (Fig. 6C). Another intriguing feature of TRPV4 signaling are Ca^{2+} oscillations that can emerge during sustained TRPV4 activation. TRPV4-dependent oscillations in trabecular meshwork cells require release of Ca^{2+} from intracellular compartments and Ca^{2+}-dependent activation of TRPM4 channels (Yarishkin et al., 2022).

10.7 Heteromerization

TRPV4 heteromerizes with TRPV1 in microvascular endothelial cells (Ma et al., 2011; O'Leary et al., 2019) and may heteromerize with TRPC1 in Müller glia (Jo et al., 2022). Heteromerization is not obligatory: TRPV1 and TRPV4 coexpress in RGCs but do not functionally interact (Lakk et al., 2018).

10.8 Stimulus-response coupling

Retinal glia, endothelial and trabecular meshwork cells respond to ECM strain with TRPV4-dependent calcium influx (Jo et al., 2022; Phuong et al., 2017; Ryskamp et al., 2016). However, corneal epithelial cells strongly express TRPV4 (Fig. 2) and respond to 1–12% strains with dose-dependent $[Ca^{2+}]_i$ elevations (Lapajne et al., 2020) yet neither the strain-evoked calcium signal nor the fraction of cell responders to strain are affected in *Trpv4*$^{-/-}$ and HC06047-treated cells. Similarly, membrane stretching in response to positive pressure activates TRPV4 in glia, trabecular meshwork cells and oocytes but not chondrocytes (Matsumoto et al., 2018; Ryskamp et al., 2016; Servin-Vences, Moroni, Lewin, & Poole, 2017). This suggests that TRPV4-mediated force transduction involves cell-type specific association with additional force-sensing lipid and protein molecules.

Hypotonicity is a universal stimulus to investigate TRPV4 activity. Indeed the the original study by Strotmann, Harteneck, Nunnenmacher, and Schultz (2000) used hypotonic stress to describe and characterize the channel. In general, TRPV4 activation tends to be evaluated with non-physiological osmotic gradients (>100 mOsm) that are not relevant to physiology or pathology and it is possible that not all expressing cell types are capable of utilizing TRPV4 for the transduction of physiological gradients (5–15 mOsm). Osmotransducers include retinal Müller cells and probably microglia (Redmon et al., 2021; Toft-Bertelsen et al., 2019), while RGCs respond to 190 mOsm but cannot sense 15 mOsm gradients (Ryskamp et al., 2011; Toft-Bertelsen et al., 2019). The differential responsiveness may reflect accessibility of the N-terminal TYR^{110}, PLA2 activation and density of AQP channels (Iuso & Križaj, 2016; Toft-Bertelsen et al., 2019; White et al., 2016).

10.9 Downstream coupling to interacting proteins

TRPV4-mediated calcium influx reflects channel's coupling to a wide array of cell membrane, ECM, and cytosolic proteins/lipids. Examples in the eye include pannexin 1 (Lapajne et al., 2020), nitric oxide synthase (Patel et al., 2021), phospholipase A2 (Ryskamp et al., 2014), the Rho-ROCK pathway (Lakk & Križaj, 2021), caveolin-1 (Lakk et al., 2021), focal adhesions (Lakk & Križaj, 2021) and the JAK/NF-κB pathway (Li et al., 2021). The agonist GSK101 depolarizes the membrane potential in endothelial and trabecular meshwork cells but hyperpolarizes microglia, presumably due to coupling with eicosanoid metabolites, nitric oxide synthase or K^+ conductances (Phuong et al., 2017; Redmon et al., 2021; White et al., 2016; Yarishkin et al., 2022) (Fig. 7).

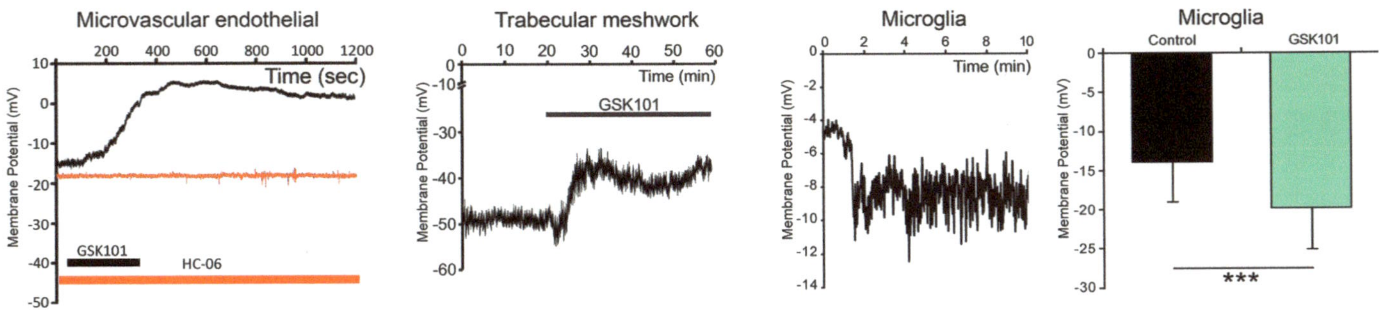

Fig. 7 Context-dependence of membrane potential regulation by TRPV4. The TRPV4 agonist GSK1016790A depolarizes trabecular meshwork and endothelial cells but hyperpolarizes microglia.

10.10 Lipid modulation

Membrane stiffness and curvature are regulated by free membrane cholesterol. β-Methyl cyclodextrin, a cholesterol-sequestering drug, inhibits TRPV4 signaling in Müller cells, enhances it in trabecular meshwork cells but has no effect on TRPV4 activity in RGCs (Lakk et al., 2017, 2021). TRPV4 associates with caveolin scaffolds that organize membrane signaling in a microvascular endothelial cell line (Goedicke-Fritz et al., 2015) but not in trabecular meshwork cells (Lakk et al., 2021).

11. Summary

TRPV4 is widely expressed across the eye, with likely functions in paradigmatic aspects of visual signaling and pathology such as corneal permeability and pain, lens accommodation and cataract, aqueous fluid production, secretion and glaucoma, regulation of retinal water content and edema, angiogenesis and barrier dysregulation, inflammation, neurodegeneration and glaucoma. While the responses to GSK1016890A and 4α-PDD accord with the universality of TRPV4 protein structure and function, agonists and physiological activation produce a remarkably diverse spectrum of activation phenotypes, sensitization, inactivation, coupling to modulatory pathways and roles in signal integration across the eye. Multi-location expression ensures that pharmacological manipulations and global knockdown will produce difficult-to-interpret phenotypes. For example, TRPV4-dependent loss of retina–blood barrier integrity (Arredondo Zamarripa et al., 2017; Phuong et al., 2017) that mirrors vascular leakage and circulatory collapse in epithelia and endothelial (Villalta, Rocic, & Townsley, 2014; Willette et al., 2008) is likely to reflect parallel effects on endothelial cells, pericytes, Müller cell endfeet, astrocyte processes and ganglion cells. Similar complexity may pertain to TRPV4 signaling at the corneal epithelial-stromal-afferent interface, and its functions in the anterior chamber that may contribute to multilevel regulation of intraocular pressure and lens accommodation but may be particularly salient in glaucoma, an irreversible blinding disease associated with dysfunction of TRPV4-expressing tissues in the anterior (ciliary body, trabecular meshwork) and posterior (retinal ganglion cells, glia, endothelial cells) eye.

Acknowledgments

Supported by: NIH (R01EY027920, R01EY031817, P30EY014800, T32EY024234), Rankin-Stauss Foundation, USAMRAA and an unrestricted grant from Research to

Prevent Blindness to the Moran Eye Center at the University of Utah. We thank Prof. Wolfgang Liedtke (Duke University) for the global TRPV4$^{-/-}$ mice and the Alan Crandall Initiative from the University of Utah for support.

References

Adapala, R. K., Katari, V., Teegala, L. R., Thodeti, S., Paruchuri, S., & Thodeti, C. K. (2021). Trpv4 mechanotransduction in fibrosis. *Cell*, *10*(11), 3053. https://doi.org/10.3390/cells10113053.

Alkozi, H. A., de Lara, M. J. P., Sánchez-Naves, J., & Pintor, J. (2017). TRPV4 stimulation induced melatonin secretion by increasing arylalkymine N-acetyltransferase (AANAT) protein level. *International Journal of Molecular Sciences*, *18*(4). https://doi.org/10.3390/ijms18040746.

Alkozi, H. A., & Pintor, J. (2015). TRPV4 activation triggers the release of melatonin from human non-pigmented ciliary epithelial cells. *Experimental Eye Research*, *136*, 34–37. https://doi.org/10.1016/j.exer.2015.04.019.

Arredondo Zamarripa, D., Noguez Imm, R., Bautista Cortés, A. M., Vázquez Ruíz, O., Bernardini, M., Fiorio Pla, A., et al. (2017). Dual contribution of TRPV4 antagonism in the regulatory effect of vasoinhibins on blood-retinal barrier permeability: Diabetic milieu makes a difference. *Scientific Reports*, 7(1), 1–16. https://doi.org/10.1038/s41598-017-13621-8.

Bagnell, A. M., Sumner, C. J., & McCray, B. A. (2022). TRPV4: A trigger of pathological RhoA activation in neurological disease. *BioEssays*, *44*(6), 2100288. https://doi.org/10.1002/bies.202100288.

Bothwell, S. W., Omileke, D., Patabendige, A., & Spratt, N. J. (2021). CSF secretion is not altered by NKCC1 nor TRPV4 antagonism in healthy rats. *Brain Sciences*, *11*(9). https://doi.org/10.3390/brainsci11091117.

Cappelli, H. C., Guarino, B. D., Kanugula, A. K., Adapala, R. K., Perera, V., Smith, M. A., et al. (2021). Transient receptor potential vanilloid 4 channel deletion regulates pathological but not developmental retinal angiogenesis. *Journal of Cellular Physiology*, *236*(5), 3770–3779. https://doi.org/10.1002/jcp.30116.

Chen, Y., Gao, J., Li, L., Sellitto, C., Mathias, R. T., Donaldson, P. J., et al. (2019). The ciliary muscle and zonules of ZinN modulate lens intracellular hydrostatic pressure through transient receptor potential vanilloid channels. *Investigative Ophthalmology and Visual Science*, *60*(13), 4416–4424. https://doi.org/10.1167/iovs.19-27794.

Chen, Y., Petrova, R. S., Qiu, C., & Donaldson, P. J. (2022). Intracellular hydrostatic pressure regulation in the bovine lens: A role in the regulation of lens optics? *American Journal of Physiology. Regulatory, Integrative and Comparative Physiology*, *322*(3), R263–R279. https://doi.org/10.1152/ajpregu.00309.2021.

Choi, H. J., Sun, D., & Jakobs, T. C. (2015). Astrocytes in the optic nerve head express putative mechanosensitive channels. *Molecular Vision*, *21*, 749–766.

Delamere, N. A., Mandal, A., & Shahidullah, M. (2016). The significance of TRPV4 channels and hemichannels in the lens and ciliary epithelium. *Journal of Ocular Pharmacology and Therapeutics*, *32*(8), 504–508. https://doi.org/10.1089/jop.2016.0054.

Diaz-Otero, J. M., Yen, T. C., Ahmad, A., Laimon-Thomson, E., Abolibdeh, B., Kelly, K., et al. (2019). Transient receptor potential vanilloid 4 channels are important regulators of parenchymal arteriole dilation and cognitive function. *Microcirculation (New York, N.Y.: 1994)*, *26*(6), e12535. https://doi.org/https://doi.org/10.1111/MICC.12535.

Donau, J., Luo, H., Virta, I., Skupin, A., Pushina, M., Loeffler, J., et al. (2022). TRPV4 stimulation level regulates Ca2+-dependent control of human corneal endothelial cell viability and survival. *Membranes*, *12*(3). https://doi.org/10.3390/membranes12030281.

Edelman, J. L., Sachs, G., & Adorante, J. S. (1994). Ion transport asymmetry and functional coupling in bovine pigmented and nonpigmented ciliary epithelial cells. *American Journal of Physiology - Cell Physiology, 266*(5 Pt 1), 35. https://doi.org/10.1152/ajpcell.1994.266.5.c1210.

Fernández-Trillo, J., Florez-Paz, D., Íñigo-Portugués, A., González-González, O., del Campo, A. G., González, A., et al. (2020). Piezo2 mediates low-threshold mechanically evoked pain in the cornea. *Journal of Neuroscience, 40*(47), 8976–8993. https://doi.org/10.1523/JNEUROSCI.0247-20.2020.

Gao, J., Sun, X., White, T. W., Delamere, N. A., & Mathias, R. T. (2015). Feedback regulation of intracellular hydrostatic pressure in surface cells of the lens. *Biophysical Journal, 109*(9), 1830–1839. https://doi.org/10.1016/j.bpj.2015.09.018.

Gao, F., Yang, Z., Jacoby, R. A., Wu, S. M., & Pang, J. J. (2019). The expression and function of TRPV4 channels in primate retinal ganglion cells and bipolar cells. *Cell Death & Disease, 10*(5), 1–12. https://doi.org/10.1038/s41419-019-1576-3.

Goedicke-Fritz, S., Kaistha, A., Kacik, M., Markert, S., Hofmeister, A., Busch, C., et al. (2015). Evidence for functional and dynamic microcompartmentation of Cav-1/TRPV4/K(ca) in caveolae of endothelial cells. *European Journal of Cell Biology, 94*(7–9), 391–400. https://doi.org/10.1016/j.ejcb.2015.06.002.

Goswami, C., Kuhn, J., Heppenstall, P. A., & Hucho, T. (2010). Importance of non-selective cation channel TRPV4 interaction with cytoskeleton and their reciprocal regulations in cultured cells. *PLoS One, 5*(7), e11654.

Gradilone, S. A., Masyuk, A. I., Splinter, P. L., Banales, J. M., Huang, B. Q., Tietz, P. S., et al. (2007). Cholangiocyte cilia express TRPV4 and detect changes in luminal tonicity inducing bicarbonate secretion. *Proceedings of the National Academy of Sciences of the United States of America, 104*(48), 19138–19143. https://doi.org/10.1073/pnas.0705964104.

Guarino, B. D., Paruchuri, S., & Thodeti, C. K. (2020). The role of TRPV4 channels in ocular function and pathologies. *Experimental Eye Research, 201*, 108257. https://doi.org/10.1016/j.exer.2020.108257.

Hochstetler, A. E., Smith, H. M., Preston, D. C., Reed, M. M., Territo, P. R., Shim, J. W., et al. (2020). TRPV4 antagonists ameliorate ventriculomegaly in a rat model of hydrocephalus. *JCI. Insight, 5*(18). https://doi.org/10.1172/JCI.INSIGHT.137646.

Iuso, A., & Križaj, D. (2016). TRPV4-AQP4 interactions 'turbocharge' astroglial sensitivity to small osmotic gradients. *Channels, 10*(3), 172–174. https://doi.org/10.1080/19336950.2016.1140956.

Jo, A. O., Lakk, M., Frye, A. M., Phuong, T. T. T., Redmon, S. N., Roberts, R., et al. (2016). Differential volume regulation and calcium signaling in two ciliary body cell types is subserved by TRPV4 channels. *Proceedings of the National Academy of Sciences of the United States of America, 113*(14), 3885–3890. https://doi.org/10.1073/pnas.1515895113.

Jo, A. O., Lakk, M., Rudzitis, C. N., & Križaj, D. (2022). TRPV4 and TRPC1 channels mediate the response to tensile strain in mouse Müller cells. *Cell Calcium, 104*, 102588. https://doi.org/10.1016/j.ceca.2022.102588.

Jo, A. O., Ryskamp, D. A., Phuong, T. T. T., Verkman, A. S., Yarishkin, O., Macaulay, N., et al. (2015). TRPV4 and AQP4 channels synergistically regulate cell volume and calcium homeostasis in retinal müller glia. *Journal of Neuroscience, 35*(39), 13525–13537. https://doi.org/10.1523/JNEUROSCI.1987-15.2015.

Kitchen, P., Salman, M. M., Halsey, A. M., Clarke-Bland, C., MacDonald, J. A., Ishida, H., et al. (2020). Targeting Aquaporin-4 subcellular localization to treat central nervous system edema. *Cell, 181*(4), 784–799. https://doi.org/10.1016/j.cell.2020.03.037.

Križaj, D. (2019). What is glaucoma? In H. Kolb, E. Fernandez, & R. Nelson (Eds.), *Webvision: The organization of the retina and visual system* (p. 1995). University of Utah Health Sciences Center.

Krizaj, D., & Copenhagen, D. R. (1998). Compartmentalization of calcium extrusion mechanisms in the outer and inner segments of photoreceptors. *Neuron, 21*(1), 249–256. https://doi.org/10.1016/s0896-6273(00)80531-0.

Krizaj, D., & Copenhagen, D. R. (2002). Calcium regulation in photoreceptors. *Frontiers in Bioscience*, 7, d2023-44. https://doi.org/10.2741/A896.

Krizaj, D., Cordeiro, S., & Strauss, O. (2022). Retinal TRP channels: Cell type-specific regulators of retinal homeostasis and multimodal integration. *Progress in Retinal and Eye Research*. In press.

Križaj, D., Ryskamp, D. A., Tian, N., Tezel, G., Mitchell, C. H., Slepak, V. Z., et al. (2014). From Mechanosensitivity to inflammatory responses: New players in the pathology of glaucoma. *Current Eye Research, 39*(2), 105–119. https://doi.org/10.3109/02713683.2013.836541.

Lakk, M., Hoffmann, G. F., Gorusupudi, A., Enyong, E., Lin, A., Bernstein, P. S., et al. (2021). Membrane cholesterol regulates TRPV4 function, cytoskeletal expression, and the cellular response to tension. *Journal of Lipid Research, 62*. https://doi.org/10.1016/J.JLR.2021.100145.

Lakk, M., & Križaj, D. (2021). TRPV4-rho signaling drives cytoskeletal and focal adhesion remodeling in trabecular meshwork cells. *American Journal of Physiology - Cell Physiology, 320*(6), C1013–C1030. https://doi.org/10.1152/ajpcell.00599.2020.

Lakk, M., Yarishkin, O., Baumann, J. M., Iuso, A., & Križaj, D. (2017). Cholesterol regulates polymodal sensory transduction in Müller glia. *Glia, 65*(12), 2038–2050. https://doi.org/10.1002/glia.23213.

Lakk, M., Young, D., Baumann, J. M., Jo, A. O., Hu, H., & Križaj, D. (2018). Polymodal TRPV1 and TRPV4 sensors colocalize but do not functionally interact in a subpopulation of mouse retinal ganglion cells. *Frontiers in Cellular Neuroscience, 12*, 353. https://doi.org/10.3389/fncel.2018.00353.

Lamb, T. D., Collin, S. P., & Pugh, E. N. (2007). Evolution of the vertebrate eye: Opsins, photoreceptors, retina and eye cup. *Nature Reviews Neuroscience, 8*(12), 960–976. https://doi.org/https://doi.org/10.1038/nrn2283.

Lapajne, L., Lakk, M., Yarishkin, O., Gubeljak, L., Hawlina, M., & Križaj, D. (2020). Polymodal sensory transduction in mouse corneal epithelial cells. *Investigative Ophthalmology and Visual Science, 61*(4), 2. https://doi.org/10.1167/iovs.61.4.2.

Li, Q., Cheng, Y., Zhang, S., Sun, X., & Wu, J. (2021). TRPV4-induced Müller cell gliosis and TNF-α elevation-mediated retinal ganglion cell apoptosis in glaucomatous rats via JAK2/STAT3/NF-κB pathway. *Journal of Neuroinflammation, 18*(1), 271. https://doi.org/10.1186/s12974-021-02315-8.

Liedtke, W., & Friedman, J. M. (2003). Abnormal osmotic regulation in trpv4$^{-/-}$ mice. *Proceedings of the National Academy of Sciences of the United States of America, 100*(23), 13698–13703. https://doi.org/10.1073/pnas.1735416100.

Loukin, S. H., Su, Z., & Kung, C. (2009). Hypotonic shocks activate rat TRPV4 in yeast in the absence of polyunsaturated fatty acids. *FEBS Letters, 583*(4), 754–758. https://doi.org/10.1016/j.febslet.2009.01.027.

Luo, N., Conwell, M. D., Chen, X., Kettenhofen, C. I., Westlake, C. J., Cantor, L. B., et al. (2014). Primary cilia signaling mediates intraocular pressure sensation. *Proceedings of the National Academy of Sciences of the United States of America, 111*(35), 12871–12876. https://doi.org/10.1073/pnas.1323292111.

Ma, X., Cheng, K. T., Wong, C. O., O'Neil, R. G., Birnbaumer, L., Ambudkar, I. S., et al. (2011). Heteromeric TRPV4-C1 channels contribute to store-operated Ca2+ entry in vascular endothelial cells. *Cell Calcium, 50*(6), 502–509. https://doi.org/10.1016/j.ceca.2011.08.006.

Mamenko, M., Zaika, O., Jin, M., O'Neil, R. G., & Pochynyuk, O. (2011). Purinergic activation of Ca2+-permeable TRPV4 channels is essential for mechano-sensitivity in the aldosterone-sensitive distal nephron. *PLOS ONE*, *6*(8), e22824. https://doi.org/10.1371/journal.pone.0022824.

Martersteck, E. M., Hirokawa, K. E., Evarts, M., Bernard, A., Duan, X., Li, Y., et al. (2017). Diverse central projection patterns of retinal ganglion cells. *Cell Reports*, *18*(8), 2058–2072. https://doi.org/10.1016/j.celrep.2017.01.075.

Matsumoto, H., Sugio, S., Seghers, F., Križaj, D., Akiyama, H., Ishizaki, Y., et al. (2018). Retinal detachment-induced müller glial cell swelling activates TRPV4 ion channels and triggers photoreceptor death at body temperature. *Journal of Neuroscience*, *38*(41), 8745–8758. https://doi.org/10.1523/JNEUROSCI.0897-18.2018.

Mergler, S., Valtink, M., Taetz, K., Sahlmüller, M., Fels, G., Reinach, P. S., et al. (2011). Characterization of transient receptor potential vanilloid channel 4 (TRPV4) in human corneal endothelial cells. *Experimental Eye Research*, *93*(5), 710–719. https://doi.org/10.1016/j.exer.2011.09.021.

Michinaga, S., Onishi, K., Shimizu, K., Mizuguchi, H., & Hishinuma, S. (2021). Pharmacological inhibition of transient receptor potential vanilloid 4 reduces vasogenic edema after traumatic brain injury in mice. *Biological & Pharmaceutical Bulletin*, *44*(11), 1759–1766. https://doi.org/10.1248/bpb.b21-00512.

Monaghan, K., McNaughten, J., McGahon, M. K., Kelly, C., Kyle, D., Yong, P. H., et al. (2015). Hyperglycemia and diabetes downregulate the functional expression of TRPV4 channels in retinal microvascular endothelium. *PLoS One*, *10*(6), e0128359. https://doi.org/10.1371/journal.pone.0128359.

Musada, G. R., Dvoriantchikova, G., Myer, C., Ivanov, D., Bhattacharya, S. K., & Hackam, A. S. (2020). The effect of extrinsic Wnt/β-catenin signaling in muller glia on retinal ganglion cell neurite growth. *Developmental Neurobiology*, *80*(3–4), 98–110. https://doi.org/10.1002/dneu.22741.

Nagle, B. W., & Burnside, B. (1984). Calmodulin-binding proteins in teleost retina, rod inner and outer segments, and rod cytoskeletons. *European Journal of Cell Biology*, *33*(2), 248–257.

Nakazawa, Y., Donaldson, P. J., & Petrova, R. S. (2019). Verification and spatial mapping of TRPV1 and TRPV4 expression in the embryonic and adult mouse lens. *Experimental Eye Research*, *186*, 107707. https://doi.org/10.1016/j.exer.2019.107707.

Netti, V., Fernández, J., Kalstein, M., Pizzoni, A., Di Giusto, G., Rivarola, V., et al. (2017). TRPV4 contributes to resting membrane potential in retinal Müller cells: Implications in cell volume regulation. *Journal of Cellular Biochemistry*, *118*(8), 2302–2313. https://doi.org/10.1002/jcb.25884.

Nisembaum, L. G., Loentgen, G., L'Honoré, T., Martin, P., Paulin, C. H., Fuentès, M., et al. (2022). Transient receptor potential-vanilloid (TRPV1-TRPV4) channels in the Atlantic Salmon, Salmo salar. A focus on the pineal gland and melatonin production. *Frontiers in Physiology*, *12*. https://doi.org/10.3389/fphys.2021.784416.

Nishimoto, R., Derouiche, S., Eto, K., Deveci, A., Kashio, M., Kimori, Y., et al. (2021). Thermosensitive TRPV4 channels mediate temperature-dependent microglia movement. *Proceedings of the National Academy of Sciences of the United States of America*, *118*(17), e2012894118. https://doi.org/10.1073/pnas.2012894118.

Okada, Y., Shirai, K., Miyajima, M., Reinach, P. S., Yamanaka, O., Sumioka, T., et al. (2016). Loss of TRPV4 function suppresses inflammatory fibrosis induced by alkali-burning mouse corneas. *PLoS One*, *11*(12). https://doi.org/10.1371/journal.pone.0167200.

Okada, Y., Sumioka, T., Ichikawa, K., Sano, H., Nambu, A., Kobayashi, K., et al. (2019). Sensory nerve supports epithelial stem cell function in healing of corneal epithelium in mice: The role of trigeminal nerve transient receptor potential vanilloid 4. *Laboratory Investigation*, *99*(2), 210–230. https://doi.org/10.1038/s41374-018-0118-4.

Okada, Y., Sumioka, T., Reinach, P. S., Miyajima, M., & Saika, S. (2022). Roles of epithelial and mesenchymal TRP channels in mediating inflammatory fibrosis. *Frontiers in Immunology*, *12*. https://doi.org/10.3389/fimmu.2021.731674.

O'Leary, C., McGahon, M. K., Ashraf, S., McNaughten, J., Friedel, T., Cincolà, P., et al. (2019). Involvement of TRPV1 and TRPV4 channels in retinal angiogenesis. *Investigative Ophthalmology and Visual Science, 60*(10), 3297–3309. https://doi.org/10.1167/iovs.18-26344.

Overby, D., & Stamer, D. (2021). Glaucoma dialogue. *International Glaucoma Review, 22*(1), 12–14. ISSN 1566-1040.

Pan, Z., Yang, H., Mergler, S., Liu, H., Tachado, S. D., Zhang, F., et al. (2008). Dependence of regulatory volume decrease on transient receptor potential vanilloid 4 (TRPV4) expression in human corneal epithelial cells. *Cell Calcium, 44*(4), 374–385. https://doi.org/10.1016/j.ceca.2008.01.008.

Pang, J. J., Gao, F., & Wu, S. M. (2021). Generators of pressure-evoked currents in vertebrate outer retinal neurons. *Cell, 10*(6), 1288.

Patel, P. D., Chen, Y. L., Kasetti, R. B., Maddineni, P., Mayhew, W., Millar, J. C., et al. (2021). Impaired TRPV4-eNOS signaling in trabecular meshwork elevates intraocular pressure in glaucoma. *Proceedings of the National Academy of Sciences of the United States of America, 118*(16). https://doi.org/10.1073/pnas.2022461118.

Phuong, T. T. T., Redmon, S. N., Yarishkin, O., Winter, J. M., Li, D. Y., & Križaj, D. (2017). Calcium influx through TRPV4 channels modulates the adherens contacts between retinal microvascular endothelial cells. *Journal of Physiology, 595*(22), 6869–6885. https://doi.org/10.1113/JP275052.

Pivonkova, H., Hermanova, Z., Kirdajova, D., Awadova, T., Malinsky, J., Valihrach, L., et al. (2018). The contribution of TRPV4 channels to astrocyte volume regulation and brain edema formation. *Neuroscience, 394*, 127–143. https://doi.org/10.1016/j.neuroscience.2018.10.028.

Preston, D., Simpson, S., Halm, D., Hochstetler, A., Schwerk, C., Schroten, H., et al. (2018). Activation of TRPV4 stimulates transepithelial ion flux in a porcine choroid plexus cell line. *American Journal of Physiology - Cell Physiology, 315*(3), C357–C366. https://doi.org/10.1152/ajpcell.00312.2017.

Rahman, M., Sun, R., Mukherjee, S., Nilius, B., & Janssen, L. J. (2018). TRPV4 stimulation releases ATP via pannexin channels in human pulmonary fibroblasts. *American Journal of Respiratory Cell and Molecular Biology, 59*(1), 87–95. https://doi.org/10.1165/rcmb.2017-0413OC.

Redmon, S. N., Yarishkin, O., Lakk, M., Jo, A., Mustafić, E., Tvrdik, P., et al. (2021). TRPV4 channels mediate the mechanoresponse in retinal microglia. *Glia, 69*(6), 1563–1582. https://doi.org/10.1002/glia.23979.

Reichenbach, A., & Bringmann, A. (2010). Müller cells in the healthy and diseased retina. *Müller Cells in the Healthy and Diseased Retina, 1–417*. https://doi.org/10.1007/978-1-4419-1672-3.

Ríos, M. O., Imm, R. N., Godínez, N. M. H., Cortes, A. M. B., Escalante, D. D. L., Liedtke, W., et al. (2019). TRPV4 inhibition prevents increased water diffusion and blood-retina barrier breakdown in the retina of streptozotocin-induced diabetic mice. *PLoS One, 14*(5), e0212158. https://doi.org/10.1371/journal.pone.0212158.

Roberts, M. W. G., Sui, G., Wu, R., Rong, W., Wildman, S., Montgomery, B., et al. (2020). TRPV4 receptor as a functional sensory molecule in bladder urothelium: Stretch-independent, tissue-specific actions and pathological implications. *FASEB Journal, 34*(1), 263–286. https://doi.org/10.1096/fj.201900961RR.

Ryskamp, D. A., Frye, A. M., Phuong, T. T. T., Yarishkin, O., Jo, A. O., Xu, Y., et al. (2016). TRPV4 regulates calcium homeostasis, cytoskeletal remodeling, conventional outflow and intraocular pressure in the mammalian eye. *Scientific Reports, 6*(1), 1–15. https://doi.org/10.1038/srep30583.

Ryskamp, D. A., Jo, A. O., Frye, A. M., Vazquez-Chona, F., MaCaulay, N., Thoreson, W. B., et al. (2014). Swelling and eicosanoid metabolites differentially gate

TRPV4 channels in retinal neurons and glia. *Journal of Neuroscience*, *34*(47), 15689–15700. https://doi.org/10.1523/JNEUROSCI.2540-14.2014.

Ryskamp, D. A., Witkovsky, P., Barabas, P., Huang, W., Koehler, C., Akimov, N. P., et al. (2011). The polymodal ion channel transient receptor potential vanilloid 4 modulates calcium flux, spiking rate, and apoptosis of mouse retinal ganglion cells. *Journal of Neuroscience*, *31*(19), 7089–7101. https://doi.org/10.1523/JNEUROSCI.0359-11.2011.

Saliez, J., Bouzin, C., Rath, G., Ghisdal, P., Desjardins, F., Rezzani, R., et al. (2008). Role of caveolar compartmentation in endothelium-derived hyperpolarizing factor–mediated relaxation. *Circulation*, *117*(8), 1065–1074. https://doi.org/10.1161/CIRCULATIONAHA.107.731679.

Sappington, R. M., Sidorova, T., Ward, N. J., Chakravarthy, R., Ho, K. W., & Calkins, D. J. (2015). Activation of transient receptor potential vanilloid-1 (TRPV1) influences how retinal ganglion cell neurons respond to pressure-related stress. *Channels*, *9*(2), 102–113. https://doi.org/10.1080/19336950.2015.1009272.

Servin-Vences, M. R., Moroni, M., Lewin, G. R., & Poole, K. (2017). Direct measurement of TRPV4 and PIEZO1 activity reveals multiple mechanotransduction pathways in chondrocytes. *eLife*, *6*. https://doi.org/10.7554/eLife.21074.

Shahidullah, M., & Delamere, N. (2022). Nitric oxide inhibits TRPV4-mediated hemichannel opening in porcine nonpigmented ciliary epithelium. *FASEB Journal: Official Publication of the Federation of American Societies for Experimental Biology*, *36*. https://doi.org/10.1096/fasebj.2022.36.S1.L7418.

Shahidullah, M., Mandal, A., & Delamere, N. A. (2012). TRPV4 in porcine lens epithelium regulates hemichannel-mediated ATP release and Na-K-ATPase activity. *American Journal of Physiology - Cell Physiology*, *302*(12), C1751. https://doi.org/10.1152/ajpcell.00010.2012.

Shahidullah, M., Mandal, A., & Delamere, N. A. (2015). Damage to lens fiber cells causes TRPV4-dependent Src family kinase activation in the epithelium. *Experimental Eye Research*, *140*, 85–93. https://doi.org/10.1016/j.exer.2015.08.013.

Shibasaki, K., Ikenaka, K., Tamalu, F., Tominaga, M., & Ishizaki, Y. (2014). A novel subtype of astrocytes expressing TRPV4 (transient receptor potential vanilloid 4) regulates neuronal excitability via release of gliotransmitters. *Journal of Biological Chemistry*, *289*(21), 14470–14480. https://doi.org/10.1074/jbc.M114.557132.

Silva, G. B., & Garvin, J. L. (2008). TRPV4 mediates hypotonicity-induced ATP release by the thick ascending limb. *American Journal of Physiology - Renal Physiology*, *295*(4). https://doi.org/10.1152/ajprenal.90365.2008.

Sonkusare, S. K., Bonev, A. D., Ledoux, J., Liedtke, W., Kotlikoff, M. I., Heppner, T. J., et al. (2012). Elementary Ca2+ signals through endothelial TRPV4 channels regulate vascular function. *Science*, *336*(6081), 597–601. https://doi.org/10.1126/science.1216283.

Strotmann, R., Harteneck, C., Nunnenmacher, K., Schultz, G., & Plant, T. D. (2000). OTRPC4, a nonselective cation channel that confers sensitivity to extracellular osmolarity. *Nature Cell Biology*, *2*(10), 695–702. https://doi.org/10.1038/35036318.

Suzuki, M., Mizuno, A., Kodaira, K., & Imai, M. (2003). Impaired pressure sensation in mice lacking TRPV4. *Journal of Biological Chemistry*, *278*(25), 22664–22668. https://doi.org/10.1074/jbc.M302561200.

Taylor, L., Arnér, K., & Ghosh, F. (2017). Specific inhibition of TRPV4 enhances retinal ganglion cell survival in adult porcine retinal explants. *Experimental Eye Research*, *154*, 10–21. https://doi.org/10.1016/j.exer.2016.11.002.

Thébault, S. (2021). Minireview: Insights into the role of TRP channels in the retinal circulation and function. *Neuroscience Letters*, *765*, 136285. https://doi.org/10.1016/j.neulet.2021.136285.

Thibodeau, M. L., Peters, C. H., Townsend, K. N., Shen, Y., Hendson, G., Adam, S., et al. (2017). Compound heterozygous TRPV4 mutations in two siblings with a complex phenotype including severe intellectual disability and neuropathy. *American Journal of Medical Genetics, Part A, 173*(11), 3087–3092. https://doi.org/10.1002/ajmg.a.38400.

Thorneloe, K. S., Sulpizio, A. C., Lin, Z., Figueroa, D. J., Clouse, A. K., McCafferty, G. P., et al. (2008). N-((1S)-1-{[4-((2S)-2-{[(2,4-dichlorophenyl)sulfonyl]amino} -3-hydroxypropanoyl)-1-piperazinyl]carbonyl}-3-methylbutyl)-1-benzothiophene-2- carboxamide (GSK1016790A), a novel and potent transient receptor potential vanilloid 4 channel agonist induces urin. *Journal of Pharmacology and Experimental Therapeutics, 326*(2), 432–442. https://doi.org/10.1124/jpet.108.139295.

Toft-Bertelsen, T. L., Križaj, D., & MacAulay, N. (2017). When size matters: Transient receptor potential vanilloid 4 channel as a volume-sensor rather than an osmo-sensor. *Journal of Physiology, 595*(11), 3287–3302. https://doi.org/10.1113/JP274135.

Toft-Bertelsen, T. L., Yarishkin, O., Redmon, S., Phuong, T. T. T., Križaj, D., & MacAulay, N. (2019). Volume sensing in the transient receptor potential vanilloid 4 ion channel is cell type-specific and mediated by an N-terminal volume-sensing domain. *Journal of Biological Chemistry, 294*(48), 18421–18434. https://doi.org/10.1074/jbc.RA119.011187.

Turovsky, E. A., Braga, A., Yu, Y., Esteras, N., Korsak, A., Theparambil, S. M., et al. (2020). Mechanosensory signaling in astrocytes. *Journal of Neuroscience, 40*(49), 9364–9371. https://doi.org/10.1523/JNEUROSCI.1249-20.2020.

Vecino, E., Rodriguez, F. D., Ruzafa, N., Pereiro, X., & Sharma, S. C. (2016). Glia-neuron interactions in the mammalian retina. *Progress in Retinal and Eye Research, 51*, 1–40. https://doi.org/10.1016/j.preteyeres.2015.06.003.

Villalta, P. C., Rocic, P., & Townsley, M. I. (2014). Role of MMP2 and MMP9 in TRPV4-induced lung injury. *American Journal of Physiology - Lung Cellular and Molecular Physiology, 307*(8), L652–L659. https://doi.org/10.1152/ajplung.00113.2014.

Weber, J., Rajan, S., Schremmer, C., Chao, Y. K., Krasteva-Christ, G., Kannler, M., et al. (2020). TRPV4 channels are essential for alveolar epithelial barrier function as protection from lung edema. *JCI Insight, 5*(20), e134464. https://doi.org/10.1172/jci.insight.134464.

White, J. P., Cibelli, M., Urban, L., Nilius, B., McGeown, J. G., & Nagy, I. (2016). TRPV4: Molecular conductor of a diverse orchestra. *Physiological Reviews, 96*(3), 911–973. https://doi.org/10.1152/physrev.00016.2015.

Willette, R. N., Bao, W., Nerurkar, S., Yue, T. I., Doe, C. P., Stankus, G., et al. (2008). Systemic activation of the transient receptor potential vanilloid subtype 4 channel causes endothelial failure and circulatory collapse: Part 2. *Journal of Pharmacology and Experimental Therapeutics, 326*(2), 443–452. https://doi.org/10.1124/jpet.107.134551.

Yarishkin, O., Phuong, T. T. T., Baumann, J. M., De Ieso, M. L., Vazquez-Chona, F., Rudzitis, C. N., et al. (2021). Piezo1 channels mediate trabecular meshwork mechanotransduction and promote aqueous fluid outflow. *Journal of Physiology, 599*(2), 571–592. https://doi.org/10.1113/JP281011.

Yarishkin, O., Phuong, T. T. T., Bretz, C. A., Olsen, K. W., Baumann, J. M., Lakk, M., et al. (2018). TREK-1 channels regulate pressure sensitivity and calcium signaling in trabecular meshwork cells. *Journal of General Physiology, 150*(12), 1660–1675. https://doi.org/10.1085/jgp.201812179.

Yarishkin, O., Phuong, T. T. T., Lakk, M., & Križaj, D. (2018). TRPV4 does not regulate the distal retinal light response. *Advances in Experimental Medicine and Biology, 1074*, 553–560. https://doi.org/10.1007/978-3-319-75402-4_67.

Yarishkin, O., Phuong, T. T. T., Vazquez-Chona, F., Bertrand, J., van Battenburg-Sherwood, J., Redmon, S. N., et al. (2022). Emergent temporal signaling in human trabecular meshwork cells: Role of TRPV4-TRPM4 interactions. *Frontiers in Immunology, 13*, 805076. https://doi.org/10.3389/fimmu.2022.805076.

Young, T. L., Whisenhunt, K., Martin, S., & Tompson, S. (2022). PIEZO1 and PIEZO2 pathogenic variants identified in primary congenital glaucoma. *Investigative Ophthalmology & Visual Science, 63*(7), 1127.

Zhao, Y., Huang, H., Jiang, Y., Wei, H., Liu, P., Wang, W., et al. (2012). Unusual localization and translocation of TRPV4 protein in cultured ventricular myocytes of the neonatal rat. *European Journal of Histochemistry, 56*(3), e32. https://doi.org/10.4081/ejh.2012.e32.

CHAPTER NINE

Role of TRPV4 in skeletal function and its mutant-mediated skeletal disorders

Rashmita Das[a,b] and Chandan Goswami[a,b,*]
[a]School of Biological Sciences, National Institute of Science Education and Research (NISER)-Bhubaneswar, Khurda, Odisha, India
[b]Homi Bhabha National Institute, Mumbai, India
[*]Corresponding author: e-mail address: chandan@niser.ac.in

Contents

Abstract

TRPV4 is a non-selective cation channel that belongs to the TRP super family. This channel can be activated by physiological temperatures and mechanical stimuli. In addition, TRPV4 is modulated by several endogenous mediators including specific lipids, cholesterol and their metabolic products. TRPV4 gene is present in all vertebrates and is widely expressed in tissues originating from ectoderm, endoderm and mesoderm. Although TRPV4 knockout is not lethal, point mutations in TRPV4 cause severe clinical phenotypes

Current Topics in Membranes, Volume 89
ISSN 1063-5823
https://doi.org/10.1016/bs.ctm.2022.07.004

with variable penetration in human population. These mutations are mostly "gain-of-function" in nature and primarily affect muscles, bones and peripheral neurons, endorsing TRPV4 as critical regulator of musculoskeletal systems. Here we critically analyze the involvement of TRPV4 in musculoskeletal system. Studies of TRPV4 mutations provide detailed information on musculoskeletal disorders at molecular, cellular and metabolic levels.

Abbreviations

ARD ankyrin repeat domains
CMT2C Charcot-Marie-Tooth neuropathy type 2C
CDSMA congenital distal spinal muscular atrophy
SPSMA scapuloperoneal spinal muscular atrophy
TRPV4 transient receptor potential vanilloid sub-type 4
VGIC voltage-gated ion channel

1. TRP channels: A brief overview

Transient receptor potential (TRP) ion channels were initially found in a *Drosophila melanogaster* mutant and have since been comprehensively examined in animals. All TRP channels are cation-selective, with some being highly selective for Ca^{2+} or Mg^{2+}. Distantly related to voltage-gated K^+, Na^+, and Ca^{2+}-superfamilies, the TRP superfamily is composed of 6 subfamilies namely TRPA (ankyrin), TRPC (canonical), TRPM (melastatin), TRPV (vanilloid), TRPP (polycystin), and TRPML (mucolipin) (Himmel & Cox, 2020; Montell, 2005). All TRP channel subunits have six putative transmembrane segments and a "pore-forming loop" located between the 5th and 6th transmembrane segments. A functional channel, i.e., the homo-tetramer, is formed by four TRP channel subunits. The amino- (N) and carboxyl- (C) termini are located intracellularly and the cytoplasmic region contains several domains and motifs required for channel function.

Several members of the TRP super family contain an N-terminal ankyrin-like repeat domain (ARD) and several other domains/motifs, including TRP-box, a coiled-coil domain, enzymatic domain, endoplasmic reticulum (ER) retention-signal domain, and Ca^{2+}-binding motif (EF-hand) in the C-terminus (Méndez-Reséndiz et al., 2020; Zheng, 2013). TRP channels are widely expressed in various tissue types. While complete deletion of certain TRP gene/s is not lethal, mutations in these TRP channels give rise to several disorders in specific organs where the proteins are expressed (discussed later). TRP channels are modulated by membrane potential, temperature, mechanical force, pH, lipids and steroids, monoterpenes, toxins, anoxia and hypoxia, ultraviolet (UV) light, Nicotinamide adenine dinucleotide (NAD^+), and its metabolites, nucleotide phosphates, several inorganic ions like Mg^{2+}, Zn^{2+}, phosphates such as pyrophosphate and pyrotriphosphate, polyvalent inorganic cations like La^{3+} and Gd^{3+} and many other mediators (Zheng, 2013; Zhu, 2011). In many cases, heavy metals can block the TRP channels effectively.

In this chapter, we will focus on TRPV4, a member that belongs to the TRPV subfamily. The TRPV subfamily has six members, TRPV1–6, that can be further divided into two groups, TRPV1–4, and TRPV5/TRPV6. The TRPV1–4 members are weakly Ca^{2+}-selective and are "heat-activated" whereas TRPV5/TRPV6 are highly Ca^{2+}-selective but are not heat-activated channels (Song & Yuan, 2010; Venkatachalam & Montell, 2007).

2. About TRPV4

The recent analysis of genomic data suggests that at least one copy of the TRPV4 gene is present in all vertebrates (Kumari et al., 2015). Higher invertebrates (such as *Drosophila*, *C elegans*) also have genes (such as *nunchung* and *Osm 9*) that are functionally homologus to mammalian TRPV4. Notably, mutation in Osm9 leads to impaired mechanosensation in *C elegans* that can be rescued by human TRPV4 (Liedtke et al., 2000, 2003). Single-cell organisms such as yeast also have a functional homologue of TRPV4 (Denis & Cyert, 2002). These findings suggest that TRPV4 function has a strong evolutionary history and it likely started from single cell organism, ~700 million years ago.

Structure of TRPV4: At protein level, TRPV4 mainly forms a homotetramer. The tetrameric structure of TRPV4 is similar to that of other TRPV channels (Fig. 1). The tetrameric structure of TRPV4 is composed of six transmembrane domains from segments 1–6. Both the N-terminal and C-terminal face the cytoplasm and contain different functional domains and motifs. A proline-rich domain (PRD) follows an N-terminal ARD consisting of 6 ankyrin repeats. A TRP-box, a MAP7-binding domain and a calmodulin (CaM)-binding site are present at the C-terminus (Nilius & Voets, 2013).

Recently a 3.8-Å resolution TRPV4 cryo-EM structure from *Xenopus tropicalis* revealed the overall architecture of the channel in great detail (Deng et al., 2018). The structure used for the Cryogenic electron microscopy (cryo EM) study was made by removing some portions of N- and

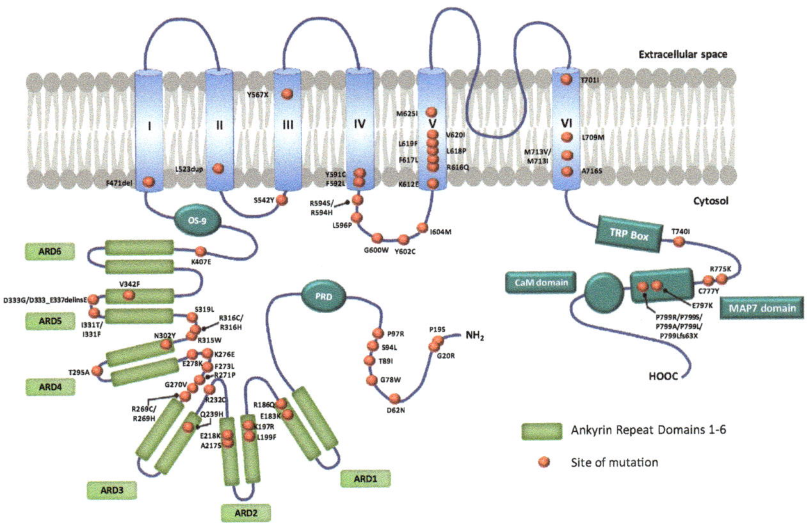

Fig. 1 TRPV4 mutation and hotspots. The TRPV4 protein in humans has approximately more than 60 different mutations (red dots) spread over several domains in N- and C-terminals as well as in the transmembrane regions.

C-terminal sections and generating a shortened construct containing amino acids 133–797. For better X-ray diffraction, a site for glycosylation was eliminated by adding the point mutation N647Q. The subsequent TRPV4 construct (133–797 N647Q) permitted near-atomic resolution cryo-EM structure. TRPV4 follows similar tetrameric architecture as its close homologues TRPV1, TRPV2, and TRPV6. A structural unit is made by first four helices, the S1–S4 domain, analogous to the voltage-sensor domain in Voltage-Gated Ion Channel (VGICs). A central ion-permeation pathway is formed by S5-S6 domains and four S1-S4 domains from four identical subunits surround them. The S4–S5 linker adopts an "ordered loop-structure" (unlike TRPV1 and TRPV2, where an α-helix structure is formed by the S4–S5 linker). In VGICs, the S4–S5 linker usually forms an α-helical segment functioning as a "mechanical lever" to couple "voltage-sensor activation" with "pore-opening," and the absence of this α-helical coupler results in different gating behavior for TRPV4. A single constriction in the lower gate was observed inside the ion permeation pathway of TRPV4. The narrowest point is formed by side chains of M714, measuring 5.3 Å in diameter, which is inadequate for ion passage, thus indicating that the structure signifies a closed conformation of the channel.

Regulation of TRPV4: TRPV4 is activated by an array of physical and chemical stimuli. For example, cell swelling, temperature, low pH, and mechanical stimuli can activate TRPV4 (Chen & Sonkusare, 2020; Plant & Strotmann, 2007; White et al., 2016). Different chemical compounds such as 4 alpha-phorbol esters (4α-PDD), endocannabinoids, arachidonic acid, and citrate can also activate TRPV4. Epoxyeicosatrienoic acid metabolites (EETs) directly activate TRPV4 via the EET-binding pocket on the S2-S3 linker (Berna-Erro et al., 2017; Vriens et al., 2004). Phorbol esters like 4α-PDD, 4α-PDH, phorbol 12-myristate 13-acetate, and synthetic agonist GSK1016790A activate TRPV4 channel and induce ionic currents through the channel pore (Toft-Bertelsen & Macaulay, 2021). The potent compound in cannabis, Δ^9-tetrahydrocannabinol (Δ^9-THC) can also activate TRPV4 and few other TRP channels. Most abundant endocannabinoids, Anandamide (N-arachidonoylethanolamine; AEA) and 2-arachidonoyl-glycerol (2-AG), act on TRPV4 and activate it. Certain *N*-acyl amides also act as TRPV4 modulators (Muller et al., 2019). Activation of TRPV4 (by physical stimuli or chemical activators) induces Ca^{2+}-influx, triggering different Ca^{2+}-dependent signaling events.

TRPV4 interactome: TRPV4 interacts with several membrane proteins (like PACSIN 3, OS-9, AIP4, Caveolins, AKAPs), which can affect the localization and function of TRPV4 (Daneva et al., 2021; Goretzki et al., 2018; Ottolini et al., 2020; White et al., 2016). Some, such as tubulin, actin, microtubule-associated proteins (MAP), and neurofilament proteins, are cytoskeletal proteins that directly interact with TRPV4 (Goswami et al., 2010). On the contrary, β-arrestin and other ion channels (for example, TRPP2, TRPC1, etc.) interact with TRPV4 and modulate TRPV4-mediated signaling (White et al., 2016). Some of the interacting partners are specified below (Fig. 2). Recent reports indicate the involvement of TRPV4 in mitochondrial regulation as TRPV4 interacts with various mitochondrial proteins, namely Mfn1, Mfn2, Hsp60, and Cyt C (Das et al., 2022; Kumar et al., 2018). These associations are to a great extent characteristic of the significance of TRPV4 in mitochondrial fission, fusion, and other functions.

TRPV4 also interacts with specific membrane components such as Phosphatidylinositol 4,5-bisphosphate (PIP_2), $PI(4.5)P_2$, cholesterol and other similar hydrophobic compounds such as fat soluble vitamins and steroids (Kumari et al., 2015; Takahashi et al., 2014). These interactions are direct, mostly confined to a specific region of TRPV4. These lipid–protein interactions are functionally important and can often vary in different TRPV4 mutants (Das et al., 2022; Das & Goswami, 2019). Interaction of TRPV4 with Vitamin D is also indicated for proper bone formation (Masuyama, 2014).

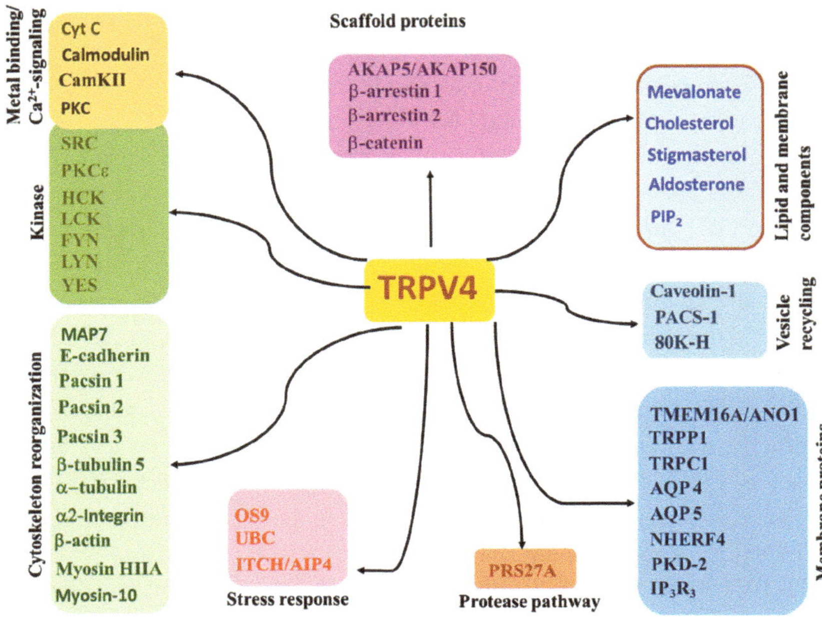

Fig. 2 TRPV4 interactome. Several proteins and lipids can interact with TRPV4. This partial list suggests involvement of TRPV4 in different cellular functions.

TRPV4 interaction with cholesterol is well-established. The TM4-loop-TM5 region of TRPV4 remains highly conserved throughout the vertebrate evolution, and interacts with cholesterol, its precursor mevalonate, and metabolic derivatives such as stigmasterol and aldosterone (Kumari et al., 2015). Thus, the TM4-Loop-TM5 region of TRPV4 can act as a molecular sensor for lipid/cholesterol metabolism. This concept is also supported by the findings that TRPV4 gene is flanked by two other genes, namely MVK (Mevalonate kinase) and GLTP (Glycolipid transfer protein), and this organization is conserved in all vertebrates. Collectively, these findings suggest that TRPV4 has co-evolved with cholesterol bio-synthesis pathway (Kumari et al., 2015).

The findings discussed above thus indicate the involvement of TRPV4 in Ca^{2+}-signaling, cytoskeletal remodeling, kinase pathways, expression of specific genes and subcellular organelle regulation (Fig. 3). Therefore, TRPV4 mutation-mediated pathophysiology can go beyond Ca^{2+}-signaling and should also be analyzed in the context of other functions of TRPV4.

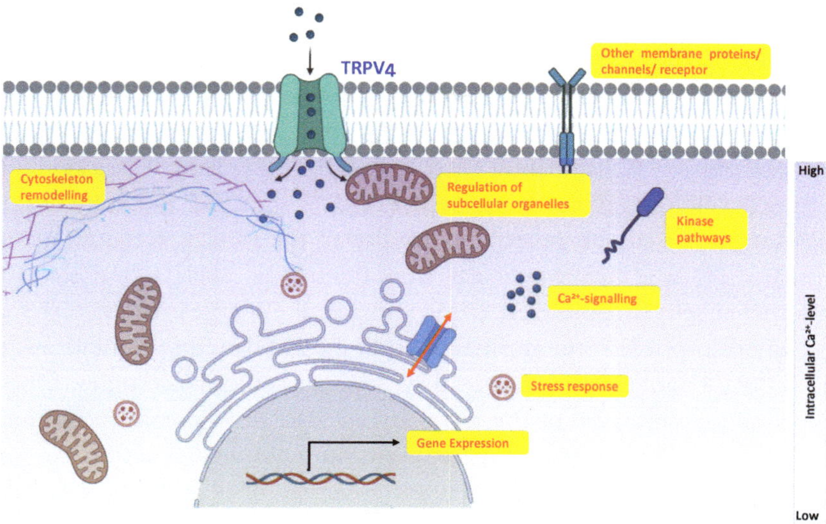

Fig. 3 Overview of TRPV4 function in cell. TRPV4 is believed to regulate several functions inside the cell, such as Ca^{2+}-homeostasis, cytoskeletal remodeling, stress response, regulation of different kinase pathways, regulation of other membrane proteins, expression of specific genes and subcellular organelle regulation.

3. TRPV4 expression in tissues with mesodermal lineage and other tissues

In the initial phase of TRPV4 research, especially when it was cloned as a homologous gene of TRPV1, it was thought that its expression is limited in peripheral neurons. Later, its expression was detected in various tissue types including brain, kidney, eyes, skin, pancreas, lungs, urinary systems, epidermal cells, and many other cell types suggesting its role in important physiological functions. Current literature indicates the expression of TRPV4 in many tissues that are originated from endoderm, ectoderm and mesoderm. However, abnormal expression of TRPV4 is closely linked to the development of many diseases including gastric, lung, and colorectal cancer (Falcón et al., 2019; Lawhorn et al., 2020; Nilius & Owsianik, 2011; Nilius & Voets, 2013; Watanabe et al., 2002). Notably, TRPV4 mutation-mediated channelopathies primarily affect skeletal muscle, cardiac muscle and most smooth muscles tissues, all of which are mesodermal in origin, suggesting a lineage-specificity.

3.1 Muscle and fibroblasts

Majority of the extracellular matrix (ECM) of the skeletal muscle is maintained by the fibroblasts, and they provide the skeletal muscle with structural support. Fibroblasts play a major role in movement generation and transmission (Chapman et al., 2016). TRP ion channels are reported in muscle cells to specifically play a part in Ca^{2+} signaling and homeostasis. TRPV4 plays a significant role in regulating the ionic concentrations in muscle cells, specifically vascular smooth muscle, skeletal muscles and endothelial cells (Brinkmeier, 2011; Filosa et al., 2013). It is also reported that any disruption in the normal physiology and structure (mutations) of the TRPV4 ion channel can lead to various pathophysiologies, which portrays its significance in the physiology of the muscle cells (Nishimura et al., 2012). Involvement of TRPV4 in mitochondrial function can also be very relevant for muscle function (Kumar et al., 2018; Woolums et al., 2020).

3.2 Bone

Bones depend on an adequate supply of Ca^{2+} to maintain their structural and mechanical properties. Ca^{2+}-signaling has been associated with osteoblast differentiation, osteoclastogenesis, and chondrocyte differentiation (all required for proper bone homeostasis). Critical role of TRPV4 in bone formation and resorption, mechanical stimulation of bones, in joint inflammation and pain has been studied by various research groups. Mutation or deletion in TRPV4 leads to a wide range of bone-related disorders, commonly termed as skeletal dysplasias and arthropathies, which further support the importance of TRPV4 in bone health. Similarly, TRPV4 knockout animals show bone defects (Hu et al., 2019; McCray et al., 2020). TRPV4-mediated intracellular Ca^{2+}-signaling is suggested to play a significant role in the events that regulate bone health (Masuyama et al., 2008; Muramatsu et al., 2007; Suzuki et al., 2013) (Fig. 4).

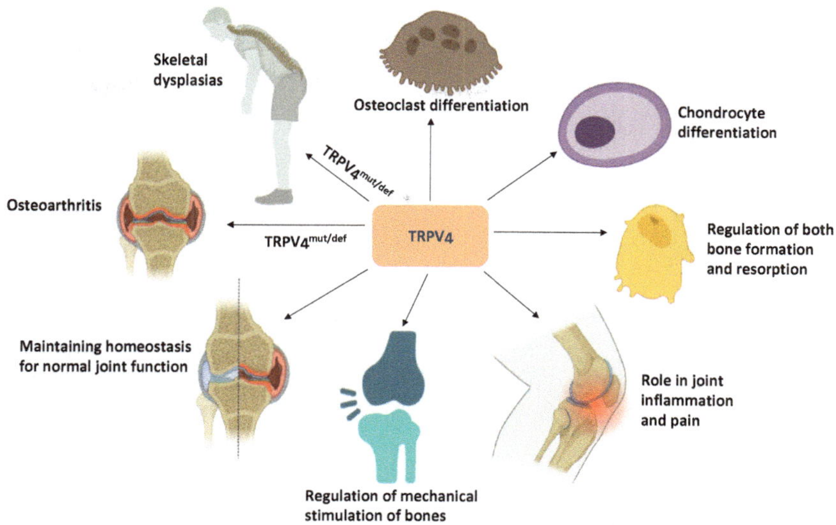

Fig. 4 TRPV4-mediated regulation of bone function. TRPV4 promotes bone formation and resorption. Deficiency or mutation of TRPV4 may lead to various bone disorders like skeletal dysplasia and osteoarthritis. TRPV4 also regulates mechanical stimulation and homeostasis required for proper joint functions.

4. Clinical characteristics of TRPV4 mutation-mediated disorders

The diseases caused by TRPV4 mutations are inherited in an autosomal dominant manner (McCray et al., 2020). Following simple Mendelian genetics, the mutations are transferred to offspring and develop into a series of diseases (Table 1). The autosomal dominant TRPV4 disorders are classified either into neuromuscular or skeletal dysplasias. Overlap between these two disorders is significant in a large number of patients (Unger et al., 2011). *TRPV4* pathogenic variants mostly cause "gain-of-function" of the ion channel as indicated from functional studies (such as Ca^{2+}-imaging, electrophysiological measurements). In some cases, like familial digital arthropathy-brachydactyly, "loss-of-function" of the ion channel is evident (Krakow, 2009; Mah, 2016; Nilius & Voets, 2013; Rock et al., 2008).

4.1 Autosomal dominant neuromuscular disorders

TRPV4 has emerged as a critical player in genetic neuropathies. Disorders caused by a heterozygous TRPV4 pathogenic variant are inherited in an

Table 1 Mutation list of TRPV4.

Sl. no.	Mutation	Residue	Disease	Effects on ion conductivity	Reference(s)
1	–	P19S	Hyponatremia (reduced response to hypotonic stress)	Loss of channel function	Tian et al. (2009)
2	58G > A	G20R	Distal hereditary motor neuropathy (dHMN)		Fawcett et al. (2012)
3	184G > A (exon 2)	D62N	Phenotype is compatible with CMT type 2C and presented with intra-cytoplasmic basophilic inclusions besides the neurogenic changes.		Evangelista et al. (2015)
4	–	G78W	Dual phenotype of metatropic dysplasia and fetal akinesia		Unger et al. (2011)
5	C366 > T (Exon 2)	T89I	Neonatal lethal metatrophic dysplasia	Elevated basal Ca^{2+}	Camacho et al. (2010)
6	281C > T	S94L	Congenital spinal muscular atrophy and arthrogryposis (CSMAA)	Increase in the channel activity	Velilla et al. (2019)
7	290C > G (Exon 2)	P97R	Distal muscular atrophy (dSMA) and vocal cord paralysis	Loss-of-function	Fiorillo et al. (2012)
8	547G > A (exon 3)	E183K	SEDM-PM2		Nishimura et al. (2010)
9	557G > A	R186Q	Charcot–Marie–Tooth disease type 2C	Slight increase in Ca^{2+} levels	Landouré et al. (2012)

10	A590 > G	K197R	Metatropic dysplasia		Camacho et al. (2010)
11	597G > C (exon 4)	L199F	Metatropic dysplasia		Dai et al. (2010)
12	649G > T	A217S	SMDK and a neurological phenotype compatible with scapulo-humeral spinal muscular atrophy		Cho et al. (2012)
13	652G > A	E218K	CMT2 with Complex neurological syndrome with demyelinating neuropathy		Fawcett et al. (2012)
14	694C > T (Exon 4)	R232C	HMN with vocal fold paralysis, SPSMA, HMSN2		Zimoń et al. (2010)
15	717G > C (Exon 5)	Q239H	Metatropicdysplaisa		Andreucci et al. (2011)
16	805C > T (exon 5)	R269C	HSMN type IIC, with SPSMA and also with distal congenital non progressive SMA.		Evangelista et al. (2015)
17	G806A (exon 5)	R269H	SMA, CMT2C, SPSMA	Gain-of-function	Deng et al. (2010)
18	c.809G > T (exon 5)	G270V	Familial digital arthropathy-brachydactyly (FDAB)	Reduced response to agonist	Lamandé et al. (2011)
19	c.812G > C (exon 5)	R271P	Familial digital arthropathy-brachydactyly (FDAB)	Reduced response to agonist	Lamandé et al. (2011)

Continued

Table 1 Mutation list of TRPV4.—cont'd

Sl. no.	Mutation	Residue	Disease	Effects on ion conductivity	Reference(s)
20	c.819C > G (exon 5)	F273L	Familial digital arthropathy-brachydactyly (FDAB)	Reduced response to agonist	Lamandé et al. (2011)
21	–	K276E	Dual phenotype of metatropic dysplasia and fetal akinesia		Unger et al. (2011)
22	832G > A (exon 5)	E278K	SMDK		Dai et al. (2010)
23	883A > G (Exon 6)	T295A	Metatropic dysplasia		Dai et al. (2010)
24	904A > T	N302Y	Charcot–Marie–Tooth disease type 2C (CMT2C)		Fawcett et al. (2012)
25	943C > T (Exon 6)	R315W	CMT2C		Chen et al. (2010)
26	C946T (exon 6)	R316C	HMSN2C	Gain-of-function	Deng et al. (2010)
27	G947 > A (exon 6)	R316H	Charcot–Marie–Tooth disease type 2C	Increased intracellular Ca^{2+}	Klein et al. (2011)
28	c.956C > T	S319L	Isolated scapular winging		Brown and Nguyen (2017)
29	A991 > T (Exon 6)	I331F	Severe Metatropic dysplasia	Constitutive open channels	Camacho et al. (2010)
30	992T > C (Exon 6)	I331T	Metatropic dysplasia		Dai et al. (2010)
31	cA992 → G (Exon 6)	D333G	SMDK	Increased Ca^{2+}-influx	Dai et al. (2010)

32	1024G > T (Exon 6)	V342F	Metatropic dysplasia		Dai et al. (2010)
33	1219A > G (Exon 7)	K407E	Metatropic dysplasia (MD)		Andreucci et al. (2011)
34	1412_1414del (Exon 8) 471	F471del	Metatropic dysplasia (MD)		Camacho et al. (2010) and Dai et al. (2010)
35	1566_68dup; p. L523dup (Exon 9)	L523dup	SMDK		Andreucci et al. (2011)
36	1625C > A (Exon 10)	S542Y	CMT2C		Chen et al. (2010)
37	1701C > A	Y567X	Charcot–Marie–Tooth disease type 2C		Fawcett et al. (2012)
38	1772A > G (Exon 11)	Y591C	Brachyolmia, SMDK		Andreucci et al. (2011)
39	1780C > A (Exon 11)	R594S	Metatropic dysplasia (MD)		Andreucci et al. (2011)
40	1774 T > C (Exon 11)	F592L	Metatropic dysplasia		Dai et al. (2010)
41	1781G > A (Exon 11)	R594H	SMDK, Parastremmatic dysplasia	Increased Ca^{2+}-influx	Nishimura et al. (2010)
42	c.1787 T > C (Exon 11)	L596P	SMDK		Dai et al. (2010)
43	1798G > T (Exon 11)	G600W	SMDK		Dai et al. (2010)
44	c.1805A > G (Exon 11)	Y602C	SEDM-PM2		Nishimura et al. (2010)

Continued

Table 1 Mutation list of TRPV4.—cont'd

Sl. no.	Mutation	Residue	Disease	Effects on ion conductivity	Reference(s)
45	C1812 > G (Exon 11)	I604M	Metatropic dysplasia		Camacho et al. (2010)
46	1834A > G (Exon 12)	K612E	Metatropic dysplasia		Narayanan et al. (2016)
47	1847G > A (Exon 12)	R616Q	Brachyolmia	Increased Ca^{2+}-influx	Rock et al. (2008)
48	C1851 > A (exon 12)	F617L	Metatropic dysplasia		Camacho et al. (2010)
49	T1853 > C (Exon 12)	L618P	Metatropic dysplasia		Camacho et al. (2010)
50	1855C > T	L619F	Metatropic dysplasia (accelerated chondrogenic differentiation of dental pulp stem cells)	Enhanced intracellular Ca^{2+}-level	Nonaka (2019)
51	1858G > A (Exon 12)	V620I	Brachyolmia	Increased Ca^{2+}-influx	Rock et al. (2008)
52	1875G > A (Exon 12)	M625I	SMDK		Dai et al. (2010)
53	2102C > T	T701I	Charcot–Marie–Tooth disease type 2C		Fawcett et al. (2012)
54	2125C > A (Exon 13)	L709M	SMDK/BO		Dai et al. (2010)
55	–	M713V/I	Central Giant cell lesions of the jaw (GCLJ)	Increased channel activity	Gomes et al. (2018)

56	c.G2146 → T (Exon 13)	A716S	SMDK	Same as WT	Krakow et al. (2009)
57	–	T740I	Dual phenotype of metatropic dysplasia and fetal akinesia		Unger et al. (2011)
58	2324G > A (Exon 14)	R775K	Metatropic dysplasia		Dai et al. (2010)
59	2330G > A (Exon 14)	C777Y	SMDK		Dai et al. (2010)
60	c.2389G > A (Exon 15)	E797K	Spondylo-epimetaphyseal dysplasia Maro- teaux—pseudo-Morquio type 2 (SEDM-PM2), SMDK and Metatropic dysplasia		Camacho et al. (2010), Dai et al. (2010), and Nishimura et al. (2010)
61	2396C > G (Exon 15)	P799R	Metatropic dysplasia		Dai et al. (2010)
62	2395C > T (Exon 15)	P799S	Metatropic dysplasia		Dai et al. (2010)
63	2395C > G (Exon 15)	P799A	Metatropic dysplasia		Dai et al. (2010)
64	C2396 > T (Exon 15)	P799L	Metatropic dysplasia, SEDM-PM2	Constitutive open channels	Camacho et al. (2010), Nishimura et al. (2010), and Dai et al. (2010)
65	999_1010del (Exon 6)	D333_E337delinsE	Mild Metatropic Dysplasia		Camacho et al. (2010)
66	2396_2412del17	Pro799Lfs63X	SEDM-PM2		Nishimura et al. (2010)

autosomal dominant manner. While the exact mechanisms behind disease pathogenesis and progression are not well-established, most of the complications occur due to "gain-of-function" nature of TRPV4 mutants. The downstream consequences of TRPV4 "gain-of- function" could be activation of Ca^{2+}-sensitive kinases and proteases, variations in gene transcription, interference in axonal transport and cytoskeletal remodeling or altered neuronal excitability (McCray et al., 2020). The majority TRPV4 mutations causing neuromuscular disease are mostly located in ankyrin repeats domains and result in increased Ca^{2+}-influx in the cells (Nilius & Voets, 2013; Verma et al., 2010).

4.1.1 Charcot-Marie-Tooth neuropathy type 2C (CMT2C)

CMT2C is an autosomal dominant peripheral neuropathy. The disease phenotype resembles CDSMA (congenital distal spinal muscular atrophy) and SPSMA (scapuloperoneal spinal muscular atrophy). At least 3 mutations located in the N-terminal region of TRPV4 (R269H, R315W and R316C) induce this pathophysiology (Auer-Grumbach et al., 2010). CMT2C is associated with laryngeal dysfunction, loss of sensorineural hearing, respiratory dysfunction, and joint contractures. Disease onset can occur anytime from birth to adulthood and seriousness of the disease symptoms varies from patient to patient.

4.1.2 Scapuloperoneal spinal muscular atrophy (SPSMA)

Specific mutations in TRPV4 (R232C, R269C, and R269H) cause this pathophysiological condition (Deng et al., 2010; Evangelista et al., 2015; Zimoń et al., 2010). Clinical features of SPSMA include laryngeal dysfunction and sensorineural hearing loss similar to CMT2C, rare cases of sensory deficits and kyphoscoliosis, weakness of muscles and atrophy in the shoulder, and neck flexor weakness. In serious cases, the absence of muscle or muscle weakness are evident from the birth itself (Delong & Siddique, 1992). The age of onset is anytime between birth-adulthood.

4.1.3 Neuronopathy, distal hereditary motor, type VIII; HMN8

HMN8 is an autosomal dominant disease, symptoms of which include proximal and distal muscle weakness, atrophy of distal leg muscles, and clubfoot. Extensive fatty atrophy in the thigh and calf muscles with preservation of the biceps femoris in the lateral thighs and the medial gastrocnemius in the posteromedial calves. Additional features include scoliosis, pes cavus, vocal cord paresis, scapular winging, and increased serum creatine kinase.

Phenotypes overlap with HMSN2C and SPSMA. The age of onset is prenatal (Astrea et al., 2012; Echaniz-Laguna et al., 2014). Mutations in TRPV4 that are linked with this condition are R269H and R315W (Chen et al., 2010; Evangelista et al., 2015).

4.2 Autosomal dominant skeletal dysplasias

Various missense mutations, frameshift and deletions of TRPV4 gene can cause skeletal dysplasia. The mutations are spread throughout the TRPV4 gene but two hot spots at residues Pro799 and Arg594 have been observed, commonly causing increased Ca^{2+}-influx inside the cells. The precise downstream events are still unidentified. The most extreme TRPV4 skeletal diseases can be lethal in utero or during childhood, and parents of the ailing child are unaffected most of the times as the variant appears for the first time in family (*de novo*). The diagnosis of TRPV4 Skeletal Dysplasias can be mild, intermediate or severe type.

4.2.1 Familial digital arthropathy-brachydactyly (FDAB)

The transmission pattern of FDAB (MIM 606835) is autosomal dominant. Patients with this disease appear normal at birth, with no evidence of developmental skeletal dysplasia. During the first decade of their life, relative shortening of the middle and distal phalanges and swelling and decreased range of motion of the interphalangeal joints become apparent. Later, other joints of the hands and feet become affected by deforming and painful osteoarthritis. Specific mutations in TRPV4 that lead to FDAB include G270V, R271P, and F273L (Lamandé et al., 2011).

4.2.2 Brachyolmia type 3; BCYM3

Brachyolmia is derived from the Greek word for "short trunk." Brachyolmias are a heterogeneous group of skeletal dysplasias that affect primarily the spine. Four types of Brachyolmias are found in nature, among which only Type 3 Brachyolmia (BCYM3; 113500) is caused by a mutation in the TRPV4 gene (OMIM 605427). It is inherited in an autosomal dominant form and characterized with severe kyphoscoliosis and flattened irregular cervical vertebrae. The other three types of Brachyolmia are inherited in an autosomal recessive fashion and do not involve TRPV4. Mutations like Y591C, R616Q, and V620I often cause this pathogenicity (Andreucci et al., 2011; Rock et al., 2008).

4.2.3 Spondylometaphyseal dysplasia, Kozlowski type

SMDs are a group of short-stature disorders illustrated by irregularities in the vertebrae and the metaphyses of the tubular bones. SMD Kozlowski (SMDK) is characterized by substantial scoliosis and minor metaphyseal abnormalities in the pelvis. The vertebrae display platyspondyly and over-faced pedicles. Inheritance of SMDK is demonstrated to be autosomal dominant and caused by variety of TRPV4 mutations like A217S, E278K, D333G, L523dup, and L596P (Dai et al., 2010; Krakow et al., 2009; Verma et al., 2010).

4.2.4 Spondyloepiphyseal dysplasia, Maroteaux type

The disease is categorized by short-trunk dwarfism and brachydactyly. While the height of the patient at birth is normal, in early childhood short stature is observed with poor growth and a short trunk. Other features like osteoporosis, kyphoscoliosis and genu valgum, etc. are also observed. Several TRPV4 mutations that lead to this phenotype are E183K, E797K and P799L (Nishimura et al., 2010).

4.2.5 Parastremmatic dwarfism

Parastremmatic dwarfism was named Parastremmatic from the Greek term "twisted". It is the rarest autosomal dominant skeletal dysplasia of TRPV4 and the causal mutation include R594H (Nishimura et al., 2010). Clinical features include deformities of the legs with severe genu valgum, bowing of the long bones, twisted thighs, short trunks, kyphoscoliosis, and multiple contractures of major joints that worsen with time.

4.2.6 Metatropic dysplasia

Metatropic dysplasia: The term metatropic dysplasia is derived from the Greek word metatropos, meaning "changing patterns", as an inversion of body proportion is seen from birth to childhood. The disease could be lethal in the prenatal or perinatal period. Often, patients have shortened limbs and narrow chest at birth, which evolves into severe kyphoscoliosis, frequently compromising lung function (Krakow et al., 2009). Numerous mutations in TRPV4 lead to this phenotype. Some of the prominent mutations are L199F, T295A, I604M, and L618P (Camacho et al., 2010; Dai et al., 2010).

5. Molecular and cellular characteristics of TRPV4 mutants

TRPV4 represents one of the most interesting candidates that harbors a series of naturally occurring point mutations which cause clinical symptoms, mainly musculo-skeletal disorders. No other TRPV channels demonstrate this level of mutation-related disorders. Close analysis of these mutations, their pathophysiology, and cellular and molecular properties indicate few important issues which are discussed in detail.

5.1 TRPV4 structure-function relationship and mutational hotspots

Conservation analysis of different structural domains of TRPV4 from all vertebrate sequences reveals interesting facts about this channel. On statistical evaluation, it is evident that some of the regions on TRPV4 are well-conserved, some domains and motifs are semi-conserved and some are less-conserved. Regions like TM4, Loop 4, TM5 and TRP-box are highly conserved, whereas PRD, Loop 3 and Loop 6 are divergent. Interestingly, a large number of mutations causing channelopathies in human population are found in conserved regions of TRPV4 (Kumari et al., 2015). Notably, point mutations in certain residues or hotspots can have prominently defective phenotypes. These phenotypes are primarily manifested by the disorders of muscles, bones, and occasionally cardiac tissue. The variability observed in different mutations is also suggestive of different effects on channel function.

5.2 Penetration of the TRPV4 mutation phenotypes

With limited genome sequencing data available so far, multiple TRPV4 mutations involved in diseases have been identified. Therefore, it can be predicted that in future, more disease-associated mutations in TRPV4 will be catalogued through genome sequencing. However, these mutations are expected to be tolerable, low-penetration, and thus low-lethality. It is highly likely that such naturally occurring channelopathies will be discovered through more genome sequencing in future. In contrast, most deleterious mutations may have embryonic or neonatal lethality. Such mutations are likely to have strong penetration but low frequency, and thus may be difficult and less-likely to be documented. For example, L618P mutation of

TRPV4 is embryonic lethal (Camacho et al., 2010). These investigations, connecting TRP channels to illness, are expected to become a high priority in biomedical sciences in future. In this context, there are strong indications that different mutations in TRPV4 can cause cancer/tumour. However, these mutations are "somatic in nature" and thus, do not propagate to the next generation.

6. Lessons learned from genetic animal models and gene diversity in human population

The involvement of TRPV4 in physiology is more complex than it appears. This is mainly due to the paradoxical findings from different model systems. The findings from genetic animal models and genetic variation in humans are analyzed below.

6.1 Consequences of TRPV4 gene deficit in animals

Although TRPV4 mutations result in a variety of hereditary bone disorders, only irregular osmotic regulation in case of TRPV4 knockout mouse has been reported (Suzuki et al., 2003). It is unclear why the phenotype of TRPV4 knockout mice is vastly dissimilar to that of mice with point mutations in TRPV4. While TRPV4 mutations result in a series of musculoskeletal disorders ranging from mild to very severe, TRPV4 knockout mouse does not demonstrate the phenotypes seen in human population with mutations (Gevaert et al., 2007; Tabuchi et al., 2005). A recent report by Nishimura et al. showed that TRPV1 and TRPV4 double knockout mice (DKO) have higher bone mass, less multinucleated osteoclasts, and lower bone resorption than WT mice (Nishimura et al., 2020). While these knockout animals are generated and maintained in controlled environments (in other words, a system that offers "no competition for survival"), it is worth mentioning that in nature, not a single vertebrate species is known where TRPV4 gene is absent. On the contrary, more than one copy of TRPV4 is present in some species. For example, in *Xenopus*, several copies of TRPV4 with minute variations are present (Kumari et al., 2015). Overall, these observations in the literature suggest that TRPV4 is not essential for survival (as knockout animals survive), at least in a controlled environment, and that TRPV4 probably offers additional benefits required for survival in evolutionary time scale.

6.2 TRPV4 gene variation in human population

With the availability of abundant human genome data at present, it is anticipated that many variations in the human genome will be identified based on ethnicity, gender, class, food habits, etc. In this regard, recent work analyzed the human genome data (2504 genomes), identifying several mis-sense variations in TRPV channels in human population (Ghosh et al., 2016). It also suggests the evolutionary stability of TRPV's in human population (Ghosh et al., 2016). Using SIFT, PolyPhen and Grantham Deviation scores, they evaluated the amino acid substitutions that are deleterious in human population. The prediction is in agreement with approximately 60 pathogenic mutations that are identified in TRPV4 termed as "channelopathies." According to the findings, TRPV4 is predicted to have the highest number of deleterious mutations among the thermosensitive ion channels of TRPV subfamily, suggesting that mutations in TRPV4 may not be tolerable, or the mutation does not continue for more number of generations.

With approximately 60 mutations identified, TRPV4 is probably one of the most disease-causing proteins known so far. Study based on 2504 human genome data suggests that the mutations in TRPV4 are clustered in specific "hotspots" on TRPV4 that are prone to mutation. One such mutational hotspot is the TM4-Loop4-TM5 region, which has almost 14 mutations that give rise to several pathophysiological conditions.

6.3 Comparisons of TRPV4 mutation-induced channelopathy with other TRPV mutations

While discussing TRPV4 mutation-induced channelopathies, relationship and comparison with other TRPV channels is obvious. Ironically, TRPV1 is expressed in several tissues and is a widely studied ion channel with many missense variations detected in the human population (Ghosh et al., 2016); yet, there is no naturally occurring TRPV1 mutation linked with a disease. Thus, mis-sense variations in TRPV1 are probably "more tolerable" in the physiological context. Similarly, no naturally occurring mutations that induce serious pathological disorder are known for TRPV2. In comparison, multiple mis-sense variations have been reported in TRPV3, and a few of these mutations have been linked with *Olmested Syndrome*, characterized by localized disorders of skin and bones (Yadav & Goswami, 2017). With advancements in genome sequencing techniques, more individuals will be sequenced in future. Currently, the number of naturally occurring pathogenic mutations stands highest for TRPV4, followed by TRPV3 and least for TRPV1 and TRPV2.

7. Conclusion and future outlook

A whole range of recent research work demonstrates that TRPV4 is major ion channel expressed in cells and tissues that collectively represent the musculoskeletal system. Recent progress in genome sequencing suggests that in human population, there is a broad spectrum of mis-sense variations for TRPV4 and not all of these variations are lethal or pathogenic. However, mutations in specific hotspots on TRPV4 are most-likely to result in clinical phenotypes with different penetration levels. In fact, TRPV4 seems to regulate a range of biological pathways including Ca^{2+} signaling, Ca^{2+}-buffering by subcellular organelles, cell death regulation, gene expression, cytoskeletal remodeling, etc. It is plausible that certain TRPV4 mutations affect the Ca^{2+}-independent effects of TRPV4. Accordingly, different point mutations in TRPV4 can have distinct effects on the development and/or functions of specific cells/tissues. Overall, TRPV4 is a potential molecular candidate for treating different forms of musculoskeletal disorders. Involvement of TRPV4 in musculoskeletal disorders can be genetic (thus transmissible to the next generation) or somatic (not transmittable to next generation). Further studies are needed to elucidate the precise pathological mechanisms for musculoskeletal disorders involving TRPV4 channels.

Acknowledgments

All the authors acknowledge the intellectual input and scientific discussion on this topic. All authors acknowledge the intramural support from NISER, DAE. We thank Ranabir Chakraborty, Ritesh Dalai and Nilesh Kumar Das for helping us with the preparation of some figures. Some of the figures are prepared using https://biorender.com. The authors thank Dr. Swapnil K. Sonkusare for his critical comments and manuscript correction.

References

Andreucci, E., et al. (2011). TRPV4 related skeletal dysplasias: A phenotypic spectrum highlighted by clinical, radiographic, and molecular studies in 21 new families. *Orphanet Journal of Rare Diseases*, *6*, 1–8.

Astrea, G., et al. (2012). Muscle MRI in TRPV4-related congenital distal SMA. *Neurology*, *78*, 364–365.

Auer-Grumbach, M., et al. (2010). Alterations in the ankyrin domain of TRPV4 cause congenital distal SMA, scapuloperoneal SMA and HMSN2C. *Nature Genetics*, *42*, 160–164.

Berna-Erro, A., et al. (2017). Structural determinants of 5′,6′-epoxyeicosatrienoic acid binding to and activation of TRPV4 channel. *Scientific Reports*, 7, 1–11.

Brinkmeier, H. (2011). TRP channels in skeletal muscle: Gene expression, function and implications for disease. *Advances in Experimental Medicine and Biology*, *704*, 749–758.

Brown, K. A., & Nguyen, T. P. (2017). A rare case of scapular winging due to a novel trpv4 mutation. *Journal of Genetic Disorders and Genetic Medicine*, *1*, 2–3.

Camacho, N., et al. (2010). Dominant TRPV4 mutations in nonlethal and lethal metatropic dysplasia. *American Journal of Medical Genetics Part A*, *152*, 1169–1177.

Chapman, M. A., et al. (2016). Skeletal muscle fibroblasts in health and disease. *Differentiation*, *92*, 108–115.

Chen, Y. L., & Sonkusare, S. K. (2020). Endothelial TRPV4 channels and vasodilator reactivity. *Current Topics in Membranes*, *85*, 89–117.

Chen, D. H., et al. (2010). CMT2C with vocal cord paresis associated with short stature and mutations in the TRPV4 gene. *Neurology*, *75*, 1968–1975.

Cho, T. J., et al. (2012). TRPV4-pathy manifesting both skeletal dysplasia and peripheral neuropathy: A report of three patients. *American Journal of Medical Genetics Part A*, *158 A*, 795–802.

Dai, J., et al. (2010). Novel and recurrent TRPV4 mutations and their association with distinct phenotypes within the TRPV4 dysplasia family. *Journal of Medical Genetics*, *47*, 704–709.

Daneva, Z., et al. (2021). Caveolar peroxynitrite formation impairs endothelial TRPV4 channels and elevates pulmonary arterial pressure in pulmonary hypertension. *Proceedings of the National Academy of Sciences*, *118*, 1–12.

Das, R., & Goswami, C. (2019). TRPV4 expresses in bone cell lineages and TRPV4-R616Q mutant causing Brachyolmia in human reveals "loss-of-interaction" with cholesterol. *Biochemical and Biophysical Research Communications*, *517*, 566–574.

Das, R., et al. (2022). Cytochrome C interacts with the pathogenic mutational hotspot region of TRPV4 and forms complexes that differ in mutation and metal ion-sensitive manner. *Biochemical and Biophysical Research Communications*, *611*, 172–178.

Delong, R., & Siddique, T. (1992). A large New England kindred with autosomal dominant neurogenic scapuloperoneal amyotrophy with unique features. *Archives of Neurology*, *49*, 905–908.

Deng, H. X., et al. (2010). Scapuloperoneal spinal muscular atrophy and CMT2C are allelic disorders caused by alterations in TRPV4. *Nature Genetics*, *42*, 165–169.

Deng, Z., et al. (2018). Cryo-EM and X-ray structures of TRPV4 reveal insight into ion permeation and gating mechanisms. *Nature Structural & Molecular Biology*, *25*, 252–260.

Denis, V., & Cyert, M. S. (2002). Internal Ca2+ release in yeast is triggered by hypertonic shock and mediated by a TRP channel homologue. *The Journal of Cell Biology*, *156*, 29–34.

Echaniz-Laguna, A., et al. (2014). Phenotypic spectrum and incidence of TRPV4 mutations in patients with inherited axonal neuropathy. *Neurology*, *82*, 1919–1926.

Evangelista, T., et al. (2015). Phenotypic variability of TRPV4 related neuropathies. *Neuromuscular Disorders*, *25*, 516–521.

Falcón, D., et al. (2019). TRP channels: Current perspectives in the adverse cardiac remodeling. *Frontiers in Physiology*, *10*, 1–12.

Fawcett, K. A., et al. (2012). Comprehensive analysis of the TRPV4 gene in a large series of inherited neuropathies and controls. *Journal of Neurology, Neurosurgery, and Psychiatry*, *83*, 1204–1209.

Filosa, J. A., et al. (2013). TRPV4 and the regulation of vascular tone. *Journal of Cardiovascular Pharmacology*, *61*, 113–119.

Fiorillo, C., et al. (2012). TRPV4 mutations in children with congenital distal spinal muscular atrophy. *Neurogenetics*, *13*, 195–203.

Gevaert, T., et al. (2007). Deletion of the transient receptor potential cation channel TRPV4 impairs murine bladder voiding. *The Journal of Clinical Investigation.*, *117*, 3453–3462.

Ghosh, A., et al. (2016). Why individual thermo sensation and pain perception varies? Clue of disruptive mutations in TRPVs from 2504 human genome data. *Channels*, *10*, 339–345.

Gomes, C. C., et al. (2018). TRPV4 and KRAS and FGFR1 gain-of-function mutations drive giant cell lesions of the jaw. *Nature Communications*, *9*, 1–8.

Goretzki, B., et al. (2018). Structural basis of TRPV4 N terminus interaction with Syndapin/PACSIN1-3 and PIP 2. *Structure*, *26*, 1583–1593.

Goswami, C., et al. (2010). Importance of non-selective cation channel TRPV4 interaction with cytoskeleton and their reciprocal regulations in cultured cells. *PLoS One*, *5*, 1–16.

Himmel, N. J., & Cox, D. N. (2020). Transient receptor potential channels: Current perspectives on evolution, structure, function and nomenclature. *Proceedings of the Royal Society B*, *287*, 1–9.

Hu, F., et al. (2019). Regulation of intracellular Ca^{2+}/CaMKII signaling by TRPV4 membrane translocation during osteoblastic differentiation. *Biophysical Reports*, *5*, 254–263.

Klein, C. J., et al. (2011). TRPV4 mutations and cytotoxic hypercalcemia in axonal Charcot-Marie-Tooth neuropathies. *Neurology*, *76*, 887–894.

Krakow, D., et al. (2009). Mutations in the gene encoding the calcium-permeable ion channel TRPV4 produce spondylometaphyseal dysplasia, Kozlowski type and Metatropic dysplasia. *American Journal of Human Genetics*, *84*, 307–315.

Kumar, A., et al. (2018). TRPV4 interacts with mitochondrial proteins and acts as a mitochondrial structure-function regulator. *bioRxiv*. https://doi.org/10.1101/330993. preprint.

Kumari, S., et al. (2015). Influence of membrane cholesterol in the molecular evolution and functional regulation of TRPV4. *Biochemical and Biophysical Research Communications*, *456*, 312–319.

Lamandé, S. R., et al. (2011). Mutations in TRPV4 cause an inherited arthropathy of hands and feet. *Nature Genetics*, *43*, 1142–1146.

Landouré, G., et al. (2012). Exome sequencing identifies a novel TRPV4 mutation in a CMT2C family. *Neurology*, *79*, 192–194.

Lawhorn, B. G., et al. (2020). Recent advances in TRPV4 agonists and antagonists. *Bioorganic & Medicinal Chemistry Letters*, *30*, 1–6.

Liedtke, W., et al. (2000). Vanilloid receptor–related osmotically activated channel (VR-OAC), a candidate vertebrate osmoreceptor. *Cell*, *103*, 525–535.

Liedtke, W., et al. (2003). Mammalian TRPV4 (VR-OAC) directs behavioral responses to osmotic and mechanical stimuli in *Caenorhabditis elegans*. *Proceedings of the National Academy of Sciences of the United States of America*, *100*.

Mah, W., et al. (2016). Gain-of-function mutation in TRPV4 identified in patients with osteonecrosis of the femoral head. *Journal of Medical Genetics*, *53*, 705–709.

Masuyama, R. (2014). Role of local vitamin D signaling and cellular calcium transport system in bone homeostasis. *Journal of Bone and Mineral Metabolism*, *32*, 1–9.

Masuyama, R., et al. (2008). TRPV4-mediated calcium influx regulates terminal differentiation of osteoclasts. *Cell Metabolism*, *8*, 257–265.

McCray, B. A., et al. (2020). Autosomal dominant TRPV4 disorders. *GeneReviews®*, *1–27*. http://www.ncbi.nlm.nih.gov/pubmed/24830047.

Méndez-Reséndiz, K. A., et al. (2020). Steroids and TRP channels: A close relationship. *International Journal of Molecular Sciences*, *21*, 1–36.

Montell, C. (2005). The TRP superfamily of cation channels. *Science's STKE*, *272*, 1–24.

Muller, C., et al. (2019). Cannabinoid ligands targeting TRP channels. *Frontiers in Molecular Neuroscience*, *487*, 1–15.

Muramatsu, S., et al. (2007). Functional gene screening system identified TRPV4 as a regulator of chondrogenic differentiation. *The Journal of Biological Chemistry*, *282*, 32158–32167.

Narayanan, D. L., et al. (2016). Metatropic dysplasia with a novel mutation in TRPV4. *Indian Pediatrics, 53*, 735–737.
Nilius, B., & Owsianik, G. (2011). The transient receptor potential family of ion channels. *Genome Biology, 12*, 1–11.
Nilius, B., & Voets, T. (2013). The puzzle of TRPV4 channelopathies. *EMBO Reports, 14*, 152–163.
Nishimura, G., et al. (2010). Spondylo-epiphyseal dysplasia, Maroteaux type (pseudo-Morquio syndrome type 2), and parastremmatic dysplasia are caused by TRPV4 mutations. *American Journal of Medical Genetics, 152*, 1443–1449.
Nishimura, G., et al. (2012). TRPV4-associated skeletal dysplasias. *American Journal of Medical Genetics Part C: Seminars in Medical Genetics, 160C*, 190–204.
Nishimura, H., et al. (2020). Transient receptor potential vanilloid 1 and 4 double knockout leads to increased bone mass in mice. *Bone Reports, 12*, 1–10.
Nonaka, K., et al. (2019). Novel gain-of-function mutation of TRPV4 associated with accelerated chondrogenic differentiation of dental pulp stem cells derived from a patient with metatropic dysplasia. *Biochemistry and Biophysics Reports, 19*, 1–7.
Ottolini, M., et al. (2020). Local Peroxynitrite impairs endothelial transient receptor potential Vanilloid 4 channels and elevates blood pressure in obesity. *Circulation, 141*, 1318–1333.
Plant, T. D., & Strotmann, R. (2007). *TRPV4: A multifunctional nonselective cation channel with complex regulation. TRP ion channel function in sensory transduction and cellular signaling cascades*. Boca Raton (FL): CRC Press/Taylor & Francis (Chapter 9).
Rock, M. J., et al. (2008). Gain-of-function mutations in TRPV4 cause autosomal dominant brachyolmia. *Nature Genetics, 40*, 999–1003.
Song, M. Y., & Yuan, J. X. J. (2010). Introduction to TRP channels: Structure, function, and regulation. *Advances in Experimental Medicine and Biology, 661*, 99–108.
Suzuki, M., et al. (2003). Impaired pressure sensation in mice lacking TRPV4. *The Journal of Biological Chemistry, 278*, 22664–22668.
Suzuki, T., et al. (2013). Osteoblastic differentiation enhances expression of TRPV4 that is required for calcium oscillation induced by mechanical force. *Bone, 54*, 172–178.
Tabuchi, K., et al. (2005). Hearing impairment in TRPV4 knockout mice. *Neuroscience Letters, 382*, 304–308.
Takahashi, N., et al. (2014). TRPV4 channel activity is modulated by direct interaction of the ankyrin domain to PI(4,5)P2. *Nature Communications, 5*, 1–15.
Tian, W., et al. (2009). A loss-of-function nonsynonymous polymorphism in the osmoregulatory TRPV4 gene is associated with human hyponatremia. *Proceedings of the National Academy of Sciences, 106*, 14034–14039.
Toft-Bertelsen, T. L., & Macaulay, N. (2021). TRPing to the point of clarity: Understanding the function of the complex TRPV4 Ion Channel. *Cell, 10*, 1–16.
Unger, S., et al. (2011). Fetal akinesia in metatropic dysplasia: The combined phenotype of chondrodysplasia and neuropathy. *American Journal of Medical Genetics, 155*, 2860–2864.
Velilla, J., et al. (2019). Homozygous TRPV4 mutation causes congenital distal spinal muscular atrophy and arthrogryposis. *Neurology Genetics, 5*, 1–9.
Venkatachalam, K., & Montell, C. (2007). TRP channels. *Annual Review of Biochemistry, 76*, 387–417.
Verma, P., et al. (2010). TRPV4-mediated channelopathies. *Channels (Austin, Tex.), 4*, 319–328.
Vriens, J., et al. (2004). Cell swelling, heat, and chemical agonists use distinct pathways for the activation of the cation channel TRPV4. *Proceedings of the National Academy of Sciences of the United States of America, 101*, 396–401.
Watanabe, H., et al. (2002). Activation of TRPV4 channels (hVRL-2/mTRP12) by phorbol derivatives. *The Journal of Biological Chemistry, 277*, 13569–13577.

White, J. P. M., et al. (2016). Molecular conductor of a diverse orchestra. *Physiological Reviews*, *96*, 911–973.
Woolums, B. M., et al. (2020). TRPV4 disrupts mitochondrial transport and causes axonal degeneration via a CaMKII-dependent elevation of intracellular Ca2+. *Nature Communications*, *11*, 1–17.
Yadav, M., & Goswami, C. (2017). TRPV3 mutants causing Olmsted syndrome induce impaired cell adhesion and nonfunctional lysosomes. *Channels*, *11*, 196–208.
Zheng, J. (2013). Molecular mechanism of TRP channels. *Comprehensive Physiology*, *3*, 221–242.
Zhu, M. X. (2011). *TRP channels*. CRC Press/Taylor & Francis.
Zimoń, M., et al. (2010). Dominant mutations in the cation channel gene transient receptor potential vanilloid 4 cause an unusual spectrum of neuropathies. *Brain*, *133*, 1798–1809.

FSC
www.fsc.org
MIX
Papier | Fördert
gute Waldnutzung
FSC® C083411

Kolibri 360 GmbH
Ferdinand-Jühlke-Straße 7
99095 Erfurt, Deutschland
produktsicherheit@kolibri360.de